Skeletal Dysplasias & Congenital Malformations

Surender Singh Yadav

Skeletal Dysplasias & Congenital Malformations

Springer

Surender Singh Yadav
Department of Orthopaedics
Mirpur Institute of Medical Sciences
Rewari, Haryana, India

ISBN 978-981-99-5681-4 ISBN 978-981-99-5679-1 (eBook)
https://doi.org/10.1007/978-981-99-5679-1

This Springer imprint is published by the registered company Springer Nature Singapore Pte Ltd.
The registered company address is: 152 Beach Road, #21-01/04 Gateway East, Singapore 189721, Singapore

Paper in this product is recyclable.

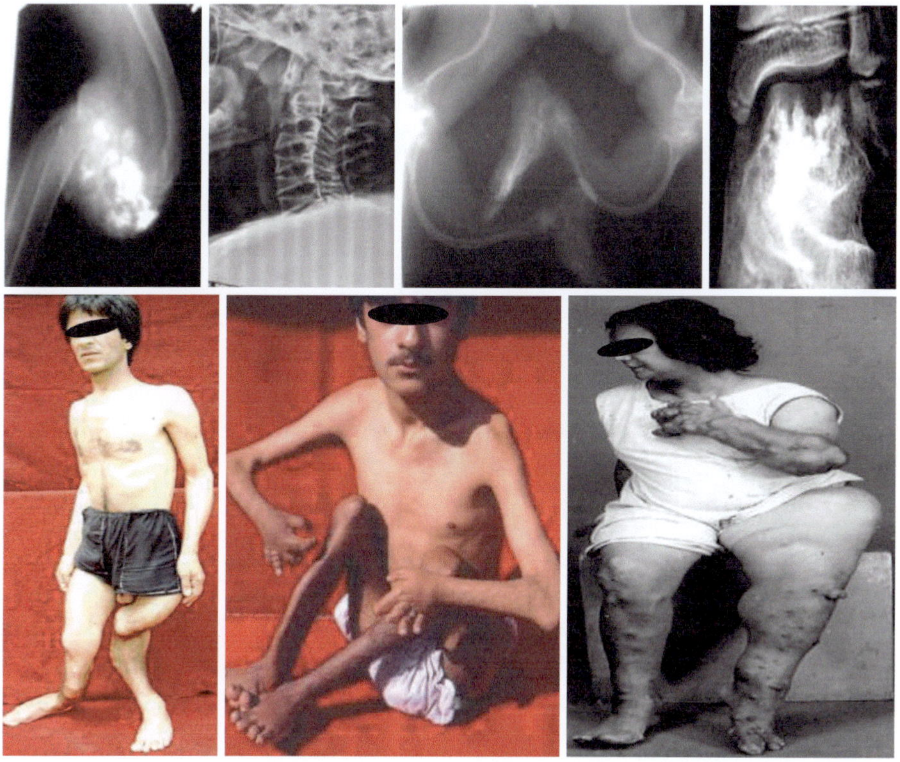

Preface

There are a number of 'rare' genetic disorders in human beings, classified as such because very few people are affected by them. While the number of people suffering from any one of these disorders is quite low, put together they account for a large number of patients. In India alone, a good number of people are estimated to be suffering from these 'rare' diseases. Ideally, it should be diagnosed by the Paediatrician and Radiologist when the parents first complain of lack of growth and limb anomalies in the child. But most often, especially in our country, it is rarely detected. It is also true that many clinicians are not inclined to take up this subject. The Government support for Research in this area has also not been encouraging.

Osteochondrodysplasias are a heterogeneous group of disorders, which result in skeletal abnormalities and often result in disproportionate short stature. As per the literature, there are many distinct skeletal dysplasias that must be differentiated one from the other, both for genetic counselling and specific therapy. In spite of all efforts it may be difficult to separate them into well-defined categories as in many cases the features often overlap. We must acquire adequate knowledge for correct identification and management of such anomalies. The orthopaedic community must stay informed about the understanding of the Clinical and Radiological observations. Histopathological assessment followed by genetic and molecular basis of these diseases is also important.

My interest in the subject grew more when I got more and more of such cases during my posting at JIPMER, Pondicherry, for nearly 14 years. This might have been because consanguineous marriages are more prevalent in South India. This was later followed in M.A.M.C., New Delhi. Since I have been identifying and managing such cases, my interest grew further and more and more patients started visiting us. Sincere effort has been made to manage these patients in a need-based fashion to bring them back to the mainstream of the society. Some of them have been able to avail the benefit of such an approach and have been listed in the text. It is also possible that such patients could not be identified and remain neglected.

Since there was a good collection of the cases over a long period, it was decided to present this material in the form of a book, which may prove useful particularly for paediatricians, radiologists, orthopaedic surgeons and other clinicians of the next generation.

The reasons for this lack of interest from treating clinicians could be many; skeletal dysplasias have not gathered much attention. As a result, there are only a few publications/books on this subject. To my knowledge, this may be the first such attempt in India to bring out some of the common anomalies of this nature in the form of a book. Moreover, the cases and the illustrations sited in the book are the personal collection of the author over a period of five decades. During this period efforts have been made for systematic collection of the cases and careful preservation of the illustrations. An attempt has been made to reveal the major clinical and radiological features of common and classical disorders of skeletal dysplasias.

The description of various features of the common dysplasias is intended to be adequate but not exhaustive. The book should provide an important source of information to the orthopaedic surgeons, paediatricians and radiologists. The useful information provided in the book will also prove helpful to the parents of such children for correct diagnosis and better management.

To my knowledge, this is the first book of its kind in our country, analysing and describing rare congenital ailments and deformities. The important topics sited in the text need widespread recognition to help such patients by the concerned care providers.

Rewari, Haryana, India Surender Singh Yadav

Acknowledgements

On the very outset, I express my gratitude to the Creator and my dear parents who brought me in this World and inspired me through their mentoring. My affectionate mother and two elder brothers deserve salutation for guiding me in my early life. Special thanks go to my wife, the late Dr. Shanti Yadav, who supported me in preparing the text.

I am indebted to the late Prof. P.K. Duraiswami for guiding me and playing an important role in introducing me to orthopaedics and imparting knowledge during my early training that has influenced my early professional career. Prof. S.M.Tuli deserves special thanks for reinforcing my idea for bringing out this Atlas.

It is my pleasure to acknowledge the cooperation of my affectionate patients, without whom this endeavour would not have been possible. They have generously provided the material and consented to be photographed. I always have immense belief and faith in my patients, and I believe that in the field of medicine, the patients are like true teachers.

I would like to take the opportunity to acknowledge those who inspired me through their sincere efforts. Mr. Mahesh Rakheja deserves mention for his cooperation.

The writing of this book has exhausted the patience of several secretaries named Rakesh Kumar, Miss. Nivya and Mr. Mohit Kumar. Thanks are also due to my colleagues who have been very cooperative during the course of preparing the text.

Surender Singh Yadav

Contents

About the Author

Surender Singh Yadav is a renowned orthopaedic surgeon. Founder-Director, PGIMS, Rohtak, & Director-Medical Education, Haryana. Past-President of Indian Orthopaedic Association 1992. Recipient of 18 national and 8 international academic awards, including 'Sher-e-Bangla National Gold Medal' from Dhaka (Bangladesh) and Dr.B.C.Roy (Major) Award as 'Eminent Medical Man' 1995 by the President. As a unique distinction, Prof. Yadav has been honoured by Five Hon'ble Presidents of India and honoured with the National Award of 'Padma Shri'—2008.Dr. Yadav obtained his MBBS degree from SMS Medical College, Jaipur, and MS in Orthopaedics from Delhi University, Delhi. Dr. S.S.Yadav has been a medical teacher in many postgraduate institutions. His original/innovative biological research is published in the ACE journals like the *JBJS* (Am.). World leaders have widely appreciated his work on non-vascularized fibular grafts.

Part I

Skeletal Dysplasias

The Skeletal Dysplasias

1

1.1 Introduction

The human skeletal dysplasias are a heterogeneous group of heritable connective tissue disorders associated with abnormalities in the size and shape of the limbs, trunk and/or skull that frequently result in disproportionate short stature. The skeletal dysplasias are interpreted as disorders of normal bone growth or remodelling and are classified by the potential site of the anatomic defect in the skeleton, epiphyseal, metaphyseal and diaphyseal.

Normally the epiphyseal cartilage proliferates in columnar fashion and in due course undergoes maturation, degeneration and calcification in its conversion to bone. If the chondroblasts are defective, multiple pathology may occur as a result of the failure or an irregularity in the maturation and calcification. If the failure is irregular throughout the skeleton, multiple pathological anomalies will appear, depending on the site of the abnormal chondroblasts.

If the failure of maturation and calcification is diffuse, the lack of the zone of preliminary calcification will result in a widespread retardation of epiphyseal ossification and, therefore, dwarfism and deformities occur.

A central immature chondroblast, if it continues to proliferate, will form an enchondroma, which gives the bone an apparent cystic appearance, and, if peripheral, this proliferating chondroblast will produce an ecchondroma. If ossification occurs in relation to this peripheral group of cartilage cells, exostosis results. The primitive connective tissue cells may continue to proliferate when there is an absence of calcium in the adjacent zone of preliminary calcification, and as a result fibrous tissues masses will occur. The resultant picture is a conglomeration of stunted growth, exostoses, enchondromata, ecchondromata and cysts; when one of these predominates, names have been applied to the clinical appearances.

In clinical characteristics, the age of presentation, and the occurrences of associated non-skeletal features are important to be understood. Careful physical examination and radiological assessment may lead to diagnosis in such cases. The family

S. S. Yadav, *Skeletal Dysplasias & Congenital Malformations*,
https://doi.org/10.1007/978-981-99-5679-1_1

history in every case is obligatory. Certain conditions may be defined clearly by their inheritance pattern.

Histopathology plays a useful role in indicating the nature of the tissue changes associated with disorders of skeletal development. After taking the clinical and radiological factors into account, pathological evaluation of skeletal tissues may point strongly towards a particular type of skeletal dysplasia or other developmental disorder. Histopathology is also useful in distinguishing between primary and secondary disorders of skeletal development. For pathological examination of a non-lethal skeletal dysplasia, biopsy of the iliac crest or another growth plate (e.g., tibia), to include the chondro-osseous junction, is useful.

The present nomenclature and distinction between various dysplasias are largely based on radiographic findings. A full skeletal survey is essential including skull, spine, hands, chest and pelvis. Some dysplasias are named for the segment of the limbs that shows them shortening, rhizomelic (proximal), mesomelic (middle), acromelic (distal) dysplasias. The abnormal bone development may also result in diastrophic dysplasia. Where the spine is involved, the prefix spondylo is used, i.e., spondyloepiphyseal dysplasias. Where the skull is involved, the prefix cranio is used. The limitations of morphology are great, but radiographic examination of the skeleton remains by far the most frequent initial approach and is often the final resolution of the diagnosis.

Skeletal dysplasias are of particular interest to Paediatricians, Radiologists, Orthopaedic Surgeons, Histopathologists and Geneticists. Because of the presence of deformities, the involvement of the Orthopaedic Surgeon is important for the management of patients with congenital skeletal dysplasias. In order to emphasize the importance of Orthopaedic Surgeons in the management of the respective disorders these disorders are specially identified. Furthermore, it is most important to be aware that most patients with these disorders have a normal intelligence. In spite of all efforts it may be difficult to separate them into well-defined categories as in many cases the features often overlap. The orthopaedic community must stay informed about the understanding of the genetic and molecular basis of these diseases. It is essential for treating clinicians who manage heritable disorders to be updated for the management of common skeletal dysplasias.

Epiphyseal and metaphyseal dysplasias are further subdivided on the basis of whether there is evidence of abnormal spinal development (e.g., spondyloepiphyseal dysplasia). The nomenclature of some skeletal dysplasias is also based on the recognition of specific radiological features of abnormal bone development, e.g., diastrophic (twisted bone) or campomelic (bent bone) dysplasia.

Dysplasias should be distinguished from underdeveloped or abnormal skeletal development which usually occur secondary to a known cause, i.e., endocrinal, nutritional or other deficiency diseases. As a result of such confusion skeletal dysplasias and other disorders of skeletal development do need careful assessment while making a diagnosis. A skeletal dysplasia should also be distinguished from a simple dystosis where there is malformation of one or more bones, which may be small or large without any change in its shape.

Clinically skeletal dysplasia often present with disproportionate short stature, deformities and with functional abnormality. On the other hand the secondary disorders of bone development may also result in proportionate short stature. Mild forms of skeletal dysplasia may, however, go unnoticed. It is, therefore, essential to take a good case history including the age of onset of symptoms, obstetric history and the family history. Appropriate radiographs must be taken and properly evaluated.

An attempt should be made to identify a given disorder on clinical, radiological and genetic grounds, the following yardstick may be followed:

1. Is the Skeletal affection generalized or localized?
2. Does any relative suffer from a similar or other congenital disorder?
3. Was the disorder present at birth?
4. Which part of the Skeleton is maximally involved?
5. Which anatomical area is affected? Epiphysis, Metaphysis or Diaphysis.
6. Is the shortening rhizomelic, mesomelic or acromelic?
7. Is there disorderly growth in the cartilage-bone zone?
8. Is there an abnormal density—sclerotic or lytic?
9. Is there a combination of skeletal along with other systems involved?
10. Any other related anomaly.

2 Skeletal Dysplasia

2.1 Genetics

Skeletal dysplasias result from disruptions in normal skeletal morphogenesis and development and are major contributors to short stature. The incidence is approximately 1/5000 births. The earlier reports reveal that the genetic causes of 56 skeletal disorders have been described. Most are the result of genetic defects. Skeletal dysplasia can present any time from prenatal period to adult life.

The correct and early diagnosis of these disorders is important for the management of complications and for future genetic counselling. It is important to be familiar with the clinical and radiological features of common dysplasias. They frequently present in the newborn period with disproportion, radiological abnormalities and occasionally other organic system abnormalities. Each syndrome usually involves different combination of symptoms, which can range from mild to severe. Because of the diversity, the diagnosis of a skeletal dysplasia is usually based on a combination of clinical, radiographic, morphologic and in some instances biochemical and molecular studies. This progress has provided us with more insights into the genes controlling normal skeletal development. The radiologist plays a major role in making an accurate diagnosis. However, the representative of other disciplines, including paediatricians, molecular biologists and pathologists can also provide important diagnostic information.

Prenatal diagnosis of skeletal dysplasia can be made as early as 14 weeks. However, despite recent advantages in imaging, foetal skeletal dysplasias are difficult to diagnose in utero due to a number of factors. Some of such disorders are lethal. Foetal radiographs can help in reaching a diagnosis or at least in identifying the probable lethal disorders. Many of the genetic disorders of skeletal development lead to significant morbidity and mortality in utero or in the early neonatal period. A basic knowledge of skeletal development and a structured, comprehensive approach to the history, physical examination and interpretation of radiographic studier are crucial. Most of the publications are single-patient case report describing

S. S. Yadav, *Skeletal Dysplasias & Congenital Malformations*,
https://doi.org/10.1007/978-981-99-5679-1_2

the clinical and radiological features of common dysplasias. We have tried to report the clinical and radiographic findings of common non-lethal skeletal dysplasias.

Although significant advances have been made in the elucidation of genetic defect, a lot more is essential in the correct diagnostic assessment and the management of such anomalies. An increasing variety of medical and surgical treatment options can be offered to affected individuals to try to improve their quality of life and lifespan. Research in the field is truly spectacular and these advances hold great promise for diagnostics, risk prediction and therapeutic design.

2.2 The Prevalence of Skeletal Dysplasias

Attempts have been made to estimate the number of living people with skeletal dysplasias (Osteochondrodysplasias). However, there are no figures available for their prevalence. In a country like ours where more than 70% of people live in rural areas, it is almost an impossible job. Moreover, the present day Orthopaedic Surgeons may not have much inclination towards the subject. The two disorders most regularly diagnosed in the perinatal period are achondroplasia and osteogenesis imperfecta.

Some of the common difficulties in their diagnosis are listed below:

1. Mildly affected patients may never attend hospital.
2. Early death may have occurred.
3. The Modern Medical team may not be able to diagnose.
4. In certain dysplasias the signs and symptoms do overlap.
5. It is not a lucrative field of medicine.

Although skeletal dysplasias are individually rare, the collective total may be substantial. It is, therefore, expected that a category of orthopaedic surgeons should gain some knowledge about this forgotten ailment.

However, we must recognize that many individual cases do not fit easily into a specific diagnostic category and try to avoid the temptation to make them fit. Therefore, unclassifiable skeletal disorders, presumably dysplastic in nature, can be so designated with the modifying phrase ‘resembling’.

3 Skeletal Dysplasia

3.1 Classification

It is mainly because of overlapping that it is difficult to classify the Skeletal Dysplasias. An attempt has been made to classify them in as simple fashion as possible. Attempt has been made to reveal the major clinical, radiological features of common and classical disorders of skeletal development. These have been divided into two:

1. According to the regions involved.
2. According to the tissue involved.

3.1.1 According to the Regions Involved

1. Chondroplasia with Epiphyseal involvement
    - Multiple Epiphyseal Dysplasia
    - Chondroplasia Punctata
2. Chondroplasia with Metaphyseal involvement
    - Achondroplasia
    - Hypochondroplasia
    - Metaphyseal Chondrodysplasia
    - Chondroectodermal Dysplasia
3. Chondroplasia with spinal involvement
    - Spondylo Epiphyseal Dysplasia
    - Spondylo Metaphyseal Dysplasia
    - Camptomelic Dysplasia
    - Diastrophic Dysplasia
    - Pseudoachondroplasia

S. S. Yadav, *Skeletal Dysplasias & Congenital Malformations*,
https://doi.org/10.1007/978-981-99-5679-1_3

4. Chondroplasia involving specific anatomic sites
   Cleidocraneal Dysplasia
   Mesomeilic Dysplasia
   Acromesomeilic Dysplasia
   Acromeilic Dysplasia
5. Chondroplasia involving Tumour like formation on cartilage
   Hereditary Multiple Exostosis
   Enchondromatosis
   Dysplasia Epiphysealis Hemimelica
6. Disorders of Skeletal Development with abnormal connective tissue formation
   Osteogenesis Imperfecta
   Marphan's Syndrome
   Ehlers-Danlos Syndrome
7. Disorders of Skeletal Development due to Lysosomal storage diseases
   Mucopolysaccharidosis
   Hurler's Disease
   Hunter's Disease
   Morquiou's Disease
   Gaucher's Disease
8. Disorders of Skeletal Development associated with Osteoscalerosis
   Osteopetrosis
   Pycnodysostosis
   Melorheostosis
   Progressive Diaphyseal Dysplasia
   Osteopoikilosis
   Osteopathia striata
9. Developmental conditions Associated with Ectopic Ossification
   Fibrodysplasia Ossificans Progressiva
   Progressive Osseous Hetroplasia
10. Other abnormalities of skeletal development
   Neurofibromatosis
   Congenital Pseudoarthrosis
   Polyostotic Fibrous Dysplasia
   Larsen's Syndrome
   Congenital Dislocation of Hip
   Slipped Upper Capital Epiphysis
   Idiopathic Osteolysis

Classification (Conti.)

3.1.2 According to the Tissue Involved

3.1.2.1 Dysplasias Due to Disturbances in Chondroid Production

1. **Abnormal maturation of growth plate Chondroblasts**
   (a) The Mucopolysaccharidoses (Chondro-Osteodystrophy)
   (b) Idiopathic
   - Achondroplasia
   - Cartilage-hair Hypoplasia
   - Metaphyseal dysostosis
2. **Heterotopic proliferation of epiphyseal Chondroblasts**
   (a) Enchondromatosis (Dyschondroplasia; Ollier Disease)
   (b) Osteochondromatosis (Hereditary Multiple Exostoses)
   (c) Epiphyseal hyperplasia (Dysplasia Epiphysealis Hemimelia)

3.1.2.2 Dysplasia Due to Disturbances in Osteoid Production

1. **Abnormal Epiphyseal Ossification**
   (a) Diastrophic Dwarfism
   (b) Spondyloepiphyseal Dysplasia
   (c) Multiple Epiphyseal Dysplasia (Fairbank)-including Blount's and Pseudoachondroplasia
   (d) Stippled Epiphyseal Dysplasia
2. **Abnormal Metaphyseal and Periosteal Ossification**
   (a) **Deficient Osteoid Production**
   - Osteogenesis imperfecta

   (b) **Excessive Osteoid Production or Deficient Osteolysis**
   - Osteopetrosis (Albers-Schonberg Disease)
   - Pyknodysostosis
   - Metaphyseal Dysplasia (Pyle Disease)
   - Diaphyseal Sclerosis
   - Melorheosteosis
   - Osteopathia Striata
   - Osteopoikilosis

   (c) **Abnormal Osteoid Production**
   - Polyostotic Fibrous Dysplasia
   - Neurofibromatosis
   - Congenital Pseudarthrosis

3.1.3 Miscellaneous Dysplasia

(a) Dyschondrosteosis (Madelung)
(b) Marfan's Syndrome
(c) Larsen's Syndrome
(d) Apert's Syndrome
(e) Cleidocranial Dysostosis
(f) Chondroectodermal Dysplasia

3.2 The Possible Stage of Identification

1. **Defects identifiable at birth**
   Achondrogenesis
   Firbrochondrogenesis
   Thanatophoric dysplasia
   Short rib-polydactyly syndrome
   Chondrodysplasia punctata
   Camptomelic dysplasia
   Achondroplasia
   Spondyloepiphyseal dysplasia congenita
   Mesomelic skeletal dysplasias
   Acromesomelic dysplasia
   Cleidocranial dysplasia
   Larsen's syndrome
2. **Defects identifiable in later life**
   Hypochondroplasia
   Pseudoachondroplasia
   Metaphyseal Chondrodysplasias
   Spondylometaphyseal dysplasia tarda
   Multiple epiphyseal dysplasia
3. **Disorganized development OFC and fibrous components of the skeleton**
   Dysplasia epiphysealis hemimelica
   Hereditary multiple exostosis
   Enchondromatosis (Ollier disease)
   Maffucci syndrome
   Metachondromatosis
   Polyostotic fibrous dysplasia (McCune-Albright disease)
   Monostotic fibrous dysplasia
   Neurofibromatosis
   Infantile myofibromatosis
4. **Defects of density of cortical diaphyseal structure and metaphyseal modelling**
   Osteogenesis Imperfecta
   Osteopetrosis
   Pycnodysostosis
   Osteopoikilosis (spotted bones)
   Osteopathia striata
   Melorheostosis
   Metaphyseal dysplasia (Pyle disease)
   Diaphyseal dysplasia (Camurati-Engelmann)
   Mixed sclerosing bone dystrophies

5. **Dysplaisa associated with heterotopic bone formation**
    Fibrodysplasia ossificans progressiva
    Progressive osseous heteroplasia
6. **Other developmental abnormalities**
    Idiopathic Osteolysis
    Other articular hypermobility syndromes

Suggested Readings

Rubin P. Dynamic Classification of Bone Dysplasia. Year Book Medical Publishers; 1964.

Beighton P. Inherted disorders of the skeletal. Churchill Livingstone; 1980.

Bassett GS. Orthopaedic aspects of skeletal dysplasia. Instruct Course Lecture. 1990;39:381.

Beighton P, editor. McKusick's inheritable Disorders of connective Tissue. C.V: Mosby; 1992.

Bassett GS. The osteochondroplasias. In: Morrissy RT, Weinstein SL, editors. Pediatric Orthopaedics. Philadelphia: Lippincott-Raven; 1969. p. 203–49.

Athanasou NA. Pathological basis Orthopaedic and Rheumatic disease. London: Edward Arnold. Beighton, P. (ed.). 2001.

Hall B. Approach to skeletal dysplasia. Pediatric Clin North Am. 1992;39:279.

Spranger JW, Langer LO, Wiedemann HR. 1974. Bone Dysplasias. An Atlas constitutional disorder of skeletal development. W.B: Saunders; 1974.

Kopits SE. Orthopaedic complications of dwarfism. Clin Orthop. 1976;114:153.

Lovell and Winter's, Raymond TM and Stuart LW. Pediatric Orthopaedics. (Vol-I, fourth Ed.) Lippincott-Raven; 1996.

Wynne-Davies, R, Hall, CM. and Apley, A.G; 1985.

Wynne-Davies R, Gormley J. The prevalence of skeletal dysplasias. An estimate of their minimum frequency and the number of patients requiring orthopaedic care. J Bone Joint Surg. 1985;67B:133–7.

Anderson PR Jr, Hague MC, Congenital bone dysplasia. A clinical, radiological and epidemiological survey. J. Med. Genet. 1989;26:37–44.

Pope FM, Smith R. Color Atlas of inherited connective tissue disorders. Mosby-Wolfe; 1995.

Dietz FR, Matthews KD. Current concepts review. Update on the genetic basis of disorders with orthopaedic manifestation. J. Bone Joint Surg. 1996;78A:1583–98.

Beighton P, Giedion A, Gorlin R, et al. International classification of osteochondrodysplasias. Am. J. Med. Genet. 1992;44:223–9.

Rimoin DL, Silberberg R, Hollister DW. Chonodro-osseous pathology in the chondrodystrophies. Clin. Orthop. 1976;114:137–52.

Frantz CH, O'Rhailly R. Congenital skeletal limb deficiencies. J Bone Joint Surg Am. 1961;43(8):1202.

Kiss S, Vizkelety T, Kollow K, Terebessy T, Hollnapy G, Szoke G. Common bone dysplasia and malformations. Colour atlas of clinical orthopaedics: Springer; 2009.

Miklos S, Franklin HS. Color Atlas of Clinical Orthopaedics, Springer-Verlag Berlin Heidelberg 2009.

4 Chondrodysplasia with Predominantly Epiphyseal Involvement

4.1 Multiple Epiphyseal Dysplasia

Multiple Epiphyseal Dysplasia **(Dysostostic Epiphysialis Multiplex, Fairbank's Disease)** is characterized by disturbance of endochondral ossification involving various epiphyses. The child may have a short stature and waddling gait due to coxa-vara. The joint of wrist and hand may be abnormal along with short fingers. Patients have short stubby digits of the hands and feet (Fig. 4.1a–e). At the proximal femoral epiphysis the anomaly may simulate Perthes' disease. There is enlargement of the epiphyses and sometimes flaring of the diaphyseal ends. There is an irregularity of density and contour of epiphyses, stiffness of joints, difficulty in walking, flattening of the femoral and humeral heads and tilting of the ankle joints into valgus. The fibula is relatively long and may subluxate or dislocate at the upper or lower ends (Fig. 4.2a–c). There may also be spontaneous dislocation of the head of the radius.

The lower extremity is commonly involved with pain and limping. The epiphyses get fragmented and become irregular resulting in mild to moderate shortening of the tubular bones. In with the passage of the time there may be evidence of early osteoarthrosis (Fig. 4.3a–d).

S. S. Yadav, *Skeletal Dysplasias & Congenital Malformations*,
https://doi.org/10.1007/978-981-99-5679-1_4

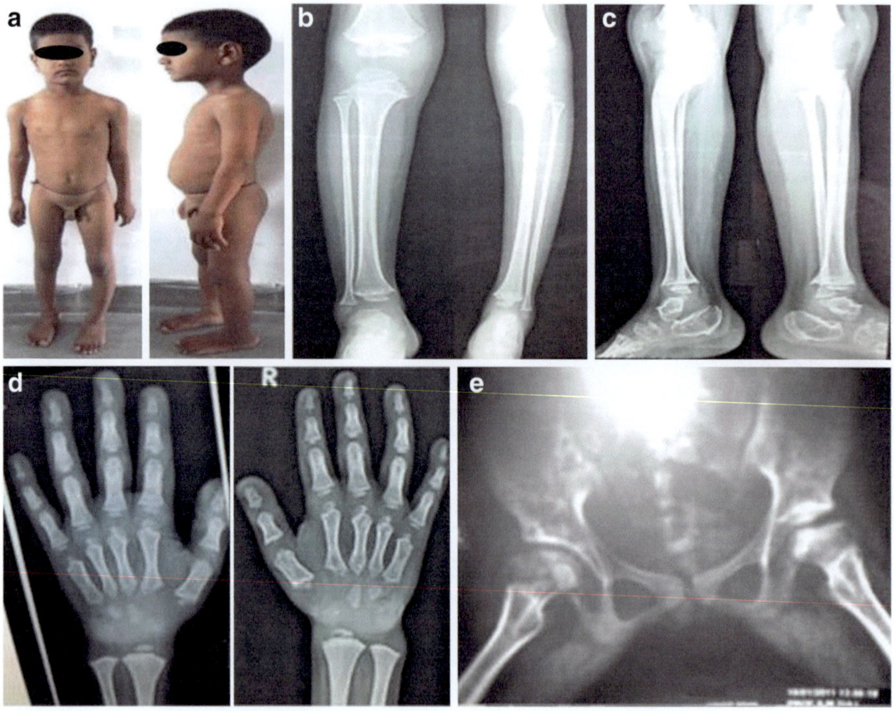

Fig. 4.1 (**a**) An obviously normal looking child has mild genu-vara on the left side. (**b**) Roentgenograms reveal evidence of multiple epiphyseal involvements at the ends of long bones. The lower end of the fibula has subluxated downwards. (**c**) Note the tibio-talar slant. (**d**) Roentgenogram of the hands revealing growth disturbances of the carpal bones. Multiple epiphyseal anomalies may be noted. (**e**) Radiological appearance of the femoral heads showing bilateral flattening

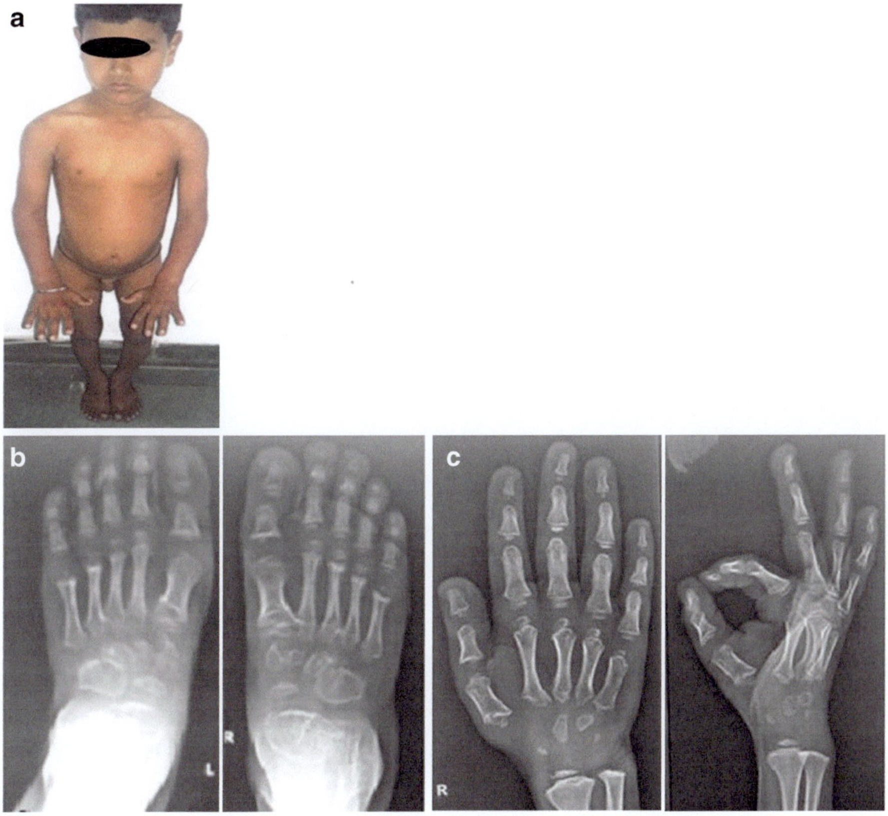

Fig. 4.2 (**a**) Clinical picture reveal short hands and feet. (**b**) Roentgenogram reveals short tubular bones of feet without significant deformity. (**c**) Roentgenogram shows short metacarpals with epiphyseal changes

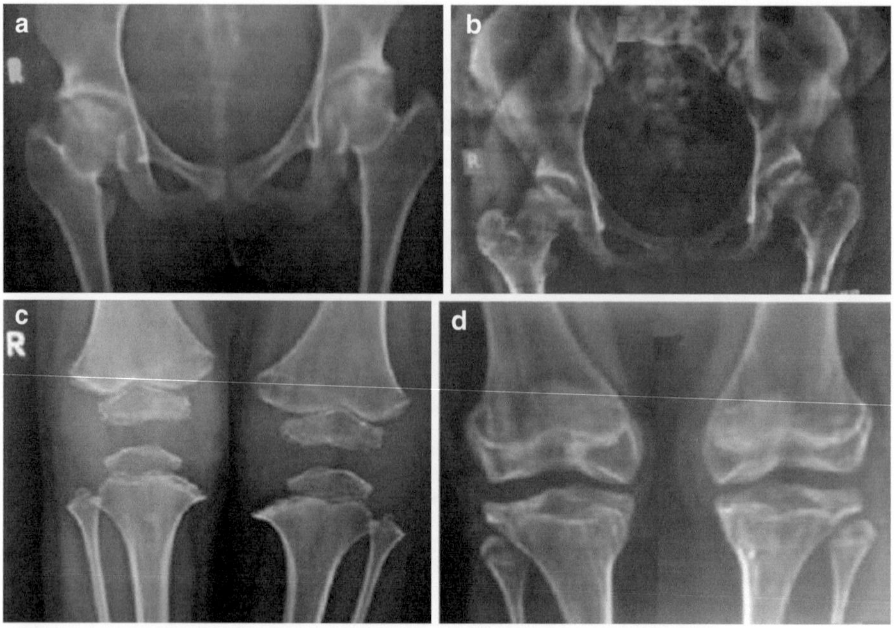

Fig. 4.3 (**a**) Roentgenogram of the hips in an older patient appears to simulate Parthe's disease. (**b**) Multiple epiphyseal dysplasia in a 13-year-old girl showing irregular femoral heads akin to bilateral Perthes' disease. (**c**) Note the epiphyseal flattening at the lower end of the femur and upper end of the tibia. (**d**) Note the articular changes in the knees with irregular epiphyses which has resulted in early osteoarthrosis

4.2 Chondrodysplasia Punctata (Chondrodysplasia Calcificans Congenita or Stippled Epiphyses)

Chondrodysplasia punctata **(stippled epiphyses)** is a rare cartilage dysplasia revealing multiple punctate areas of calcification in an unossified epiphyseal cartilage. This lethal disorder is characterized by symmetric rhizomelic shortening of the limb. Joint contractures are also observed. The face is flat with depression of the nasal bridge. Radiologically there are symmetrical multiple irregular opacities present in ends of the long bones (Fig. 4.4a–c). Vertebral anomalies may be noted in the form of scoliosis.

This uncommon condition is characterized by multiple punctuate opacities appearing in the un-ossified epiphysis cartilage at birth. There is a milder form which is not noted at birth. However, great variations may occur. Dwarfism is absent and the stippled areas merge into normal-appearing epiphyseal osseous centres. Great variability occurs in these cases. There may be calcification of the sternal margins and the respiratory cartilage.

An 18-year-old girl was confined to bed since birth. She presented with roentgenograms revealing multiple punctate areas of calcification. She was diagnosed as

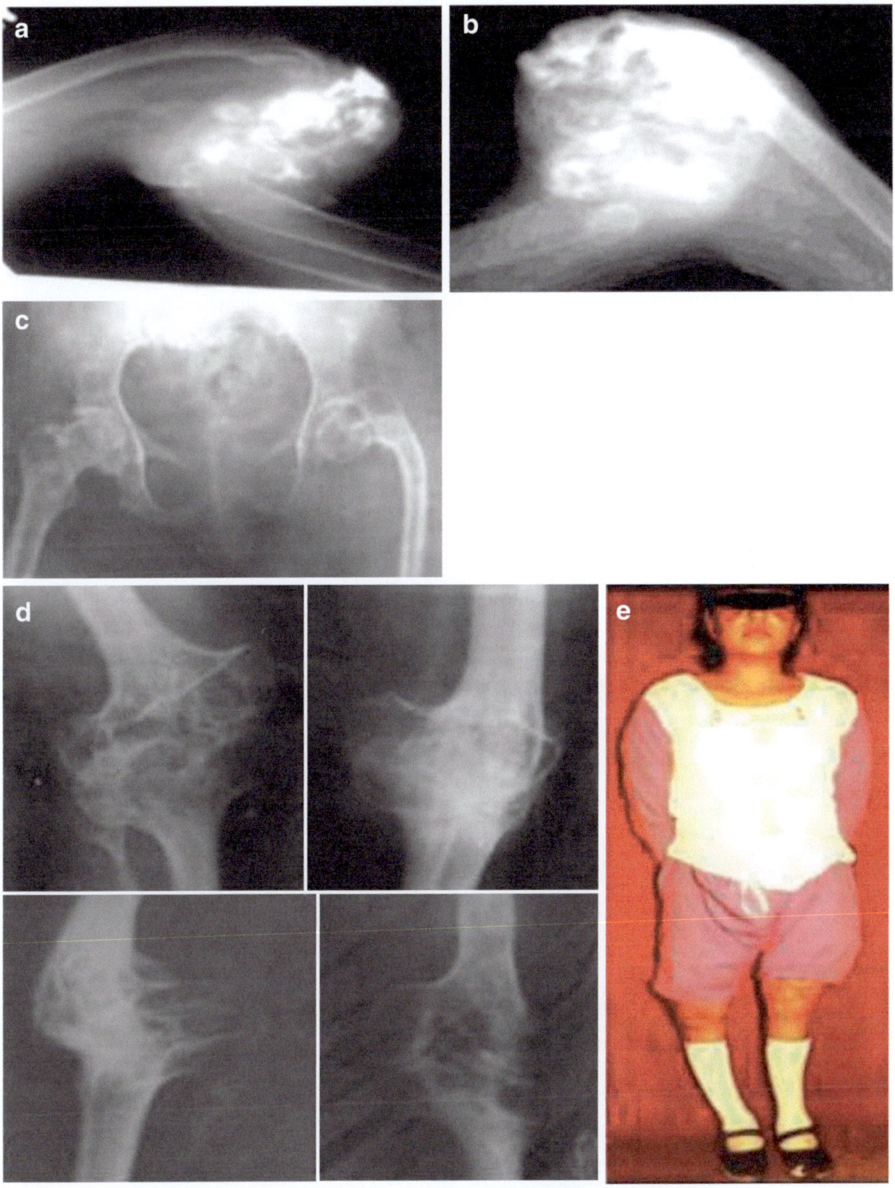

Fig. 4.4 (**a**, **b**) Roentgenogram of a rare case revealing extensive flocculent appearance and excessive calcific stippling of secondary ossification centres of lower end of the femora and upper end of the tibiae in both the knees. (**c**) Roentgenogram reveals hypoplasia of the left femur with severe coxa-vara. (**d**) Roentgenogram shows bilateral knee fusion to provide stability at the knees. (**e**) The patient stands after arthrodesis of both the knees. Note the depression of the nasal bridge

a case of Chondrodysplasia Punctata with gross deformities of the knees. Under the circumstances a difficult decision was taken to arthrodese both the knees. After a span of 1 year, the knees revealed fusion and the girl could stand and walk with support (Fig. 4.4d, e).

4.3 Kniest Dysplasia

Kniest dysplasia is a newly defined chondrodystrophy and is characterized by disproportionate short trunk dwarfism, kyphoscoliosis and significant joint stiffness and contractures. Patients may have flat faces, cleft palate, myopia and hearing loss. The skeletal changes can be observed in infancy. There may be clubfoot deformity with various types of hernia. Intelligence is normal. Along with disproportionate short trunk dwarfism there is rhizomelic involvement of the extremity. There may be stiffness of the joints of the fingers.

Radiologically there is osteoporosis along with atlantoaxial instability due to abnormalities of the odontoid. There may be platyspondyly along with anterior wedging. The metaphyses are expended and flared. The epiphyses, once ossified, show cloud-like effect on both sides of the epiphyseal line (Fig. 4.5a–c). Long bones are dumbbell shaped in the epiphyseal and metaphyseal regions about the hip knees

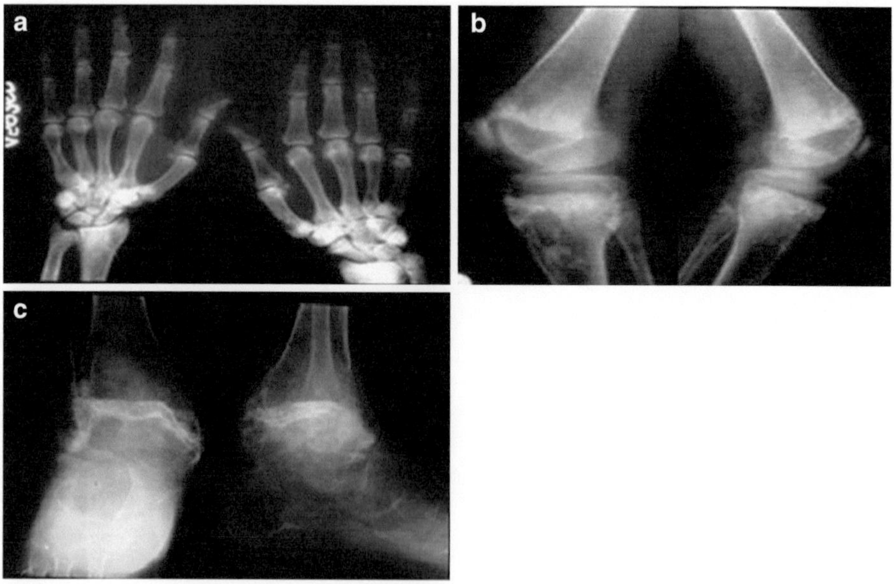

Fig. 4.5 (**a**) Roentgenogram of a 54-year-old male with kniest dysplasia. Note the flocculent calcifications within the epiphyseal and metaphyseal regions of the hand. (**b**) Roentgenogram shows extensive flocculent calcification within the epiphysio-metaphyseal regions at the distal femur and proximal tibia. Note the expended metaphyses. (**c**) Roentgenogram of the ankle reveals broadening and abnormal calcification at the distal end of the tibia and fibula. The ends of the long bones are dumble-shaped

and ankles which later show epiphyseal ossification. The joints are large and deformed.

The iliac crests are short and broad. The femoral heads ossify in late childhood and are typically deformed and subluxated. There is generalized osteoporosis and characteristic flattening and squared off epiphyses of the hands with joint space narrowing.

The differential diagnosis of kniest dysplasia includes metatropic dysplasia and spondyloepiphyseal dysplasia congenita.

Suggested Readings

Fairbank T. Dysplasia epiphysialis multiplex. Br J Surg. 1947;34:325.

Griffiths HE, Witherow PJ. Perthes' disease and multiple epiphyseal dysplasia. Postgrad Med J. 1977;53:464.

Hesse B, Kohler G. Does it always have to be Perthes' disease? What is epiphyseal dysplasia? Clin Orthop. 2003;414:219–27.

Ingram RR. The shoulder in multiple epiphyseal dysplasia. J Pediatr Orthop. 1992;12:241.

Jacobs PA. Dysplasia epiphysialis multiplex. Clin Orthop. 1968;58:117.

Spranger J. The epiphyseal dysplasias. Clin Orthop. 1976;114:46.

Andersen PE, Justesen P. Chondrodysplasia punctata: report of two cases. Skelet Radiol. 1987;16:223.

Stanescu R, Stanescu V, Muriel MP, et al. Multiple epiphyseal dysplasia. Fairbank type. Morphologic and biochemical study of cartilage. Am J Med Genet. 1993;45:501.

Apley AG, Wientroub S. The sagging rope sign in Perthes' disease and allied disorders. J Bone Joint Surg (Br). 1981;63:43.

Sheffield LJ, Dranks DM, Mayne V, Hutchinson LA. Chondrodysplasias punctata. 23 cases of a mild and relatively common variety. J Pediatr. 1976;89:916–23.

Ingram RR. Early diagnosis of multiple epiphyseal dysplasia. J Pediatr Orthop. 1992;12:241–4.

Fairbank HAT. Dysplasia epiphysealis punctata. Symptoms: stippled epiphyses, chondrodrstrophia calcificans congenital (Hunermann). J Bone Joint Surg (Br). 1949;31:114.

Lawrence JJ, Schlesinger AE, Kozowski K, et al. Unusual radiographic manifestations of chondrodysplasia punctata. Skelet Radiol. 1989;18:15.

Poulos A, Sheffield LJ, Sharp P, et al. Rhizomelic chondrodysplasia punctate. Clinical, pathologic and biochemical findings in two patients. J Pediatr. 1988;113:685–90.

Lachman RS, Rimoin DL, Hollister DW, et al. The Kniest syndrome. AJR. 1975;123:805.

Siggers DC, Rimoin DL, Dorst JP, et al. The Kniest syndrome. Birth Defects Orig Artic Ser. 1974;10:193.

Heymans HAS, Oorthuys JWE, Nelck G, et al. Rhizomelic chondrodysplasia punctate: another peroxisomal disorder. E Engl J Med. 1985;313:187.

Rittler M, Menger H, Spranger J. Chondrodysplasia punctate, tibia-metacarpal (MT) type. Am J Genet. 1990;116:88.

Treble NJ, Jensen FO, Bankier A, Rogers JG, Cole WG. Development of the hip in multiple epiphyseal dysplasia. J Bone Joint Surg Br. 1990;72:1061–104.

Chondrodysplasia with Predominantly Metaphyseal Involvement

5

5.1 Achondroplasia

Achondroplasia results in dwarfism with disproportionately short extremities. The trunk is normal, whereas the long bones of the extremities are short. The head is large with characteristic rhizomelic appearance of the extremities (Fig. 5.1a). The condition is identifiable at birth. Although the milestones are delayed the intelligence and lifespan are normal, spinal canal stenosis may be observed at various stages of life. In achondroplasia, dwarfing is presumably related to slow endochondral ossification. Bone formation since intramembranous and periosteal ossifications are normal. The spine often shows progressive kyphoscoliosis, lordosis and there may be narrowing of the vertebral canal. The pelvis is rectangular, short and wide. There is characteristic pelvic configuration (Fig. 5.1b). The femoral neck area is disproportionate in the length (Fig. 5.2a–d). The ilia are square and broad the acetabula flat with horizontal acetabular roof and the three prong pear shape of the sciatic notch. Acetabular roofs are irregular and horizontally placed (Fig. 5.2e). The long bones are short with smaller epiphyseal centres (Fig. 5.2f, g). The fibulae are relatively longer in comparison to tibia.

Radiologically the long bones are short with smaller epiphyseal centres. There is clear flaring of the metaphyses all the ends of the long bones is cupped and the small epiphyseal ossification centres fill-up the cup in a shallow ball-and-socket pattern (Fig. 5.1b). The tubular bones of the hands and feet are short and broad. Patient may have trident hands with equal length of fingers (Fig. 5.3a–c). The lumbar spine may reveal relatively smaller interpedicular distance. The cranial vault is large. The upper and lower extremities are very short in all the patients and the shortening is rhizomelic that is the arms are shorter than the forearms and the thighs are shorter than the legs. The tubular bones of the legs are bowed.

S. S. Yadav, *Skeletal Dysplasias & Congenital Malformations*,
https://doi.org/10.1007/978-981-99-5679-1_5

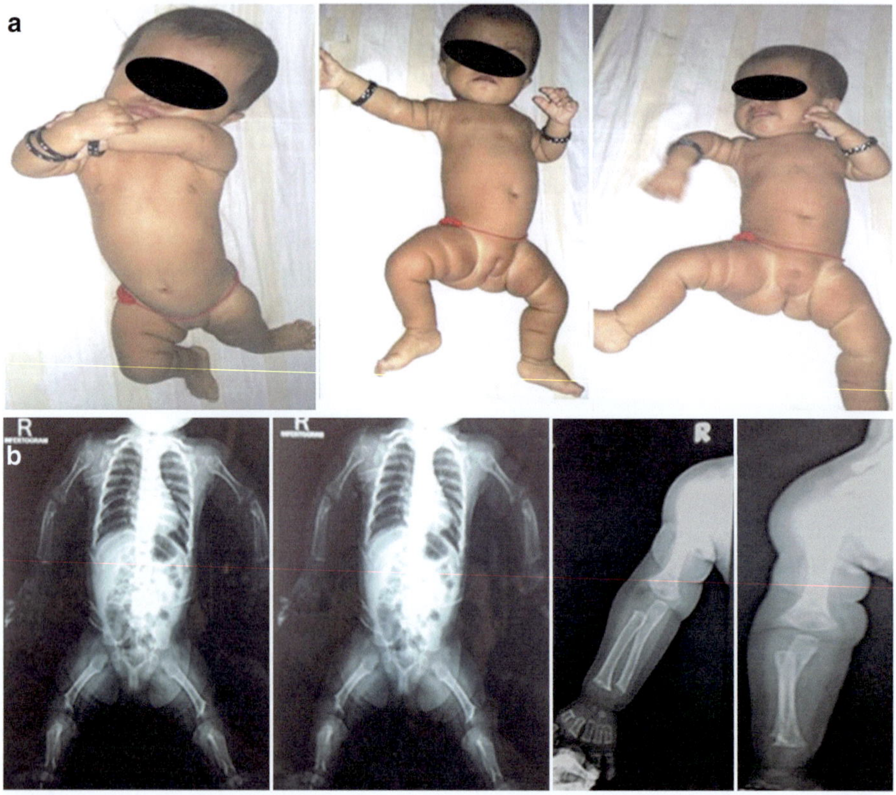

Fig. 5.1 (**a**) A 6-month-old child with achondroplasia revealing normal trunk and short extremities with skin folds and rhizomelic presentation. Note the large head. (**b**) Roentgenogram of the same patient showing normal trunk and short limbs. Note the rhizomelic shortening of the limbs

The achondroplastic is a 'Dwarfed' individual in whom:

1. Trunk length (crown to pubis) is normal.
2. Leg length (Pubis to heel) is reduced.
3. Span length (fingertip to thumbtip) is reduced.

Generally such patients do not require treatment. However, in case of spinal canal stenosis an adequate decompression may be carried out.

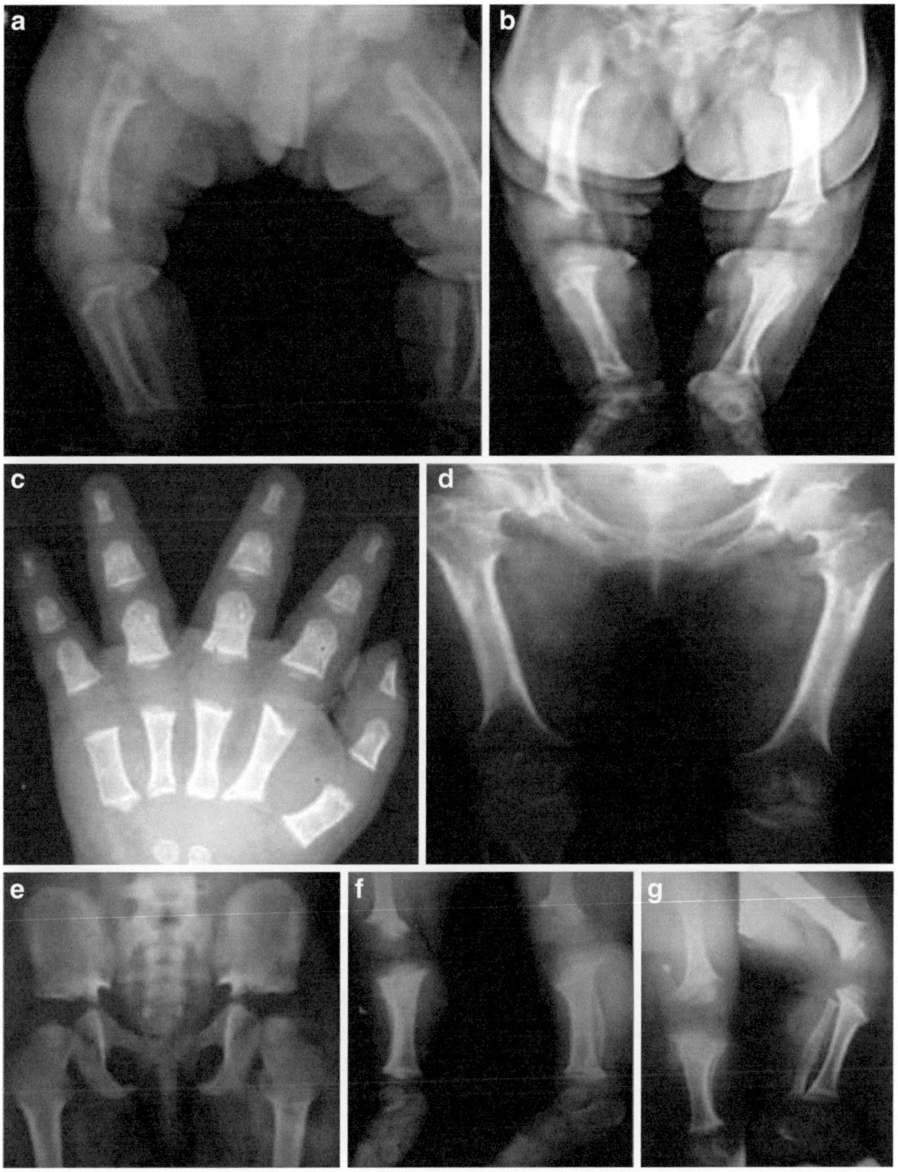

Fig. 5.2 (**a**) Roentgenograms reveal shortened diaphysis and broadened epi-metaphyses of the femur and tibia. (**b**) Note the multiple skin folds at the thigh. (**c**) Trident hand in achondroplasia showing short and thick metacarpals and phalanges. The stunting effect on the short miniature bones. (**d**) Pelvis and femora in achondroplasia -Note the marked alteration in shape and length of the femora. Note the horizontal roof of the acetabulum. (**e**) Note the characteristic pelvic configuration. The ilia are square and broad the acetabula flat with horizontal acetabular roof. Note the 'trident sign'. The three-prong pear shape of the sciatic notch. The femoral neck area is disproportionate in the length. (**f, g**) Rontgenograms of an achondroplastic knees showing sclerotic irregularities of the joint with upword and outword tilting of the femoral epiphyseal line

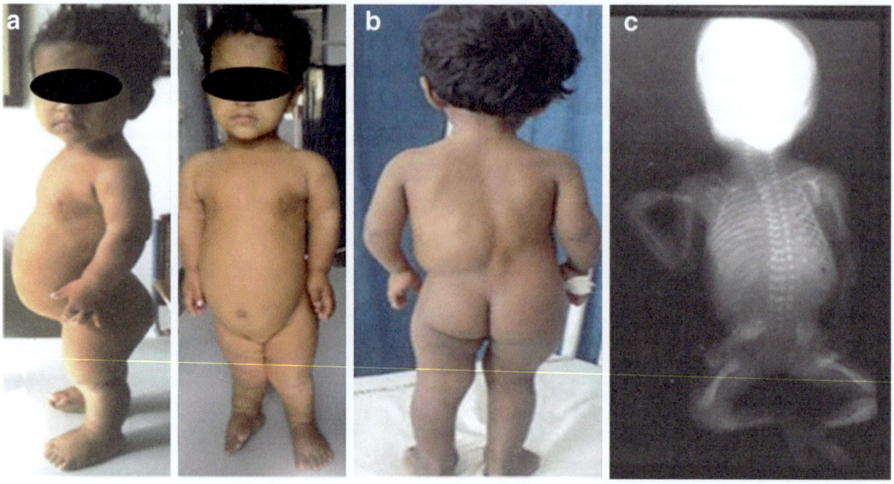

Fig. 5.3 (**a**) A typical case of achondroplasia with depression at the root of the nose. Note the rhizomelic shortening. (**b**) Posterior photograph of a 3-year-old girl with achondroplasia showing scoliosis. (**c**) Radiograph revealing classical features of achondroplasia

5.2 Hypochondroplasia

Hypochondroplasia reveals features similar to those of achondroplasia but are much milder. Often it may go unnoticed. The clinical and radiological features of this growth disturbance are less severe than achondroplasia.

5.3 Thanatophoric Dysplasia

It is a common form of lethal skeletal dysplasia. Patients reveal short-limbed dwarfism and usually die of respiratory distress. Usually head is large and there is craniofacial disproportion. The chest is narrow with hypoplastic ribs. The limbs are short and curved. Radiologically there is cupping of the metaphyses with curved and short bones of the extremities.

5.4 Metaphyseal Chondrodysplasia

Metaphyseal chondrodysplasia ***(Metaphyseal dysostosis)*** is a progressive disease resulting in growth abnormalities in the metaphyseal area. The patients are usually dwarf (Fig. 5.4a). The metaphyseal abnormalities predominantly exist in the form of bowing of the legs with shortening. There is enlargement of the metaphyses of the growing bones (Fig. 5.4b, c). There is limited joint mobility and waddling gait. Hypercalcemia and hypophosphatemia may be observed.

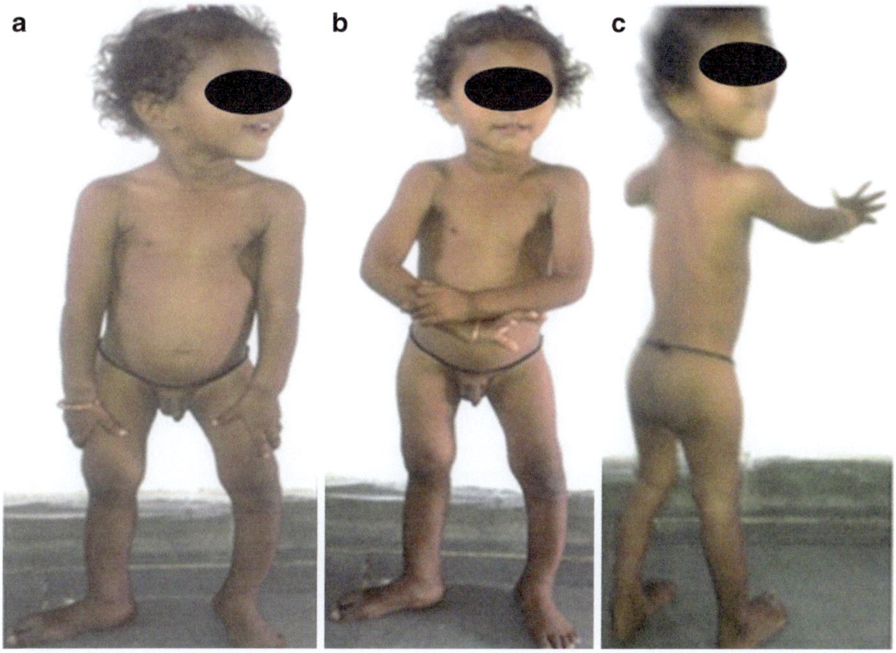

Fig. 5.4 (**a**–**c**) A 3-year-old boy with metaphyseal chondrodysplasia shows genu-vara and lateral bowing of the left leg. The joints of lower extremities and the wrists appear broad. The spine appears normal

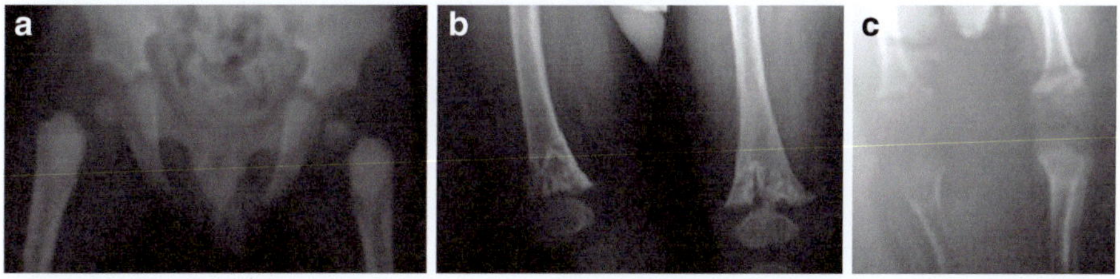

Fig. 5.5 (**a**) Radiograph reveals early coxa-vara and metaphyseal sclerosis. (**b**) Roentgenogram of the knee reveals metaphyseal chondrodysplasia with straight lines, parallel to the long axis of the bone. (**c**) Roentgenogram reveals the lower end of the femur and upper end of the tibia with metaphyseal flaring

Radiologically the growth plate is expended, cupped, irregular along with patchy mineralization. There may be coxa-vara and presence of straight lines in the metaphyseal area which are parallel to the long axis of a tubular bone (Fig. 5.5a–c). There is osteopenia and rickets like changes (Fig. 5.6a, b). There may be fragmentation of the metaphyseal zone and cystic changes may be observed. There may be hyperostosis of the base of the skull. The terminal phalanges may be hypoplastic (Fig. 5.7a–c). The epiphyseal ossification centres are smooth and appear normal.

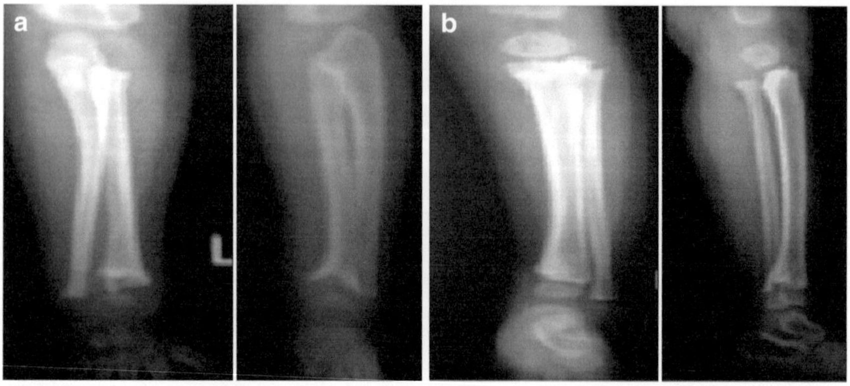

Fig. 5.6 (**a**) Roentgenogram of the forearm revealing metaphyseal changes at the wrist. (**b**) Roentgenogram of the leg showing metaphyseal changes in the tibia. Note the distal end of the fibula is migrating distally resulting in minimal varus at the ankle

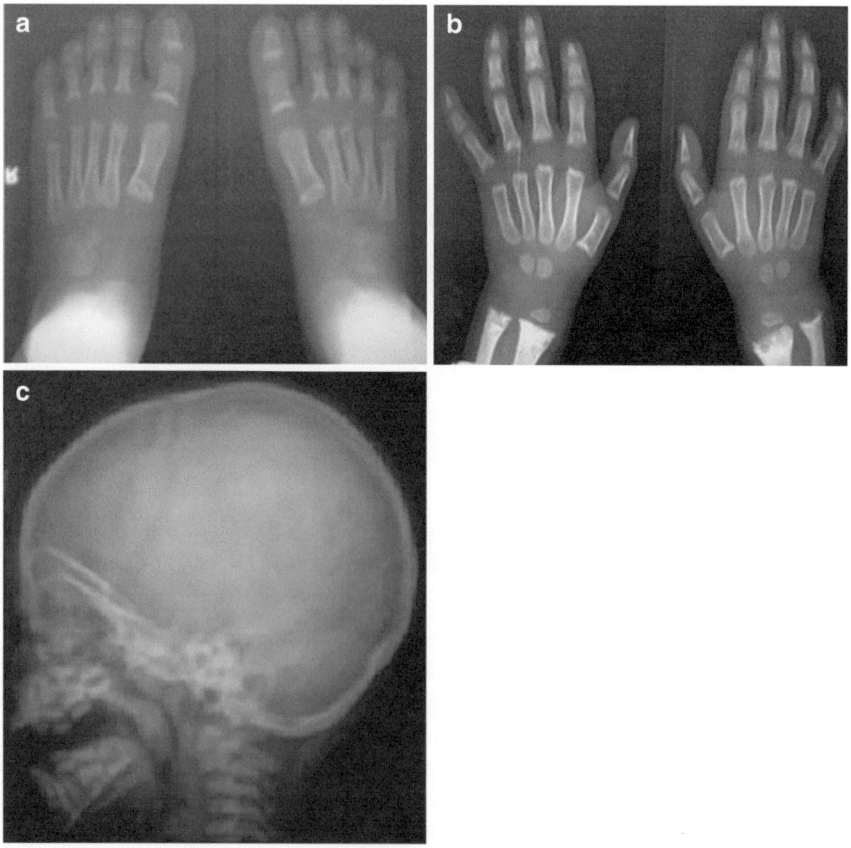

Fig. 5.7 (**a**, **b**) Radiograph showing the stunting effect on the short miniature bones with broadening and cupping of the metaphyses. The terminal phalanges are hypoplastic. (**c**) Roentgenogram of the skull reveals thickened base. The changes are progressive and increase with age

5.5 Chondroectodermal Dysplasia

Chondroectodermal dysplasia (Ellis-van Creveld syndrome) is characterized by short-limb disproportionate dwarfism with micromelia which is more marked in the peripheral skeleton (acromelic micromelia), short stature and may be postaxial polydactyly and dysplastic finger nails. Ectodermal dysplasia involving the gums, teeth and dysplastic finger nails may also be observed. There is valgus alignment of the knees which frequently progresses producing marked knock knees. Congenital heart disease may occur in 50% cases. One third (1/3) cases are stillborn or die.

Radiologically the large tubular bones are short and have club-shaped ends. The fibulae are disproportionately shorter than tibiae. The forearm may be smaller with dislocation of the head of the radius.

Ossification in the lateral portion of the proximal tibial epiphysis and metaphysis is retarded resulting in slopping of the tibial metaphysis revealing genu valgum. The patella may be dislocated superolaterally. The iliac bones are short small and square. The tubular bones may be short in the peripheral skeleton. There may be mesomelic type of appearance. Typically the capitate and hamate bones are fused. There may be metacarpal abnormalities like fusion at the basis of metacarpals and metatarsals (Fig. 5.8). There may be syndactyly with polydactyly.

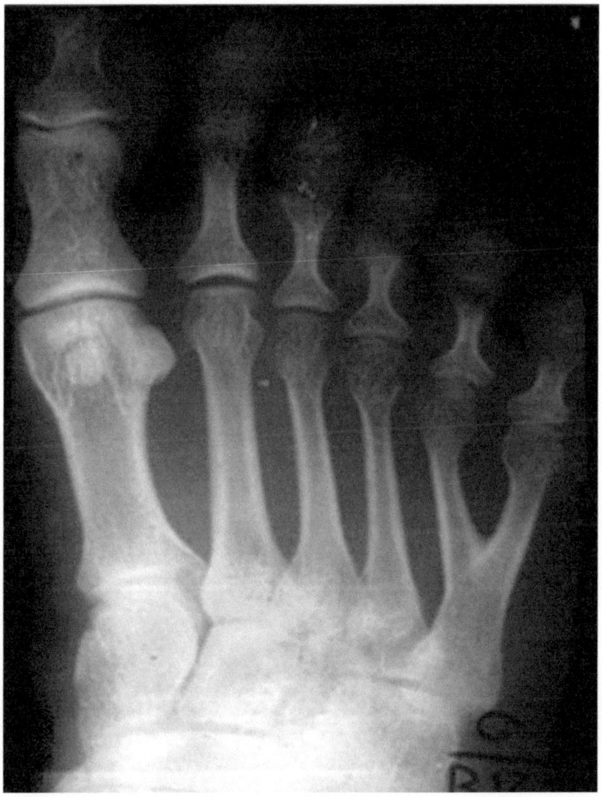

Fig. 5.8 Roentgenogram of a patient with chondroectodermal dysplasia shows fusion of the basis of the fifth and an additional metacarpal leading to polydactyly

Suggested Readings

Bailey JA. Orthopaedic aspects of achondroplasia. J Bone Joint Surg Am. 1970;52:1285.
Beals RK. Hypochondroplasia. A report of five kindreds. J Bone Joint Surg Am. 1969;51:728.
Fortuna A, Ferrante L, Santoro. Narrowing of the thoracolumbar spinal canal in achondroplasia. J Neurosurg Sci. 1989;33:185.
Hall BD, Spranger J. Hypochondroplasia: clinical and radiological aspects in 39 cases. Radiology. 1979;133:95–100.
Langer LO Jr, Baumann PA, Gorlin RJ. Achondroplasia. Am J Roentgenol. 1967;100:12.
Ponseti IV. Skeletal growth in achondroplasia. J Bone Joint Surg Am. 1970;52:701–16.
Gilbert EF, Opitz JM, Spranger JW, et al. Chondrodysplasia punctate-rhizomelic form, pathologic and radiographic studies of three infants. Eur J Pediatr. 1976;123:89.
Ponseti IV. Skeletal growth in achondroplasia. J Bone Joint Surg. 1970;52A:701–16.
Rosenbloom AI, Smith DW. The natural history of metaphyseal dysostosis. J Pediatr. 1965;66:857.
Kahanovitz N, Rimoin DL, Sillence DO. The clinical spectrum of lumbar spine disease in achondroplasia. Spine. 1982;7:137.
Sommer A, Young-Wee T, Frye T. Achondroplasia-hypochondroplasia complex. Am J Med Genet. 1987;26:949–57.
Harris R, Patton J. Achondroplasia and thanatophoric dwarfism in the newborn. Clin Genet. 1971;2:61–72.
Wynne-Davies R, Walsh K, Gormley J. Achondroplasia and hypochondroplasia. Clinical variation and spinal stenosis. J Bone Joint Surg (Br). 1981;63:508.
Wang H, Rosenbaum AE, Reid CS, et al. Pediatric patients with achondroplasia. CT evaluation of the craniocervical junction. Radiology. 1987;164:515.
Herring JA. Kyphosis in an achondroplastic dwarf. J Pediatr Orthop. 1983;3:250.
Lutter LD, Langer LO. Neurological symptoms in achondroplastic dwarfs- surgical treatment. J Bone Joint Surg Am. 1977;59:87.
Rimoin DI, Hughes GN, Kaufman RL, et al. Endochondral ossification in achondroplastic dwarfism. N Engl J Med. 1970;283:728.
Scott CI Jr. Achonroplastic and hypochondroplastic dwarfism. Clin Orthop. 1976;114:18.
Yamada H, Nakamura S, Tajima M, et al. Neurological manifestation of pediatric achondroplasia. J Neurosurg. 1981;54:49.
Nehme AM, Riseborough EJ, Tredwell SJ. Skeletal growth and development of the achondroplastic dwarf. Clin Orthop. 1976;116:8.
Carter EM, Davies JG, Raggio C. Advances in understanding etiology of achondroplasia and review of management. Orthopaedics. 2007;19(1):32–7.

Chondrodysplasia Predominantly Involving the Spine

6

6.1 Spondyloepiphyseal Dysplasia Congenita

Spondyloepiphyseal dysplasia congenita (SED Congenita) is a specific disorder that is differentiated from other SEDs including chondroplasia punctata, morquio disease, pseudoachondroplasia, and SED tarda variants. Congenital spondyloepiphyseal dysplasia is characterized by short stature associated with short trunk and spinal deformity due to growth disorders of the spine and the epiphyseal abnormalities of the limb bones evident at birth (Fig. 6.1). The principal metabolic fault appears to

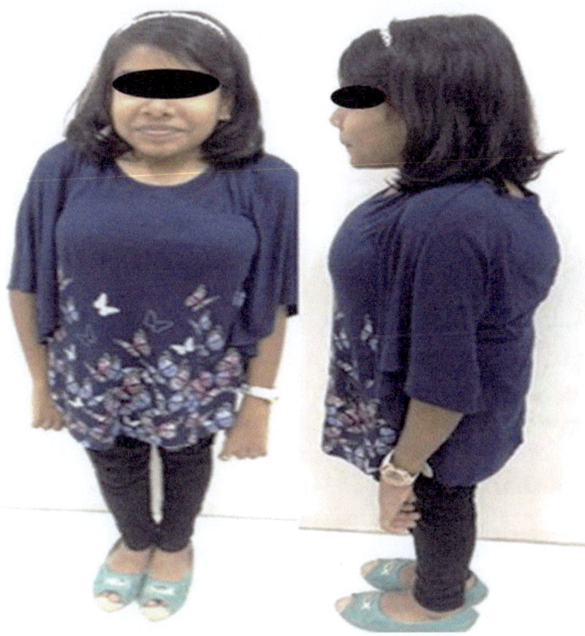

Fig. 6.1 A 14-year-old girl with Spondyloepiphyseal dysplasia congenital has short stature and severe Kyphoscoliosis

S. S. Yadav, *Skeletal Dysplasias & Congenital Malformations*,
https://doi.org/10.1007/978-981-99-5679-1_6

be inability to ossify epiphyseal centres. The face appears flat and the neck and spine are short. The neck is short and a pectus carinatum deformity occurs. The epiphyseal centres are much more irregular and often multiple with short stubby hands, vertebra plaina and saddle nose. There is severe kyphoscoliosis with other thoracic deformities. Hip dislocations and eye changes may also be observed. There may be limited extension of the elbows.

Radiologically there is kyphoscoliosis. There is evidence of platyspondyly with flattening and dorsal wedging of vertebral bodies with irregular end plates. There may be hypoplasia of the odontoid process. The short tubular bones reveal variable epiphyseal changes with delayed ossification resulting in flattening and fragmentation. There is evidence of marked hip dysplasia with coxa-vara (Fig. 6.2a–g). In later life early osteoarthritic changes are observed.

The main differential diagnosis is Morquio's disease and other congenital skeletal dysplasias. Atlantoaxial subluxation may occur and lead to spinal cord compression. There is premature osteoarthritis.

6.2 Spondyloepiphyseal Dysplasia Tarda

The skeletal changes are generally less severe and of late onset. There is moderate shortening and development of thoracic kyphoscoliosis. The chest may appear broad with severe protrusion and progressive degenerative changes in the hip and spine may result in early osteoarthrosis. Radiologically there is flattening of the vertebral bodies (Platyspondyly) and progressive kyphoscoliosis. The pelvis may be abnormal in shape. There is evidence of epiphyseal dysplasia, more so in the proximal ends of the large limb bones with evidence of flattening of the articular surface and premature development of osteoarthrosis.

6.3 Spondylo-Metaphyseal Dyschondroplasia

Spondylo-metaphyseal dyschondroplasia ***(Metaphyseal dysostosis***) reveals predominantly metaphyseal abnormalities in the form of bowing of the legs with shortening. There is enlargement of the metaphyses of the growing bones. There may be prominent short stature with unusual faces and clubfeet at birth (Fig. 6.3a–m). Hypercalcemia and hypophosphatemia may be observed. Radiologically the growth plate is expanded, cupped irregular along with patchy mineralization. There is osteopenia and rickets like changes. There may be fragmentation of the metaphyseal zone and cystic changes may be observed. There may be hyperostosis of the base of the skull. Spinal changes are seen in the form of kyphoscoliosis. There is limited joint mobility and waddling gait. Radiological changes reveal platyspondyly with tongue shaped appearance of the vertebrae.

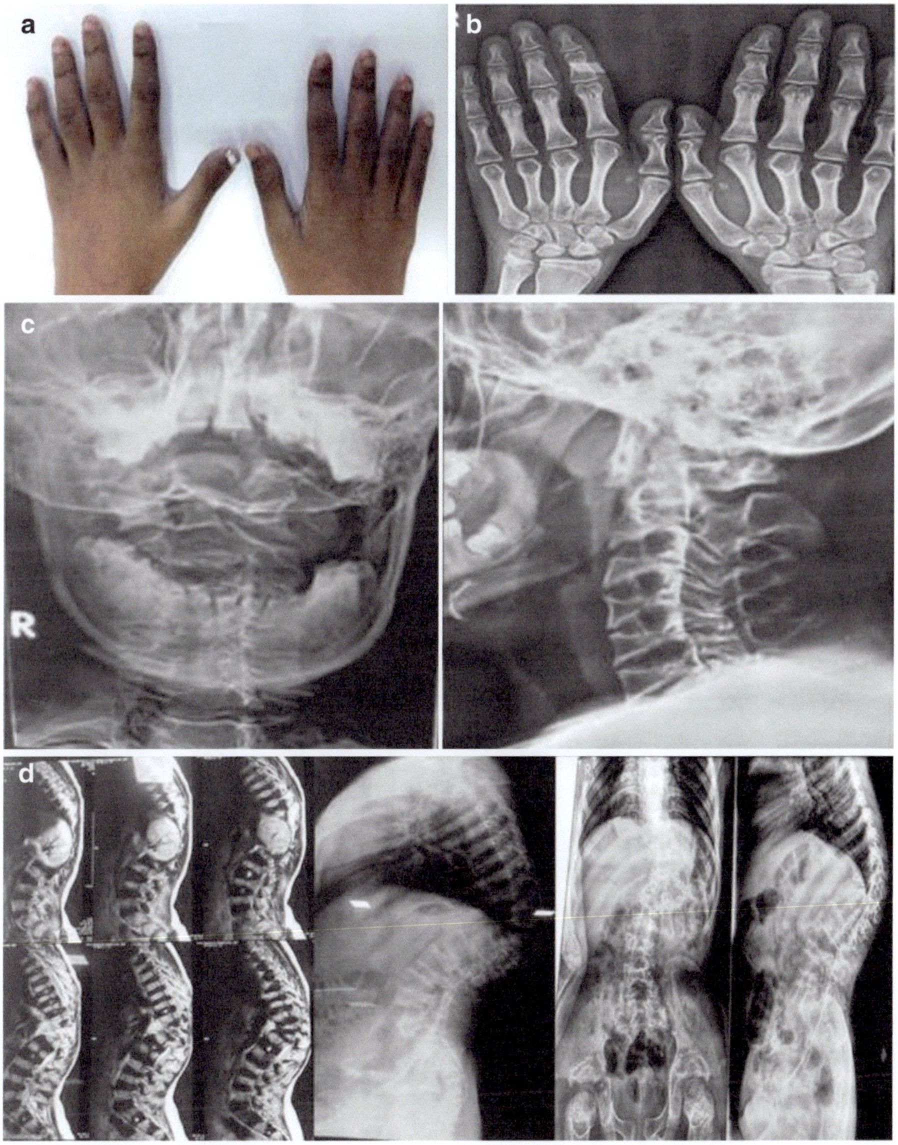

Fig. 6.2 (**a**) The hands reveal stubby fingers. (**b**) Roentgenogram shows short tubular bones of the hands with minimal broadening and shortening. The heads of the metatarsals are usually flat. The ossification centres of the carpals are delayed. (**c**) The same patient has severe posterior wedging of the cervical vertebrae and hypoplastic odontoid. (**d**) Roentgenogram of the dorso-lumbar spine reveal severe Kyphoscoliosis and increased lumbar lordosis along with generalized platyspondyly with anterior tongue like projections of the thoracic and lumbar vertebral bodies. (**e**) The roentgenogram of the pelvis of the same patient reveals short and small iliac crests with horizontal acetabular roofs and delayed ossification of the pubis. The ossification of capital femoral epiphysis is usually delayed resulting in coxa-vara. (**f**) The roentgenogram of a patient with SED congenita showing severe bilateral coxa-vara. Note the small and deformed heads of the femora due to delayed ossification of the capital femoral epiphyses. (**g**) Roentgenogram of the ankle and feet reveal delayed ossification centres of the tarsal bones

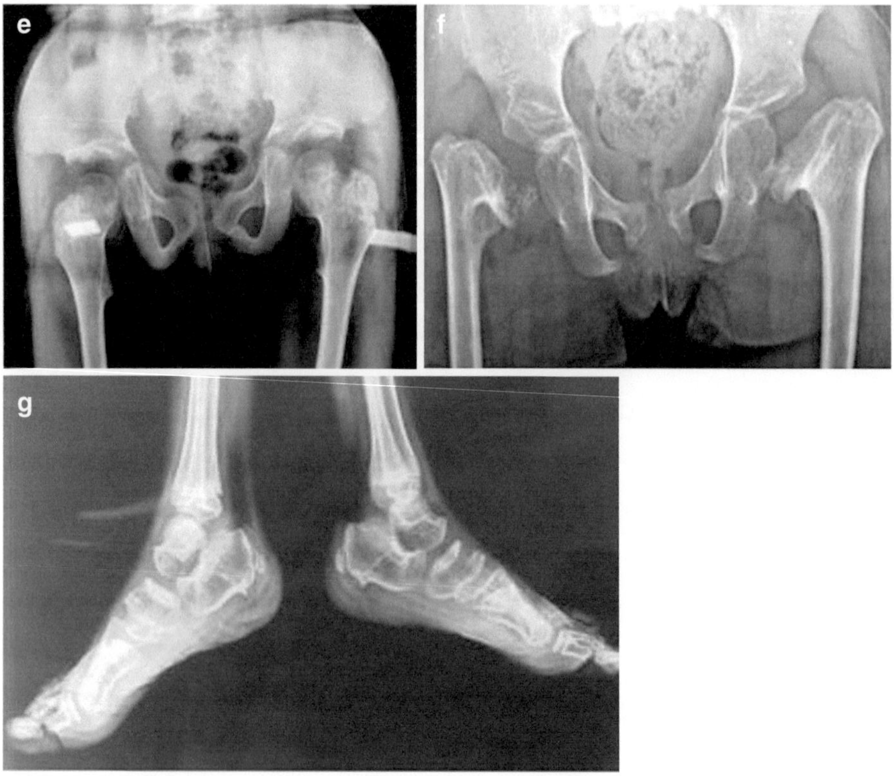

Fig. 6.2 (continued)

Fig. 6.3 (**a**) A 40-year-old dwarf male revealing broadening of the metaphyses of the long bones. The back reveals kyphoscoliosis. (**b**) The same patient reveals enlarged head with unusual face. Note the Clavicular prominence on either side. (**c**) Roentgenogram revealing sclerosis of the base of the skull. (**d**) Roentgenogram showing completely distorted pelvis due to unusual and massive chondroid tissue. The metaphyses of the proximal femora are an enlarged. However, the epiphyses appear undisturbed. (**e**) Roentgenogram reveals changes in the acetabulum and proximal femoral metaphyses with normal hip joint. (**f**) Similar changes are seen in both the ends of the radius and ulna. (**g**) Roentgenogram reveals abnormal chondroid at the proximal humeri. The movements at the shoulder are, however, normal. (**h**) Roentgenogram of the dorso-lumbar spine shows kyphoscoliotic changes with unusual presentation of the dorso-lumbar spine. (**i**) Roentgenogram of the chest revealing kyphoscoliotic changes in the spine with normal air entry in the lungs. (**j**) Roentgenogram of the leg revealing metaphyseal fraying at the either ends of tibia along with corner fractures. (**k**) Radiograph revealing broadening of both the ends of the metacarpals. The stunting effect on the short miniature bones with fused epiphyses. The terminal phalanges are hypoplastic. (**l**) Radiograph of the feet with fusiform outline of the metaphyseal expansion. The effect can be seen on the either ends of the metatarsals. The proximal end of the proximal phalynx in the left foot shows wide open physeal line. The terminal phalanges are relatively hypoplastic. (**m**) The father and the son

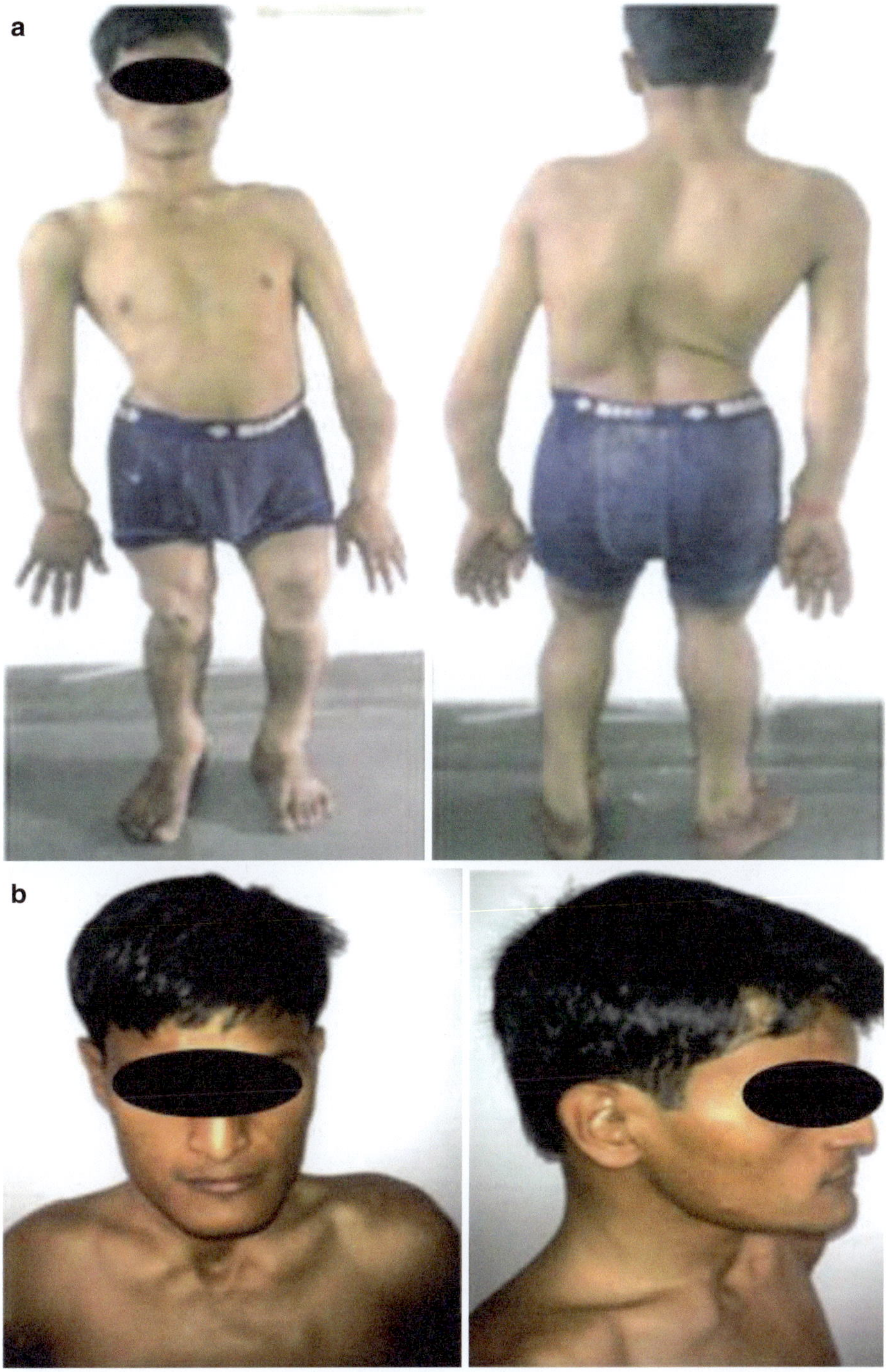
a
b

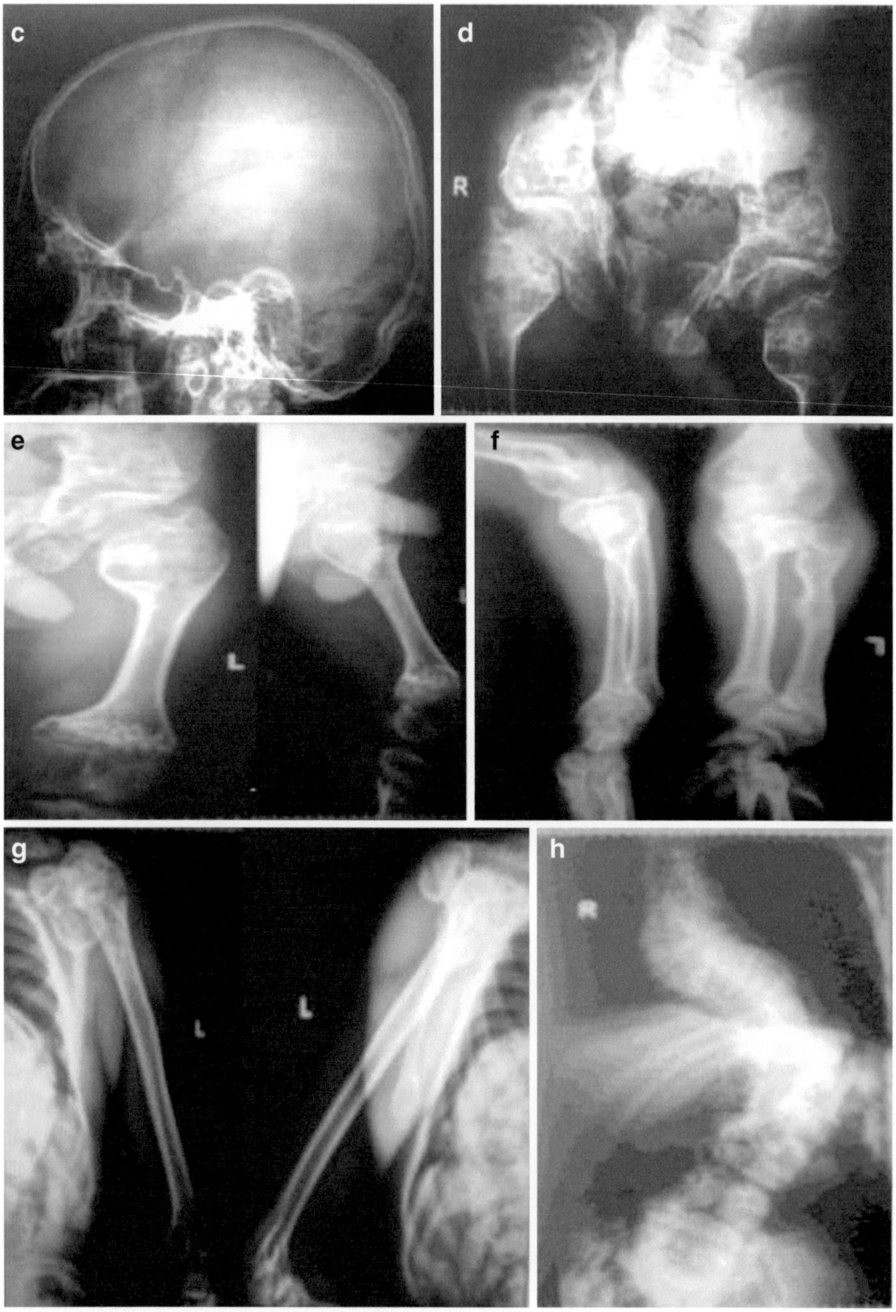

Fig. 6.3 (continued)

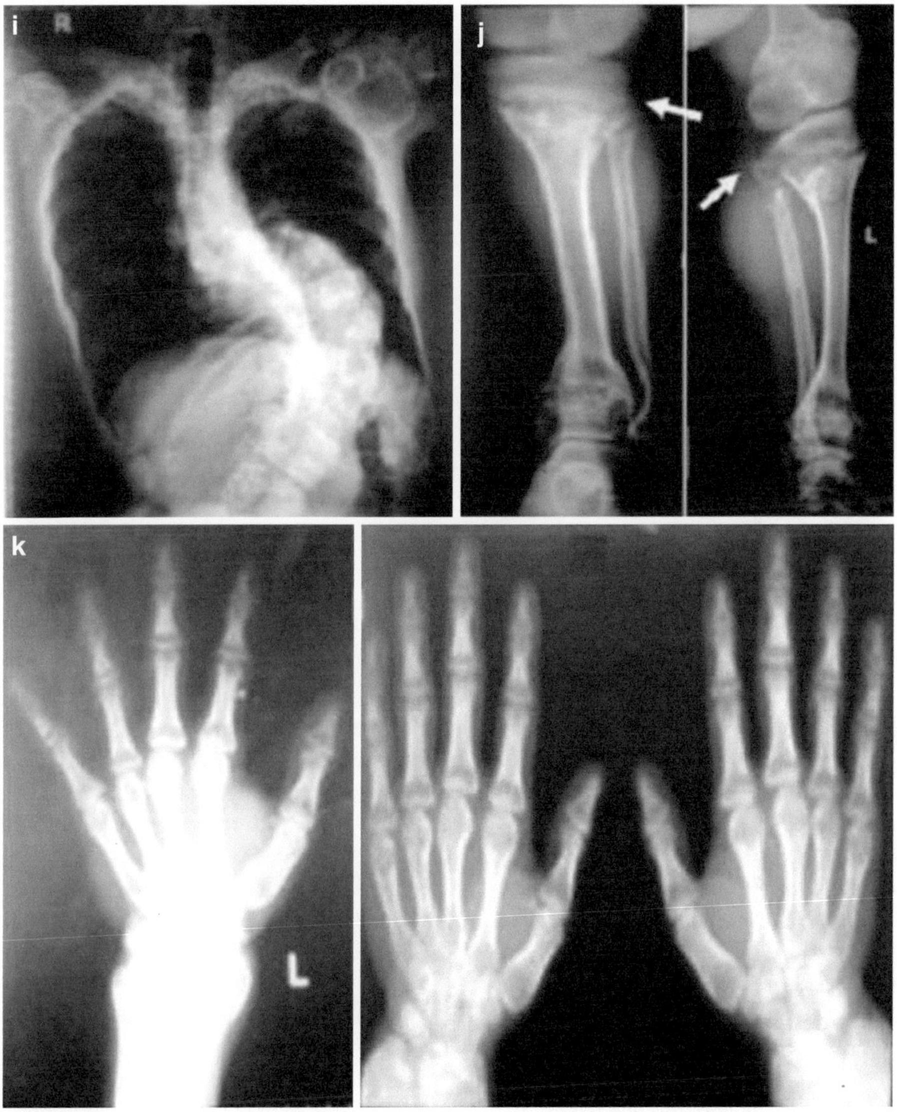

Fig. 6.3 (continued)

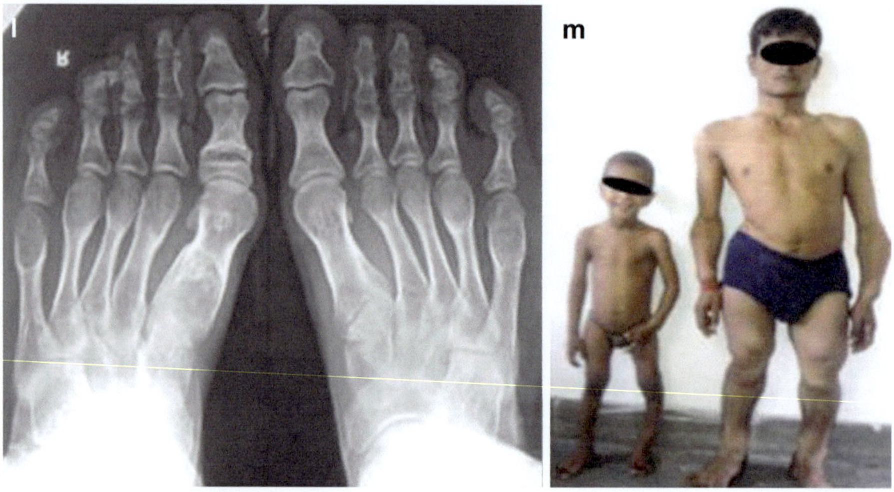

Fig. 6.3 (continued)

6.4 Diastrophic Dysplasia

Diastrophic is a Greek word meaning twist. Diastrophic dysplasia is characterized by severe micromelic dwarfism, small bones of hand and feet and the tubular bones are affected. This unusual variety of short-limb dwarfism is characterized by a triad of 'hitch-hiker' thumbs', severe foot deformities and ear calcification. Congenital dislocation of the hip may be observed. Extreme short stature with micromelia is almost diagnostic at birth. Mostly there is scoliosis and 25% of the patients may demonstrate cleft palate. The hands are broad and reveal ulnar deviation. Kyphoscoliosis may develop later in childhood. Joint movements may be limited followed by joint contracture and external ear abnormalities. Face is long with narrow nasal bridge, flaired nostrils and broad mouth withy cleft palate. However, the head is of normal size and configuration with rhizomelic pattern may exist. The intelligence is normal.

Radiologically the long bones are short and broad. The epiphyseal ossification centres appear late and are irregular flattened. There is evidence of cervical kyphosis. In 30% cases there is symphalangism of the proximal interphalangeal joints (Fig. 6.4a). The first metacarpal is usually short and ovoid (Fig. 6.4b). There may be flaring of the metaphyses with bulbous joints. The fibulae are shorter than tibiae. There may be hypoplasia of the femoral capital epiphysis (Fig. 6.5). Joint dislocations may occur. The ulnae are short and the hand may reveal ulnar deviation. There may be posterior wedging of the fifth lumbar vertebra (Fig. 6.6a, b).

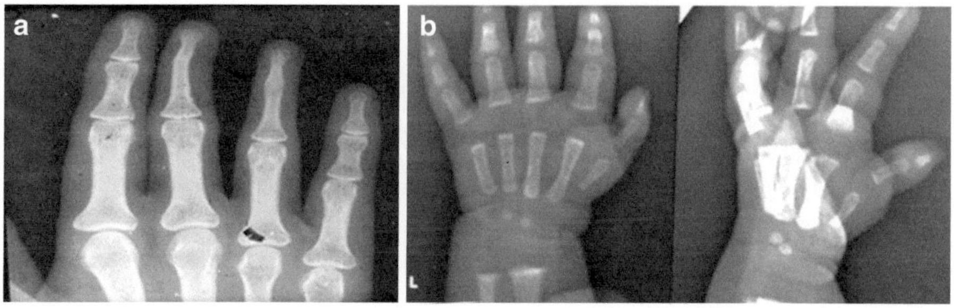

Fig. 6.4 (**a**) Congenital symphalangism of the distal phalanges of the middle and ring fingers. (**b**) Roentgenogram reveals short and ovoid first metacarpal

Fig. 6.5 The roentgenogram reveals bilateral hip deformities. Note the saucer-shaped defect of ossification and subluxation on either side

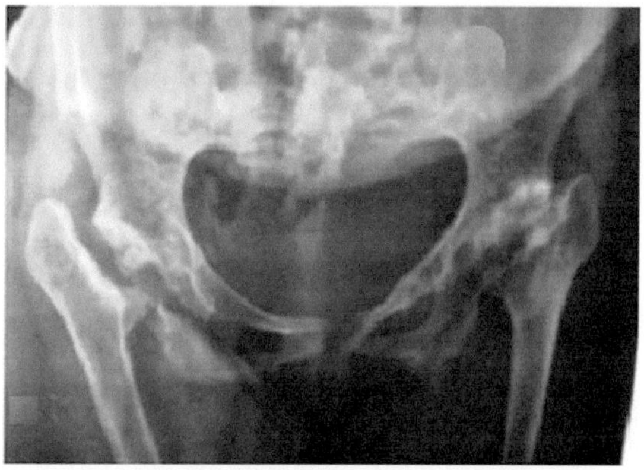

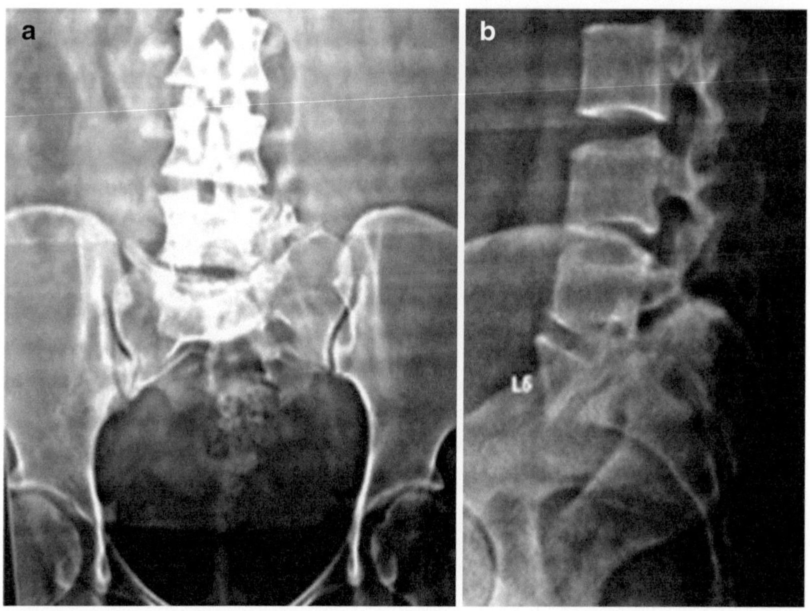

Fig. 6.6 (**a**, **b**) Roentgenogram of lumbo-sacral region revealing posterior wedging of fifth lumbar vertebra

6.5 Campomelic Dysplasia

Campomelic dysplasia is uncommon skeletal anomaly characterized by bent limbs (Campomelic). Long-limb and short-limb (Kyphomelic) types are described. There is bowing of long bones particularly in the lower limbs (Fig. 6.7a). Long-limb type affects the bones like tibia and femur. A skin dimple may be observed over the bent tibia. The short-limb type may be associated with craniosynostosis. Many congenital anomalies have been associated including clubfeet, cleft palate. Renal and cardiac defects have been described.

Radiologically the long-limb type reveals bowing of the femur and tibia (Fig. 6.7b). There may be hypoplasia of the scapula and the fibula. Pelvis is usually narrow. There may be platyspondyly. In short-limb type, the proximal long bones especially the femur is short as well as bent (Fig. 6.8a–f). The ribs may also be short and there may be evidence of narrow thorax.

Kyphomelic deformity is characterized by generalized anterior bowing of long bones (Fig. 6.9a, b). When severe, the patient may seek medical help.

6.5.1 Metatropic Dysplasia

Metatropic dysplasia is a skeletal disorder characterized by short stature (dwarfism) with other skeletal abnormalities. The term "metatropic" is derived from the Greek word "metatropos," which means "changing patterns." This name reflects the fact that the skeletal abnormalities associated with the condition change over time. In infancy the trunk is long and as the growth the proceeds the trunk become smaller resembling Morquio Disease. There may be a tail-like projection over the sacrum.

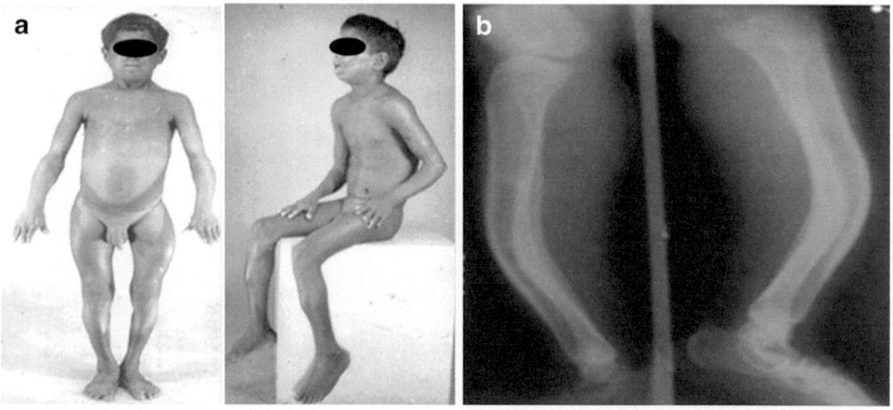

Fig. 6.7 (**a**) A patient with Camptomelia showing bent in both tibiae and the femora. Note the small hands with short fingers (Acromelic-dysplasia). (**b**) Roentgenogram of the tibia and fibula revealing the bent in the midshafts with sclerosis

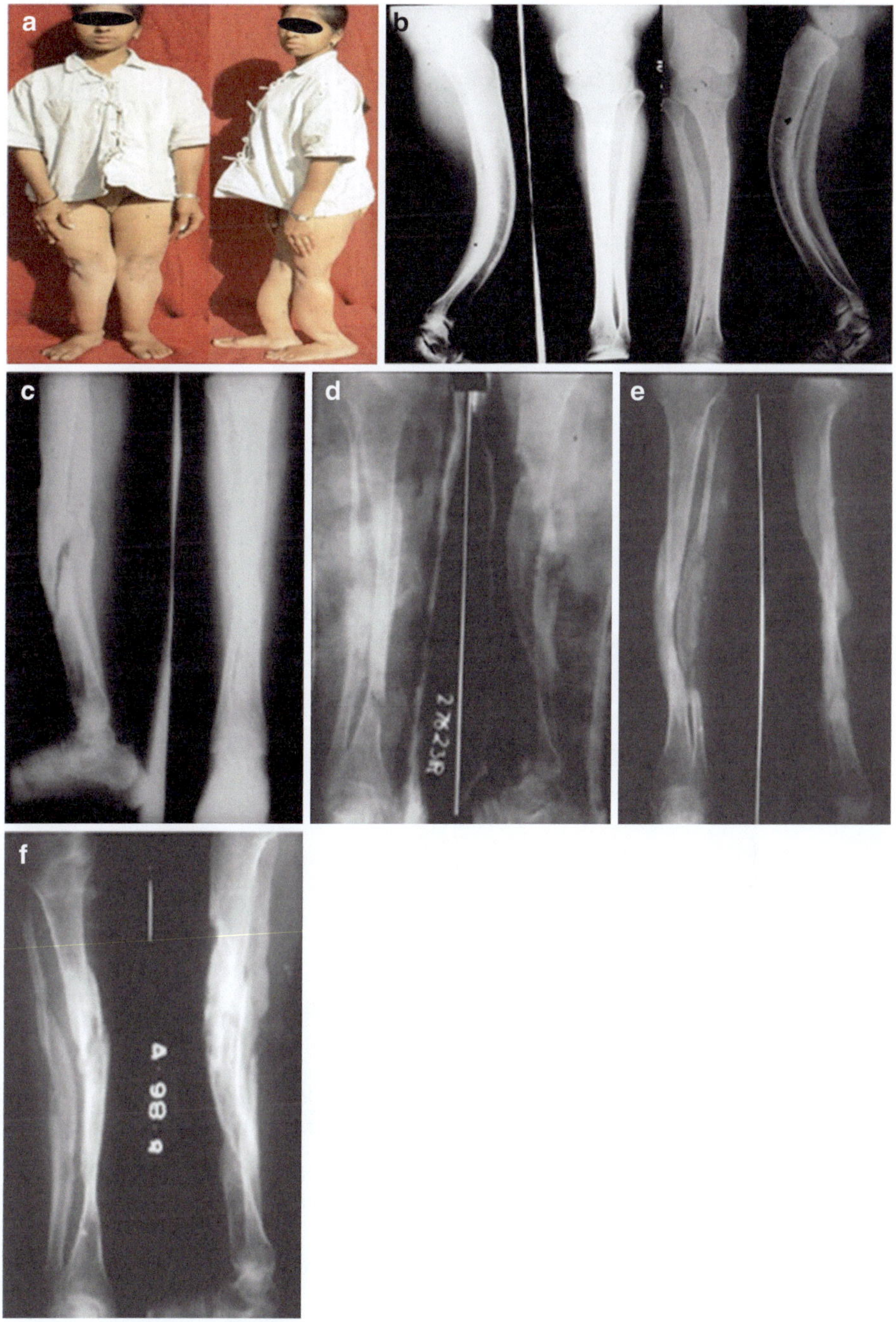

Fig. 6.8 (**a**) A 30-year-old lady showing kyphomelic deformity of both the legs. (**b**) Roentgenogram on either side reveals kyphomelic deformity of the tibia and fibula. (**c**, **d**, **e**, **f**) Roentgenograms showing surgical correction by performing DODO to straighten both the bones

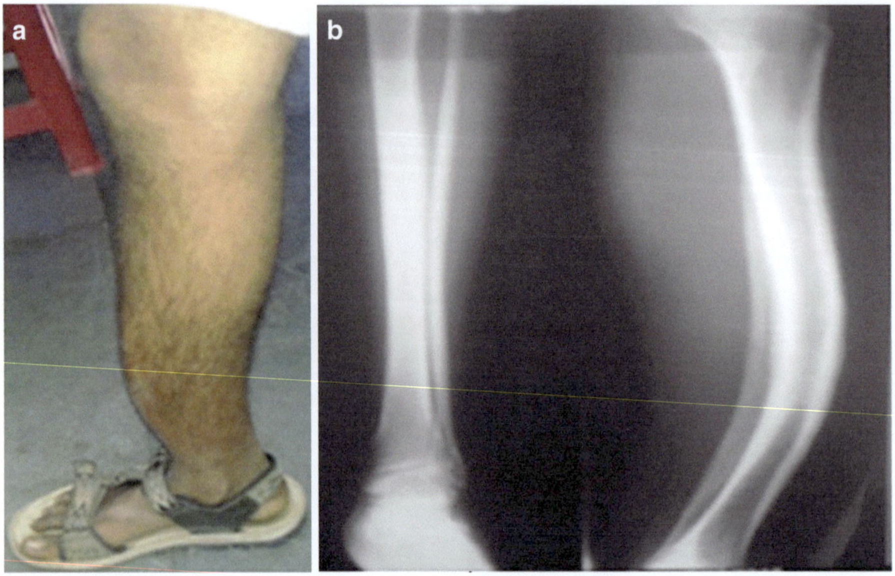

Fig. 6.9 (**a**) Right leg revealing kyphomelic changes in the tibial shaft. (**b**) Roentgenogram showing kyphomelic tibia and fibula with sclerosis in their middle third

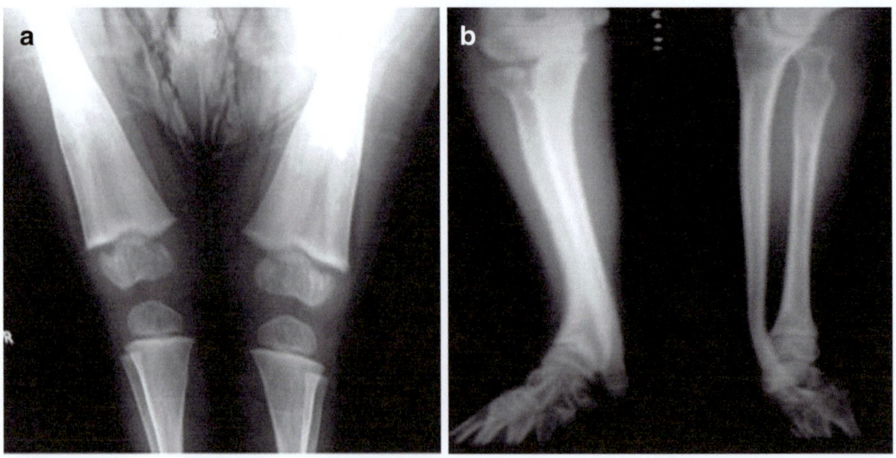

Fig. 6.10 (**a**) Roentgenogram reveals metaphysio-diaphyseal 'trumpet' shaped femur and tibia. (**b**) Roentgenogram revealing long ulna with subluxation at its distal end

The Kyphoscoliosis gradually progresses. The limbs are short with bulbous enlargement of joints with restricted motion. The head and face are normal.

Radiologically there is platyspondyly with wide spaces. There may be wedging of the vertebrae and hypoplastic odontoid process. There is metaphyseal flaring with dumble shaped bones. In infants the relationship of the metaphyses and diaphyses is like 'Trumpet' (Fig. 6.10a). There may be distortion of pelvic bones. The ulna may be longer than radius (Fig. 6.10b). There is horizontal acetabular roofs. The iliac bones are broad and square. There may be signs of premature osteoarthrosis with genu-vara. The treatment depends on the symptoms of the patient.

Suggested Readings

Wynne-Davies R, Hall C. Two clinical variants of spondyloepiphysial dysplasia congenital. J Bone Joint Surg (Br). 1982;64:435.

Diamond LS. A family study of spondyloepiphyseal dysplasia. J Bone Joint Surg Am. 1970;52:1587.

Kaibara N, Takagishi K, Katsuki I. Spondyloepiphyseal dysplasia tarda with progressive arthropathy. Skelet Radiol. 1983;10:13.

Williams B, Cranley RE. Morphologic observations on four cases of SED congenital. Birth Defects. 1974;10:75.

Kozlowski K. Metaphyseal and spondylometaphyseal chondrodysplasia. Clin Orthop. 1976;114:83–93.

Spranger JW, Langer LO. Spondyloepiphyseal dysplasia congenita. Radiology. 1970;94:313–22.

Maroteaux P, Spranger J. The spondylometaphyseal dysplasias. A tentative classification. Pediatr Radiol. 1991;21:293.

Yang SS, Chen H, Williams P, et al. Spondyloepiphyseal dysplasia congenital. Arch Pathol Lab Med. 1980;104:208–11.

Boden SA, Kaplan FS, Fallon MD, et al. Metatropic dwarfism. J Bone Joint Surg Am. 1987;69:174.

Rimoin DL, Siggers DC, Lachman RS, Silberberg R. Metatropic dwarfism, the kniest syndrome and the pseudoachondroplastic dysplasia. Clin Orthop. 1976;114:70–82.

Fielding JW, Hensinger RN, Hawkins RJ. Os odontoideum. J Bone Joint Surg Am. 1980;62:376.

Houston CS, Opitz JM, Spranger JW. The campomelic syndrome. Am J Med Genet. 1983;15:3–28.

Bethem D, Winter RB, Lutter L. Disorders of the spine in diastrophic dwarfism. J Bone Joint Surg Am. 1980;62:529.

Herring J. The spinal disorders in diastrophic dwarfism. J Bone Joint Surg Am. 1978;60:177.

Horton WA, Hall JG, Scott CI, et al. Growth curves for height for diastrophic dysplasia, spondyloepiphyseal dysplasia congenital, and pseudoachondroplasia. Am J Dis Child. 1982;136:316.

Hollister DW, Lachman RS. Diastrophic dwarfism. Clin Orthop. 1976;114:61.

Poussa M, Merikanto J, Ryoeppy S, et al. The spine in diastrophic dysplasia. Spine. 1991;16:881.

Richards BS. Atlanto-axial instability in diastrophic dysplasia. A case report. J Bone Joint Surg Am. 1991;73:614.

Ryoeppy S, Poussa M, Merikanto J, et al. Foot deformities in diastrophic dysplasia. An analysis of 102 patients. J Bone Joint Surg. 1992;74:441.

Kash IJ, Sane SM, Samaha FJ, et al. Cervical cord compression in diastrophic dwarfism. J Pediatr. 1974;78:862.

Krecak J, Starshak RJ. Cervical kyphosis in diastrophic dwarfism: CT and MRI findings. Pediatr Radiol. 1987;17:321.

McKusick VA, Goodman RM. Pinnal calcification: observations in systemic disease not associated with disordedcalcium metabolism. JAMA. 1962;179:230.

7 Chondroplasia Involving Specific Anatomical Sites

7.1 Cleidocranial Dysplasia

Cleidocranial Dysplasia (Dysostosis) is a rare condition with abnormal development of bones especially formed by intramembranous ossification. The clavicles, skull and pelvis are often involved. Due to such a presentation the shoulders are hypermobile (Figs. 7.1a–c and 7.2a–c). Radiologically there may be complete absence of clavicles of both the sides or only the medial part may be present. Recurrent dislocation of major joints may occur. There may be spina bifida and kyphoscoliosis. Widening of the symphysis pubis and the sacroiliac joints may be observed. The pubic and iliac bones may be hypoplastic. The skull has large patent fontanelles (Fig. 7.3a, b). The skull bones show lack of ossification to the extent that the skull has to be protected. Long bones are thin and various other anomalies of the hands like distal Phalangeal tapering may be observed. The proximal femora may reveal elongated femoral heads (Chefhat-sign).

S. S. Yadav, *Skeletal Dysplasias & Congenital Malformations*,
https://doi.org/10.1007/978-981-99-5679-1_7

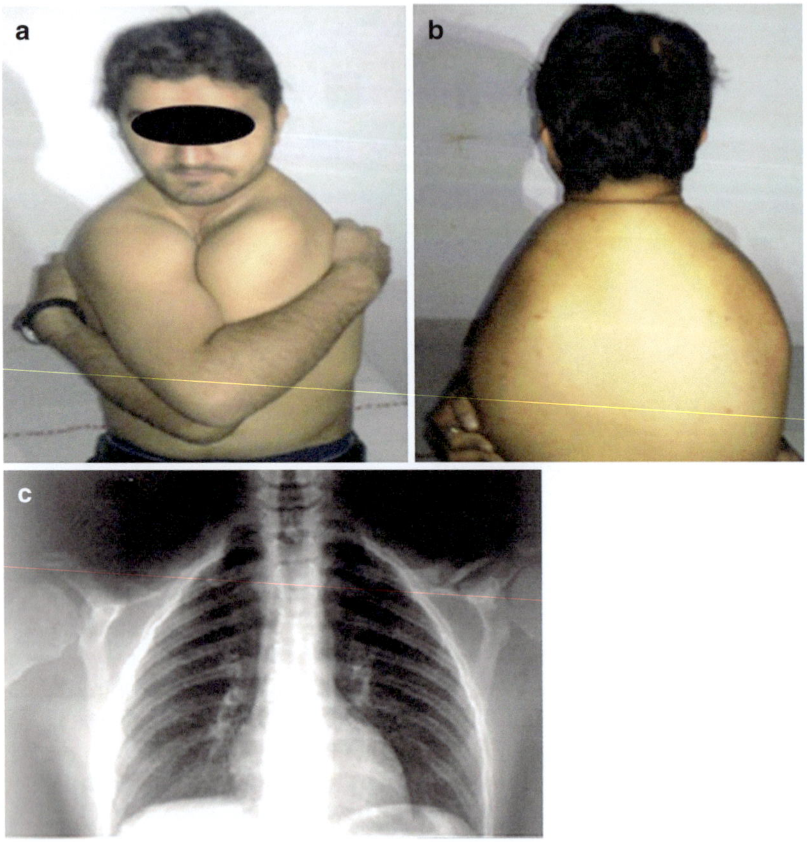

Fig. 7.1 (**a**) An otherwise well-developed 25-year-old man with cleidocranial dysplasia can approximate the shoulders anteriorly—a typical clinical feature. (**b**) Posterior view of the same patient while approximating the shoulders. Note the large head. (**c**) Anteroposterior radiograph of the chest revealing complete clavicular aplasia

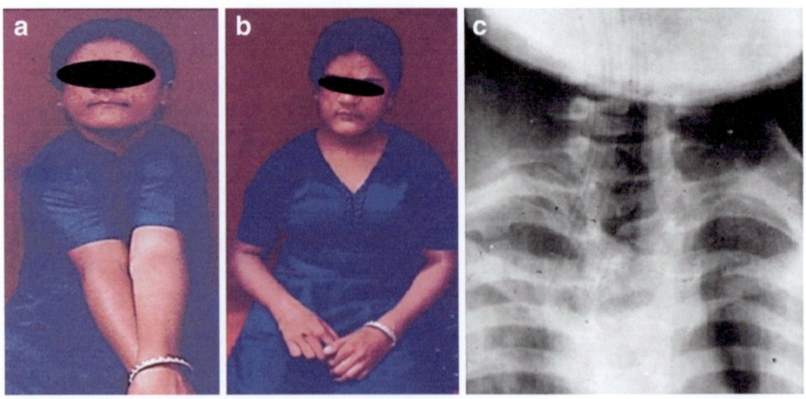

Fig. 7.2 (**a**) Patient with absence of clavicales revealing the classical clinical sign of approximating both the shoulders anteriorly. (**b**) The same patient at ease with normal appearance. (**c**) An anteroposterior radiograph revealing complete bilateral clavicular aplasia

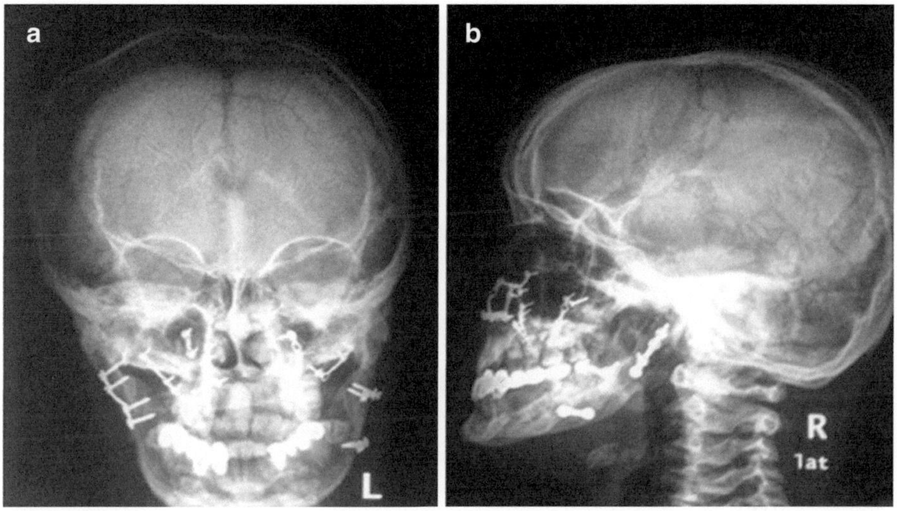

Fig. 7.3 (**a**) The anteroposterior skull radiograph of the same patient demonstrates wide sutures. Note the bossing of the perietal bones (The patient had undergone mandibular and maxillary surgery). (**b**) Lateral radiograph also reveals wide sutures with bossing of the frontal and occipital bones. Note the presence of 'Wormian bones' in the occipital region

7.2 Pseudoachondroplastic Dysplasia

Pseudoachondroplasia is relatively common and is characterized by achondroplasia-like rhizomelic short stature. The main criteria of differentiation are the normal size of the head (Fig. 7.4a–c). The infants appear normal at birth but retardation of skeletal growth is obvious by the end of first or second year of life. The trunk is long and the limbs are short and become more evident with age. The hand and feet are short and broad. There may be ligament laxity in the joints of the hands and feet. There is limited elbow extension. Hip joint movement may be restricted and there may be waddling gait. There may be varus or valgus deformity of the knees. Associated deformities include kyphoscoliosis or increased lumbar lordosis and joint laxity with secondary osteoarthritic changes. The intelligence and faces are normal.

Radiologically the pelvis is broad and squared iliac wings and narrow sacrosciatic notches. The acetabuli are dysplastic with horizontal roofs. Metaphysis of all long bones are markedly flared and irregular with deformed and fragmented epiphyses. The tubular bones are also flared. There may be hypoplasia of the capital femoral epiphysis along with evidence of dysplastic hips. Epiphyses are small irregular and often fragmented with delayed appearance particularly the femoral capital and humeral epiphyses are most effective (Fig. 7.5). Medial beaking of the femoral neck is considered as the characteristic features. The vertebral bodies appear ovoid and have anterior tongue like projection. Interpedicular distance is characteristically normal.

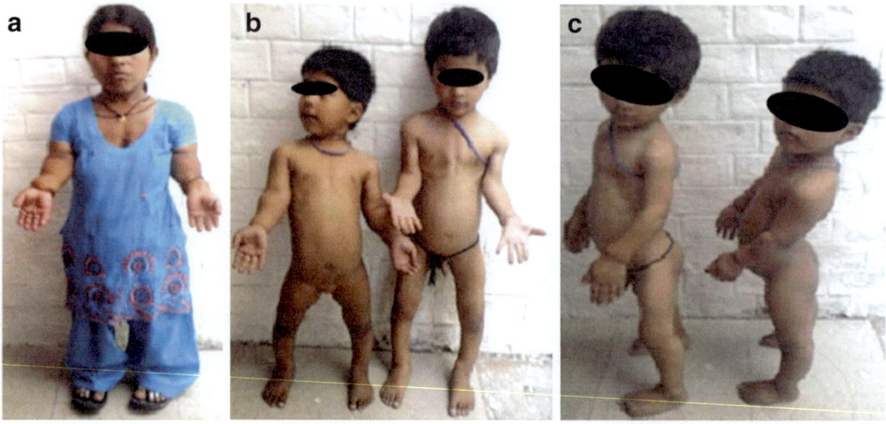

Fig. 7.4 (**a**) A family of Pseudoachondroplasia showing normal trunk with short limbs of rhizomelic type. (**b**) The children with pseudoachondroplasia also have rhizomelic shortening and normal head. (**c**) Note the exaggerated lumbar lordosis

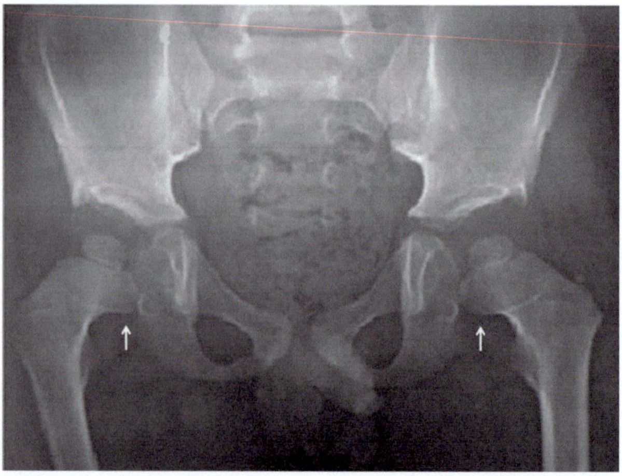

Fig. 7.5 Pelvis in a case of pseudoachondroplasia with squrish iliac wings and horizontal acetabular roofs akin to achondroplasia and the wide triradiant cartilage. Note the small heads of the femora on either side. Medial beaking of the femoral neck may also be noted

Differential diagnoses for this radiographic appearance in context of clinical findings include: achondroplasia, morquio syndrome, hypothyroidism, multiple epiphyseal dysplasia (MED) and spondyloepiphyseal dysplasia (SED) congenita.

In morquio syndrome, the spinal changes are prominent with flat vertebrae, central beaking and marked kyphosis. Metacarpals may show proximal tapering with short, wide tubular bones. Metaphyseal widening and irregularity is absent.

In hypothyroidism, epiphyseal changes may simulate pseudohypochondroplasia, but the dwarfism is symmetrical and the metaphyses are normal. The skull shows Wormian bones. The classical pelvis changes are also lacking and the general platyspondyly is absent.

7.3 Mesomelic Dysplasia

Mesomelic dysplasia is a condition with short stature associated with disproportionate limb shortening that is commonly seen in the mid portions of the limbs. Radiologically there is shortening of long tubular bones mostly in the middle segment. The severe form of mesomelic dysplasia produces prominent short stature in adulthood. There may be genuvalgum and club foot. There may be hypoplasia of mandible. Radiologically the tubular bones are predominantly short in the middle segment. There is also hypoplasia of the proximal fibula and marked shortening of the tibia.

Dyschondrosteosis is a mild but common form of mesomelic dysplasia associated with a madelung deformity with reducible dorsal subluxation of the distal ulna (Fig. 7.6a, b). The tibia and fibula are short and the elbow moments are restricted. Radiologically the radius is short in relation to the ulna with dorsal subluxation of the ulna (Fig. 7.7a, b). Excision of distal end of the ulna may be required to relieve pain and cosmetic improvement. The medial portion of the proximal tibia is prominent and the carpal bones have a triangular configuration.

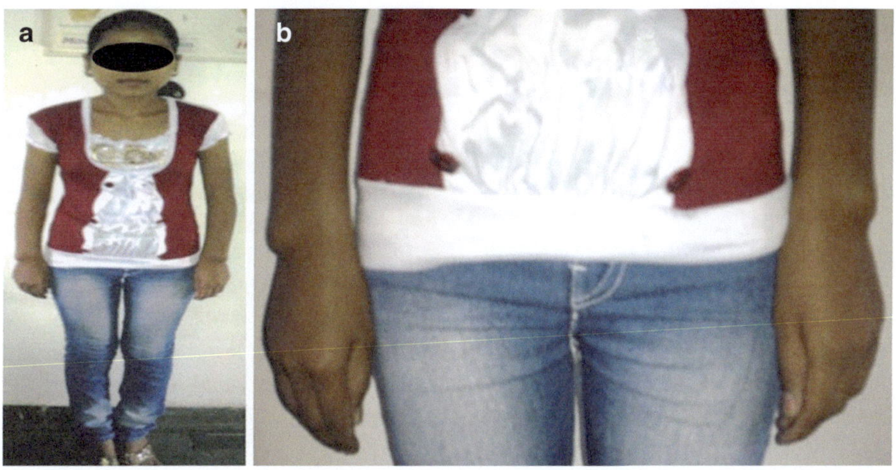

Fig. 7.6 (**a**) An 18-year-old girl showing prominent head of the ulna on both sides. (**b**) A close-up view of the anomaly

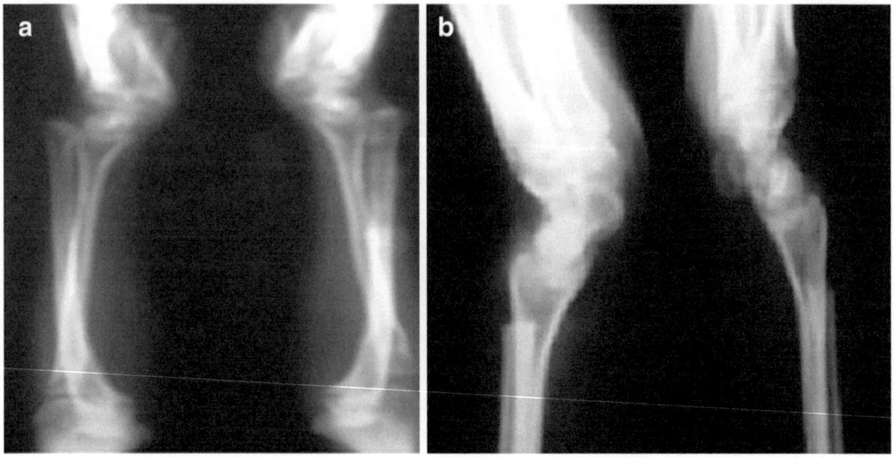

Fig. 7.7 (**a**) Roentgenogram of both the forearm revealing luxation of the head of the ulna with bowing of the radius on either side. (**b**) Roentgenogram revealing the picture after excision of heads of the ulnae. The patient was satisfied with the result

7.4 Acromelic Dysplasias

Acromelic dysplasia is characterized by shortening of the distal extremities. There may also be evidence of short stature. Severe forms of this condition have been described including ***Acrodysostosis*** in which the hand and feet are short stubby (Fig. 7.8a–d). There is short stature with facial dysmorphism with growth failure and deformities. Major features are nasal hypoplasia with flat and depressed nasal root with short stature the bones of the hands and feet are short. Mental retardation has been observed in some cases.

Radiologically the phalanges, metacarpals and metatarsal are short and wide and have cone-shaped epiphyses. The skull shows brachycephaly and thickening of the base and vault of the skull. There is nasal hypoplasia and acetabular dysplasia. Osteoarthritic changes may be observed later in life.

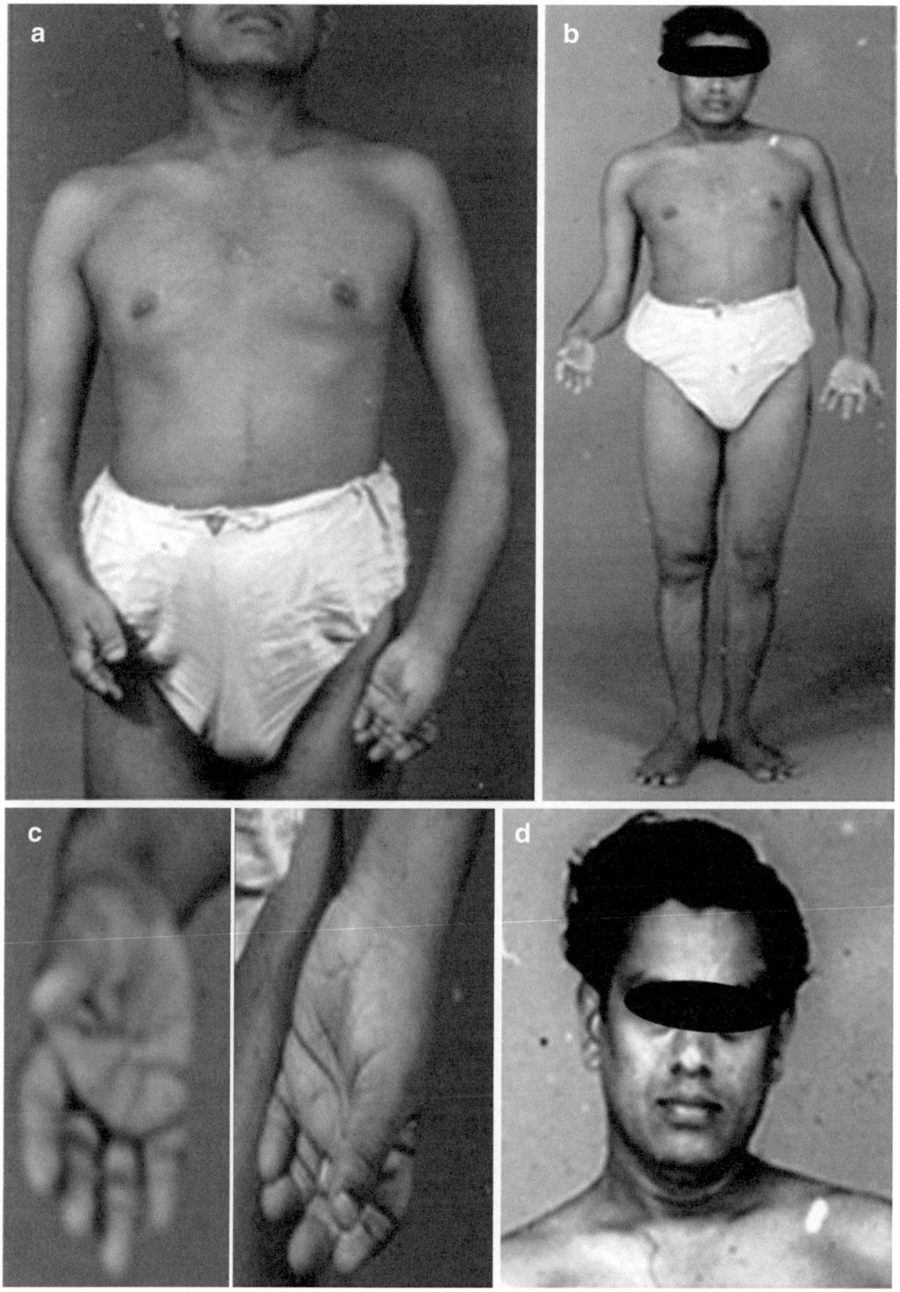

Fig. 7.8 (**a**) A case of acromelic dysplasia with smaller size of hands and feet. (**b**) Note the mesomelic shortening on the right forearm. (**c**) The finger and toes are also short and wide. (**d**) There is an evidence of nasal hypoplasia with depressed nasal root

Suggested Readings

Jarvis JL, Keats TE. Cleidocranialdysostosis: a review of 40 new cases. AJR. 1974;121:5.

Eventov I, Reider-Grosswasser I, Weiss S, et al. Cleidocranial dysplasia. A family study. Clin Radiol. 1979;30:323.

Kaitila II, Leisti JT, Rimoin DL. Mesomelic skeletal dysplasia. Clin Orthop. 1976;114:94.

Hall JG. Pseudoachondroplasia. Birth Defects. 1975;1:187.

Kopits SE, Lindstrom JA, McKusik VA. Pseudoachondroplastic dysplasia: pathodynamics and management. Birth Defects. 1974;10:341.

Bick EM, Marie P, Sainton P. The classic: on hereditary cleidocranialdysostosis. Clin Orthop. 1968;58:5.

Vickers D, Nielsen G. Madelung deformity: surgical prophylaxis (physiolysis) during the late growth period by resection of the dyschondrosteosis lesion. J Hand Surg (Br). 1992;17:401.

Fagg PS. Wrist pain in the Madelung's deformity of dyschondrosteosis. J Hand Surg (Br). 1988;13:11.

Langer LO Jr. Dyschondrosteosis, a heritable bone dysplasia with characteristic roentgenographic features. AJR. 1965;95:178.

Chondroplasia Involving Tumour-Like Formation of Cartilage

8

8.1 Osteochondromatosis (Hereditary Multiple Exostoses; Diaphyseal Aclasis; Metaphyseal Aclasis)

Hereditary multiple exostoses (***Diaphyseal Aclasis***) are characterized by the formation of multiple osteochondromas in several bones (Fig. 8.1). Masses of osteocartilaginous tissues are observed near the growing ends of the long bones. The growth of such swellings takes place till adulthood. They also result in growth disturbances with deformity and shortening of the affected bones (Fig. 8.2). Because of pressure changes the neurovascular bundle in the neighbourhood may get involved. Spinal lesions are rare. Radiologically there are multiple asymmetrical, sessile or pedunculated osteochondromas seen in the vicinity of the metaphyses. The thickness of the cartilage-cap can be assessed by MRI. The disease affects the bones or portions of bones, which are developed both from cartilage and from membrane, and it has a marked hereditary tendency. In late cases with huge tumour a CT scan may be necessary (Fig. 8.3a, b).

The most characteristic features of metaphyseal aclasis are a failure of remodelling of the metaphysis and the occurrence of multiple outgrowths or exostoses from the surface of the shaft of long bones (Fig. 8.4a–d). The membranous bones of the skeleton are immune. It is essentially a failure of maturation of the cartilage cells. In the forearm and the leg an exostosis from one of the bones may cause pressure absorption of the adjacent surface of the neighbouring bone and may attain great proportions. In the diffuse type there may be extensive distortion and deformity of the whole skeleton and the stature is short. The limb may show deformities of the nature of bowing or knock knee. If the process is inadvertently fractured, it interferes with the free play of associated tendons and act as a mechanical obstruction to joint movement and may give rise to disability (Fig. 8.5a–e). A bursa may form over the projection. The only treatment indicated is in symptomatic case by removing the exostosis which is increasing in size rapidly or giving rise to trouble by pressure or malignant change (Fig. 8.6a–d).

S. S. Yadav, *Skeletal Dysplasias & Congenital Malformations*,
https://doi.org/10.1007/978-981-99-5679-1_8

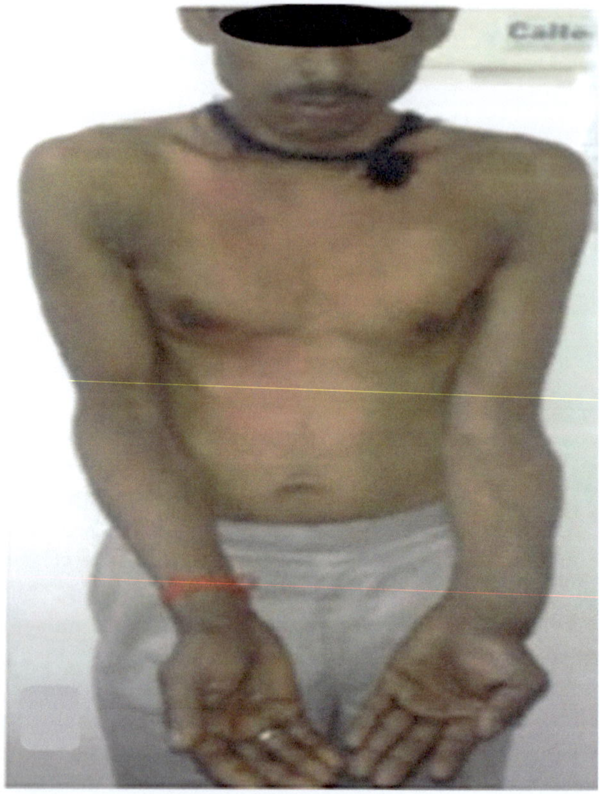

Fig. 8.1 A case of familial osteochondroma from the distal end of the ulna resulting in shortening and deformity of the left forearm

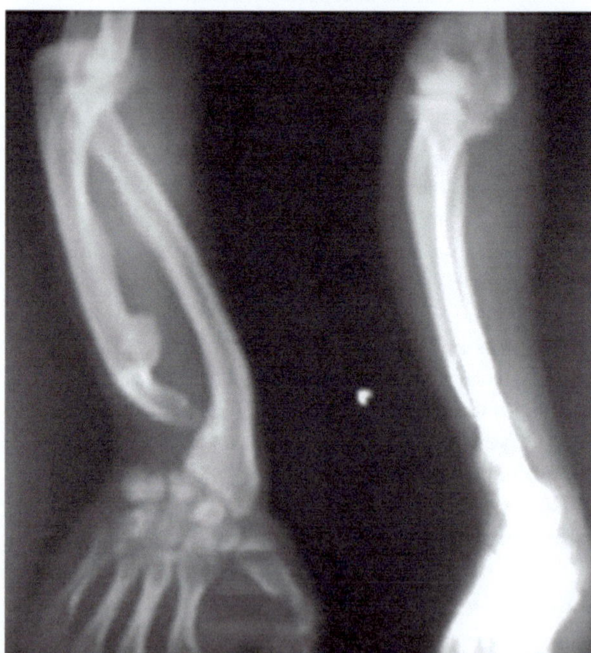

Fig. 8.2 Radiograph of left forearm with an osteochondroma near the distal end of the unla with deformity and shortening

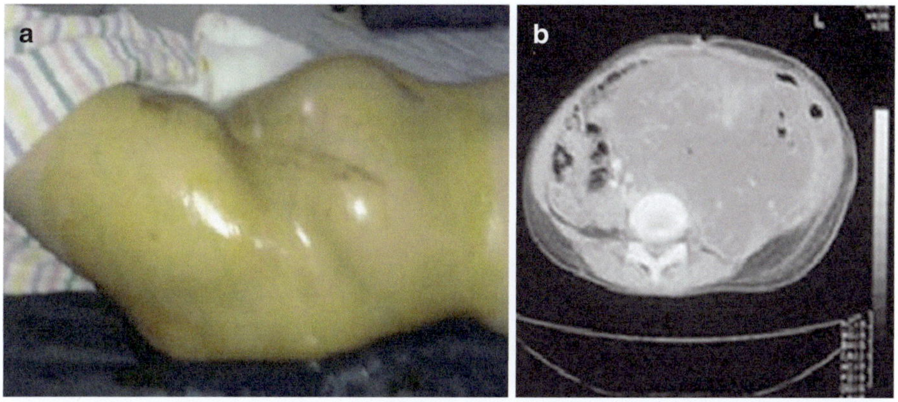

Fig. 8.3 (**a**) The daughter of the above patient, an advanced case of chondrosarcoma of the left Ilium. (**b**) CT scan reveals massive extension in the pelvis displacing the pelvic contents on the right side

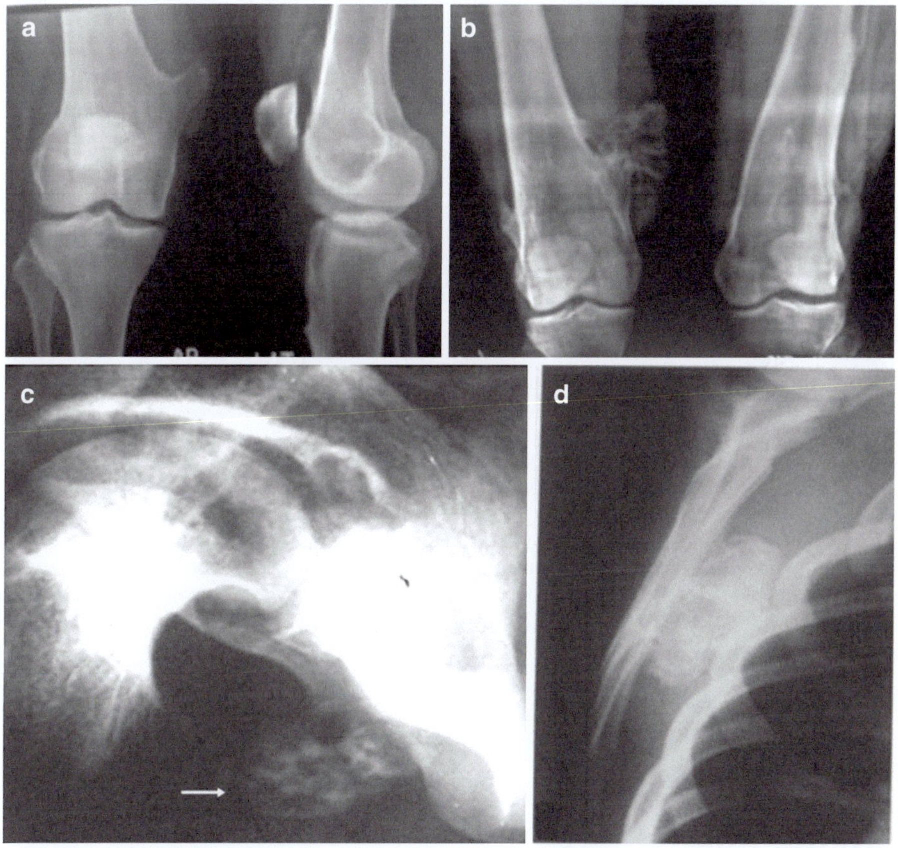

Fig. 8.4 (**a**) Roentgenogram reveals an osteochondroma arising from the medial side of the metaphyseal area of the distal femur. (**b**) An osteochondroma arising from the same site with multi-lobulated growth. (**c**) Exostosis from the ischial ramus of the Pelvis (Arrow). (**d**) A massive sessile osteochondroma from the under surface of the scapula, rubbing against the ribs and has been a cause of pain in the area, a rare site

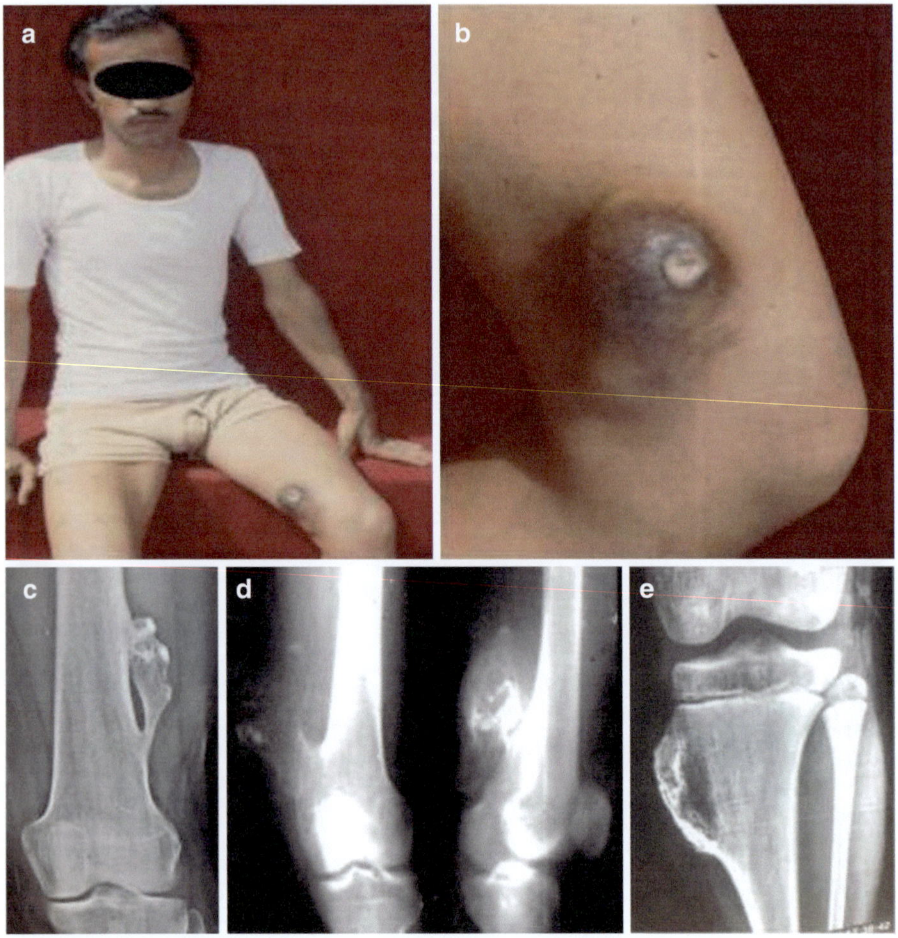

Fig. 8.5 (**a**) Osteochondroma of the distal end of the left femur in a 35-year-old man. Ulceration with dark-brown margins are seen. (**b**) Such a presentation may reveal signs of malignancy. (**c**) Radiograph of a solitary exostosis arising from the medial metaphyseal area of the distal femur. (**d**) Radiograph of the same patient showing pedunculated exostosis from the medial aspect of the distal femur on left side. The fundus of the lesion has lost the clarity of the boundaries revealing a possible malignant change. (**e**) A Sessile osteochondroma arising from proximal end of the left tibia. Note the shortening of the tibia and proximal migration of the head of the fibula

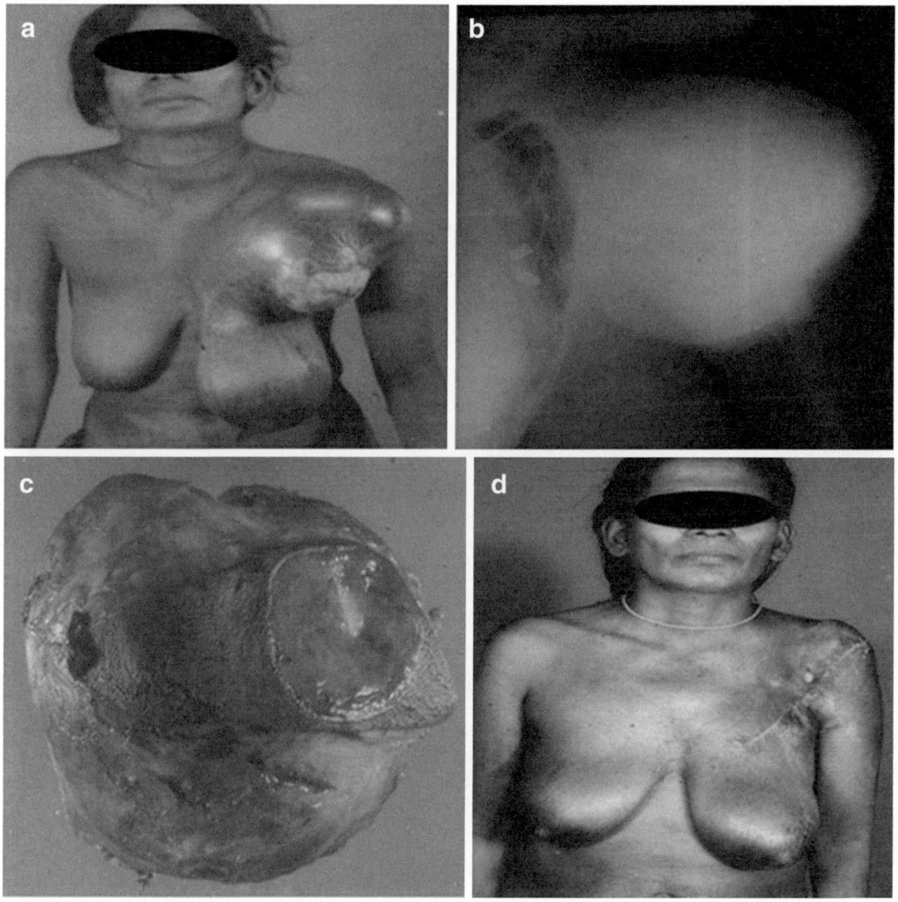

Fig. 8.6 (**a**) A huge osteochondroma from the anterior aspects of the upper rib displacing the left breast downwards. (**b**) Roentgenogram revealing the massive growth. (**c**) The excised specimen. (**d**) The patient after surgery. The breast has been restored to the original position

8.2 Enchondromatosis

Enchondromatosis ***(Dyschondroplasia, Ollier Diseases)*** is a distinct developmental disorder of endochondral ossification. It is a rare, non-hereditary disorder which results in the failure of endochondral ossification of the growth plate cartilage. Dyschondroplasia may be unilateral or bilateral and it may be confined to a single bone. The disease is one affecting proliferating cartilage in the metaphyseal region so that cartilage persists abnormally. More than 50% may go in for chondro sarcoma. It may be seen near the ends of the diaphyses in the long bones or in the short long bones of the hands and feet. The metaphysis is usually broadened and its texture as poor. Occasionally small exostoses project from the surface. These are

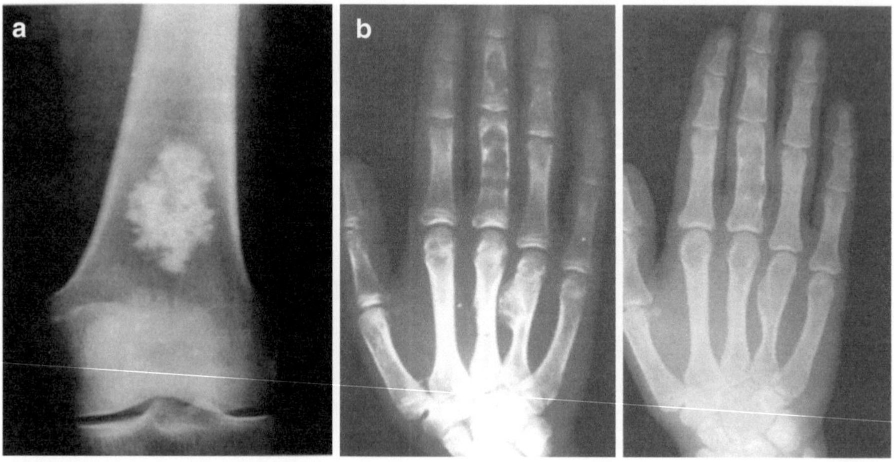

Fig. 8.7 (**a**) Roentgenogram of a pop-corn types of enchondroma in the distal end of the femur. (**b**) Multiple enchondromas of the metacarpals and phalanges

less well formed than in metaphyseal aclasis. The ischium and the pubis may appear fluffy and stippled to the presence of small rounded areas of cartilage. Stippled or ring like (pop-corn) ossification is often seen particularly in older lesions (Fig. 8.7a, b). If one side of the growth plate is affected, an angular deformity results (Fig. 8.8a–g). As the child grows, the striping of the bone is replaced by irregular mottling and speckling and the affected bones are stunted. Sometimes the deformity is so severe that it results in cosmetic disturbances (Fig. 8.9a, b).

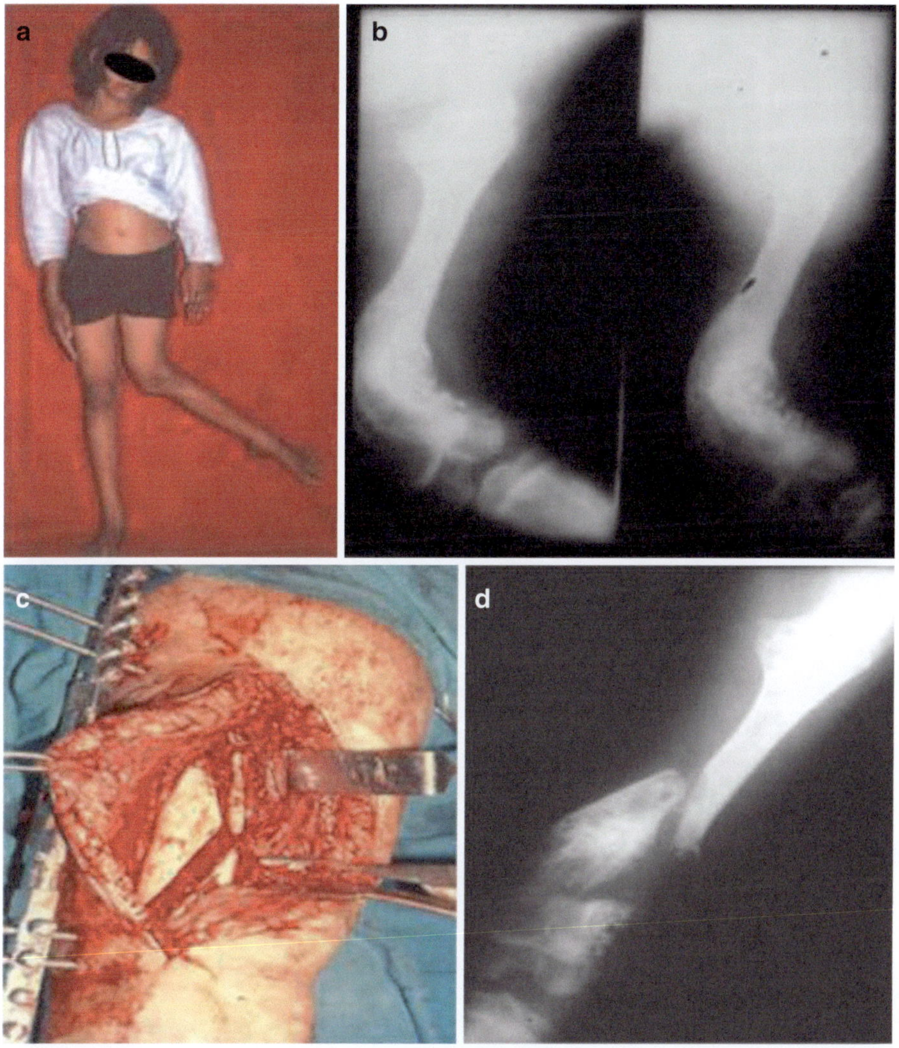

Fig. 8.8 (**a**) The clinical picture revealing a valgus deformity at the left knee. In this case the growth plate is affected only on lateral side which has resulted in an angular deformity. (**b**) Massive achondroplasia showing valgus deformity of the lower end of the femur with irregular density. (**c**) Intra-operative picture showing the Double Obligue Diaphyseal Osteotomy (DODO) stabilized with external fixator. (**d**) Postoperative radiograph revealing DODO and correction of the deformity. (**e**) Radiograph showing early healing. (**f**) Complete healing and remodelling after DODO. (**g**) Clinical picture of the patient

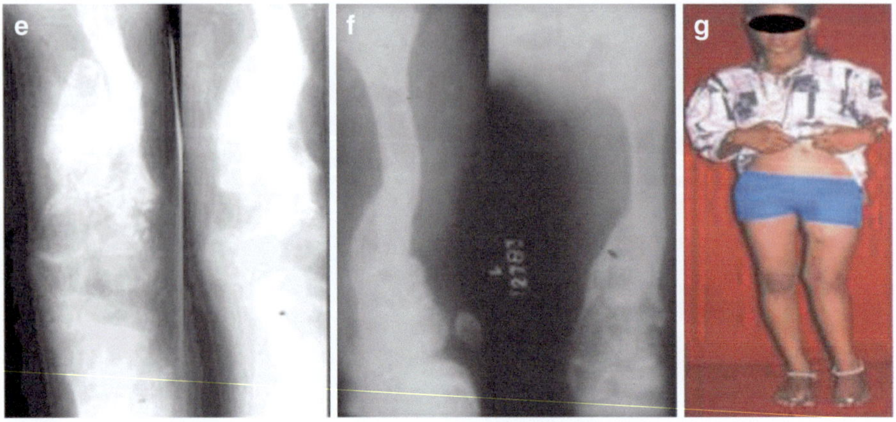

Fig. 8.8 (continued)

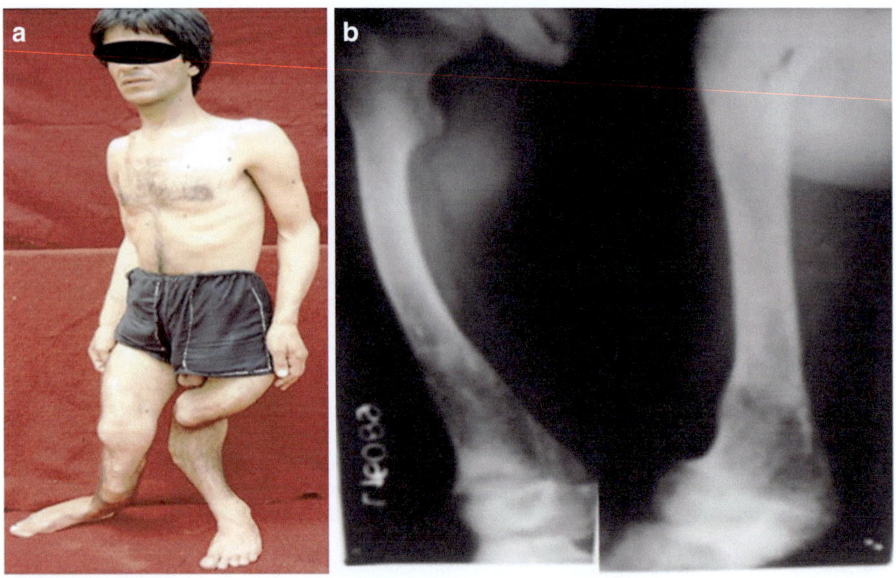

Fig. 8.9 (**a**) Massive enchondromatosis of the bones of lower extremity. Note the severe deformity of the left femur and broadening of the proximal tibia with severe valgus deformity. (**b**) Enchondromatosis of the lower end of the femur showing shortening and irregular density of the distal of the femur

8.3 Maffucci's Syndrome

It is characterized by multiple enchondromata occurring simultaneously with multiple cutaneous and soft tissue cavernous hemangiomatous and can be severely disabling (Fig. 8.10a–f). This hamartomatous proliferation is asymmetric. Although

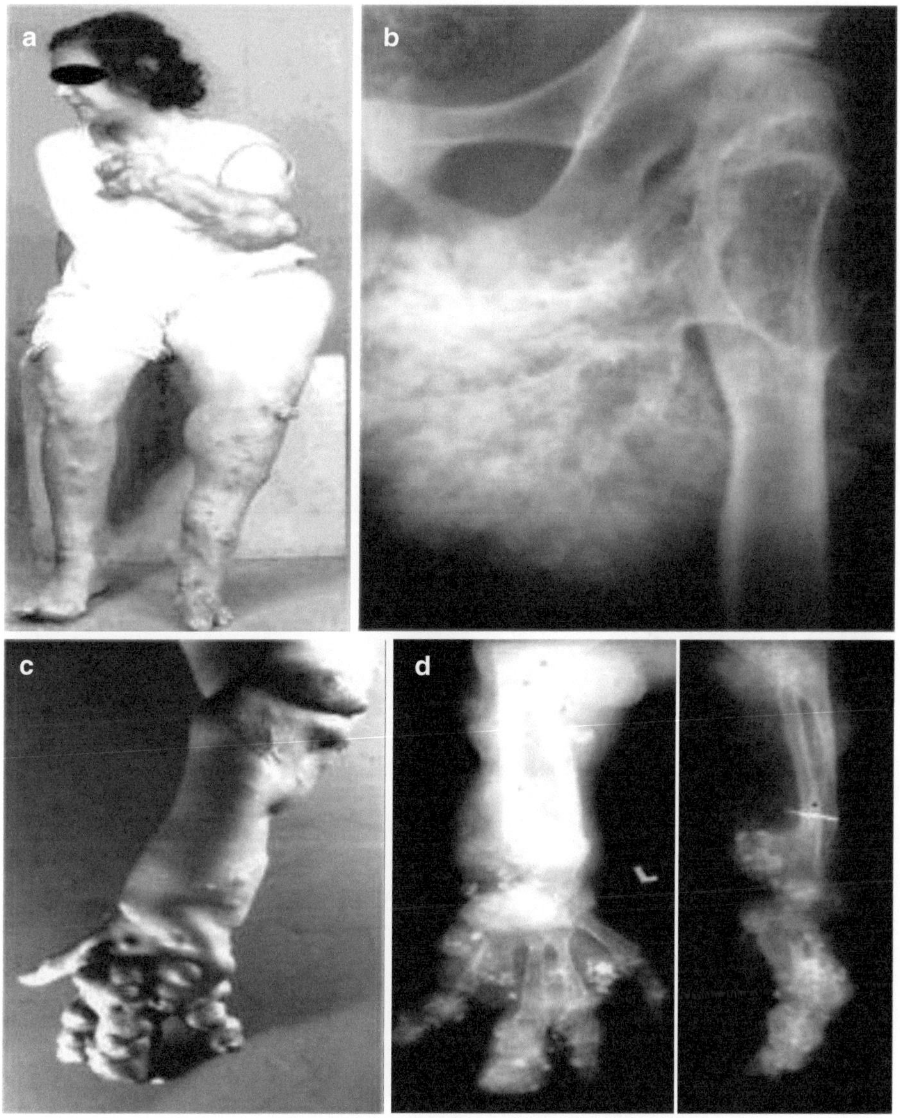

Fig. 8.10 (**a**) A rare and advanced case of Maffucci syndrome principally involving the bones on the left side of the body. (**b**) Roentgenogram reveals enlarged masses of cartilage. (**c**) The clinical picture of the left hand and the Forearm showing multiple nodules. (**d**) The roentgenogram reveals multiple cartilaginous masses in the hand and forearm. The rounded opacities shows multiple hemingomas. (**e**) The proximal end of the humerus reveals involvement with multiple cartilaginous masses. (**f**) The X-Ray reveals stippled pattern of sclerosis with rounded and elongated defects in the left ilium. The proximal end of the femur on left side is also involved

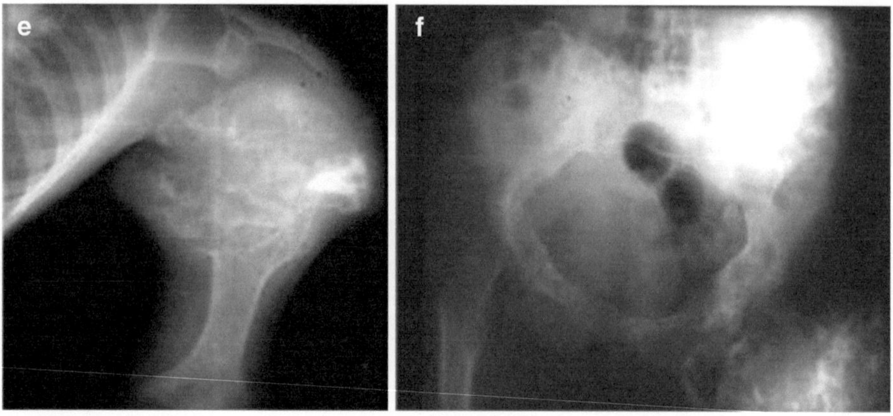

Fig. 8.10 (continued)

the child is normal at birth, nodules appear on the fingers or toes along with tumour-like presentation in other bones. The distribution is asymmetrical and may be unilateral. Bluish haemangiomata may be seen in soft tissues which gradually increase in size. By second decade the tumour growth is quiescent.

The affected long bones are usually short and bowed resulting in distorted appearance of the bone involved. The excessive size of the tumour may interfere with normal function and may need an amputation. The bones commonly involved are femur, tibia, ilium followed by short bones of the hand and feet. The vascular component is a diffuse cavernous haemangioma with phlebectasia, and frequent thrombi and phleboliths. In about 30% cases there is a definite possibility of malignant transformation in the form of chondrosarcomata. The patient otherwise has normal intelligence and no familial cases have been reported.

8.4 Dysplasia Epiphysealis Hemimelia

Dysplasia epiphysealis hemimelica is a developmental disorder of childhood in which there is asymmetrical cartilaginous overgrowth of one, or more than one, epiphysis or of a tarsal or carpal bone. This disorder is usually limited to either the medial or lateral half of a single extremity with pain, deformity and restricted motion. Pain is not common but when present it is in the ankle, more frequently than in the knee. The roentgenographic appearance of dysplasia epiphysealis hemimelica is usually characteristic (Fig. 8.11a–c). The mature lesion appeared as an irregular, frequently lobulated, osseous from the epiphysis or from a tarsal or carpal bone.

This unusual growth disturbance produces overgrowth of long bone epiphyses, forming an anterior, posterior or side projection–osteochondroma which affects epiphysis instead of the diaphysis. It affects a single epiphysis, the great majority being of the talus, distal femur or proximal tibia (Fig. 8.12a, b). The protruding epiphyseal mass interferes with joint function. The epiphyseal osseous nucleus results in an excessive growth.

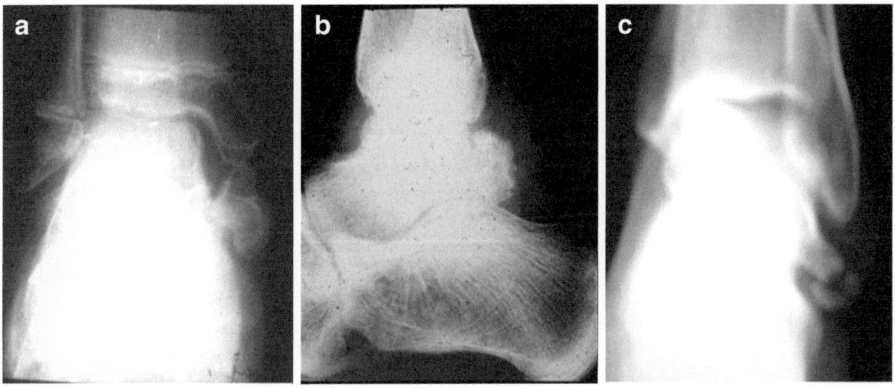

Fig. 8.11 (**a–c**) Dysplasia epiphysealis hemimelica involving the medial, posterior and the lateral aspects of the talusin three different cases, respectively

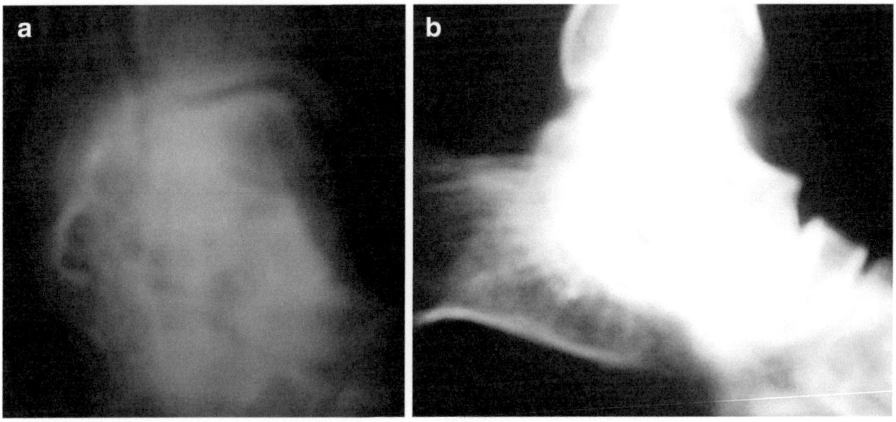

Fig. 8.12 (**a**) Roentgenogram reveals an abnormal growth from the lateral aspect of the talus. (**b**) Roentgenogram shows dysplasia epiphysealis hemimelica from the neck of the talus

Suggested Readings

Peterson HA. Multiple hereditary osteochondromata. Clin Orthop. 1989;239:222–30.

Liu J, Hudkins PG, Swee RG, et al. Bone sarcomas associated with Ollier's disease. Cancer. 1987;59:1376–85.

Milgram JW. The origin of osteochondromas and enchondromas. Clin Orthop. 1983;174:264.

Dorfman GD, Czerniak B. Bone tumour. Mosby; 1998.

Keret D, Spatz DK, Caro PA, Mason DE. Dysplasia epiphysialis hemimelia. Diagnosis and treatment. J Pediatr Orthop. 1992;12:365–72.

Kettlekamp DB, Campbell CJ, Bonfiglio M. Dysplasia epiphysialis hemimelica. J Bone Joint Surg. 1966;48A:746–66.

Rosero VM, Kiss S, Terebessy T, Kollo K, Szoke G. Dysplasia epiphysealis hemimelica (Trevor's disease): 7 of our own cases and a review of the literature. Acta Orthop. 2007;78: 856–61.

Disorders of Skeletal Development with Abnormal Connective Tissue Formation

9

9.1 Osteogenesis Imperfecta

(Osteitis fragilis, fragilitas ossium congenital, osteopsathyrosis idiopathica and brittle bones) It is a hereditary often congenital and familial disease with blue sclera. There is a genetic damage to primitive mesenchymal cells. Osteoblasts are in large numbers, they are imperfect in their function and the normal development of mature bone is disturbed. There are deformities and the stature is short (Fig. 9.1a). Osteoporosis is proportional to the severity. The diaphyses and metaphyses are usually very thin with a pencil line cortex. The epiphysis is usually normal but may be cystic with a prominent vertebral trabeculae and kyphoscoliosis. These depend on the age of the patient and the severity of the condition. The cortex appears as a pencil-line and bones are stunted, diffusely rarefied and may have expanded club-shaped extremities. Old deformities following fractures are often apparent and ultimately the long bones may reveal concertina deformities even with non-union at certain sites (Fig. 9.1b). The biochemistry is variable.

S. S. Yadav, *Skeletal Dysplasias & Congenital Malformations*,
https://doi.org/10.1007/978-981-99-5679-1_9

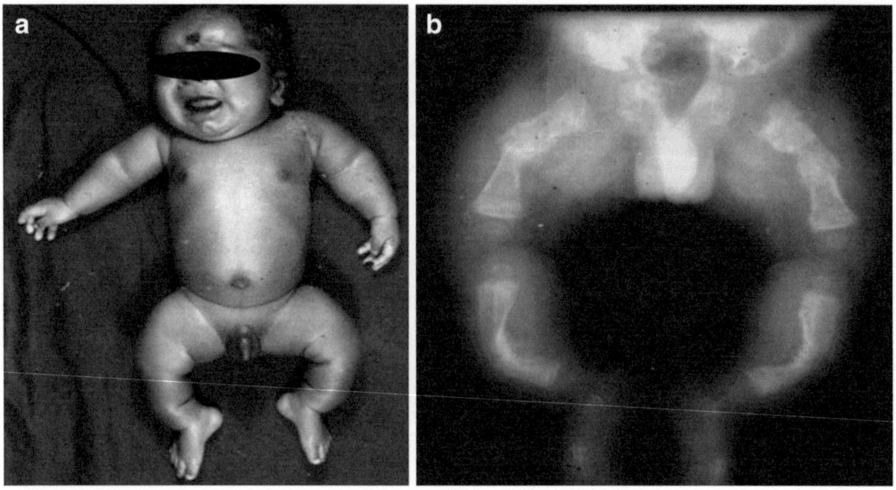

Fig. 9.1 (**a**) A newly born male child of osteogenesis imperfecta involving all the four limbs. (**b**) Roentgenogram reveals contracted pelvis and typical concertina deformity of the limb bones with multiple fractures. There is beading of the ribs

9.2 Osteogenesis Imperfecta Type I

This condition is inherited autosomal dominant and accounts for about 60% of all cases of OI. Approximately one-third of cases are due to new mutations, most commonly in the gene. Patients typically have blue sclerae, a feature which is especially prominent in childhood. Patients may reach normal height or be slightly short. These patients have relatively few fractures; most occur in long bones and tend to heel normally (Fig. 9.2a, b). Fractures rarely occur in infancy but commence when the child begins to walk. Early onset adult hearing loss may occur. There may also be joint hypermobility.

The skeleton usually shows generalized osteopenia and there may be evidence of previous fracture and associated deformity. Wormian bones may be present in cranial sutures (Fig. 9.3).

9.3 Osteogenesis Imperfecta II

This is a severe and lethal form of OI. It is usually due to a new dominant mutation and accounts for approximately 20% of all cases of OI. Affected neonates are short; they are usually stillborn or die in the early neonatal period due to respiratory failure or intracranial haemorrhage. They have numerous fractures and exhibit severe deformities. There is beading of the ribs and deformity of the limb bones as well as poor ossification of the cranial vault and platyspondyly.

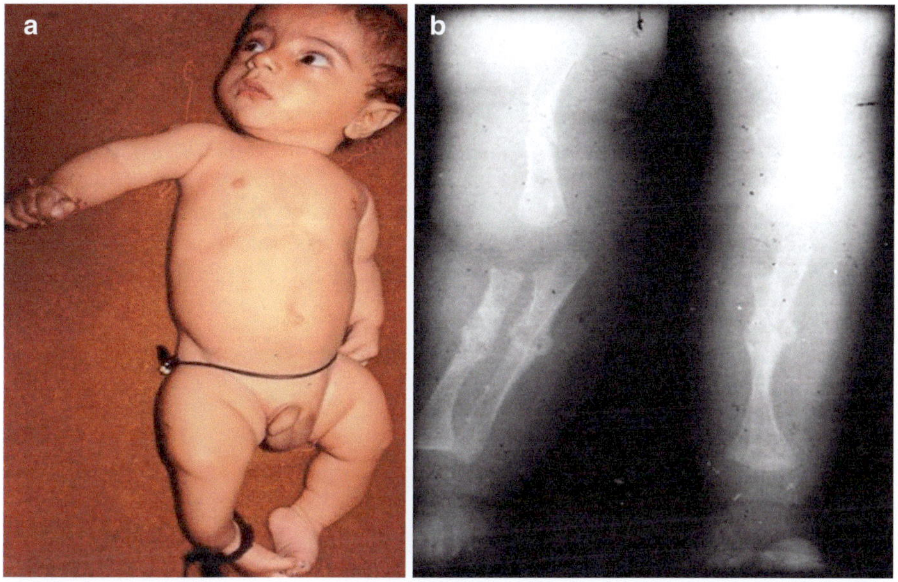

Fig. 9.2 (**a**) A male child with severe deformities of the Limbs. Note the blue sclerae. (**b**) Roentgenogram revealing diaphyseal fractures of the long bones of the upper limb in a case of osteogenesis imperfecta type I

Fig. 9.3 Skull X-ray reveals Wormian bones in a case of Osteogenesis imperfecta

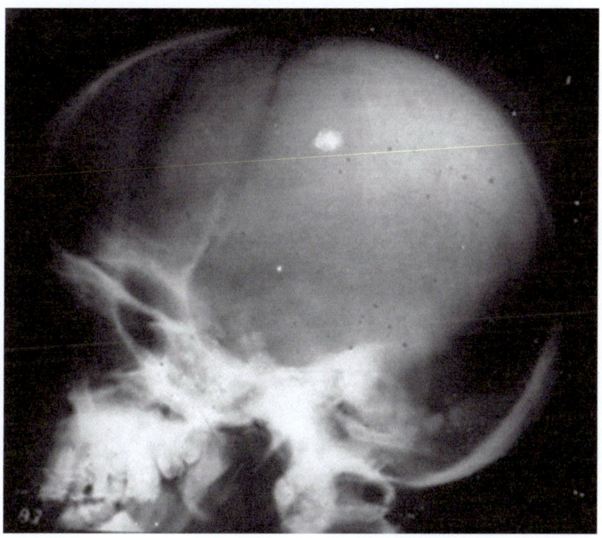

9.4 Osteogenesis Imperfecta III

This is a rare form of OI which results in severe progressive disease. Fractures are commonly present at birth or appear during infancy and continue to occur throughout childhood. This results in progressive skeletal deformity with growth retardation (Fig. 9.4a–f). Short-limbed dwarfism may be noted at birth. The sclerae are usually blue-grey but may be white later in life. There is marked joint laxity. Hearing loss and dentinogenesis imperfecta may occur. There is marked osteopenia with numerous fractures and severe deformity (Fig. 9.5). The calvarium is partly

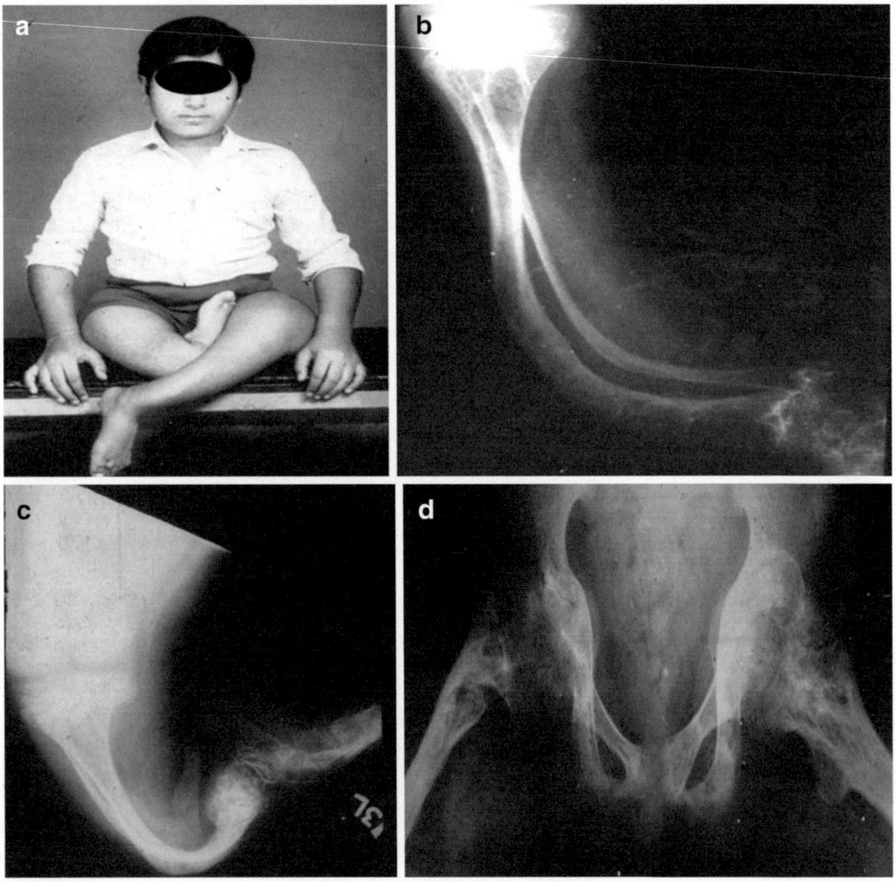

Fig. 9.4 (**a**) A 13-year-old boy showing deformities of the lower extremities. (**b**) The radiograph of the left leg reveal deformity of the distal tibia and fibula in a case of OI type III. Cystic changes are seen in the involved tibia. (**c**) The radiograph of the right leg reveal severe kyphotic deformity of the right leg. (**d**) The radiograph reveals protrusio acetabuli and grossly involved proximal femora with coxarthrosis of the hips. (**e**) Roentgenoram reveals correction and healing of the left leg deformity after multiple osteotomies and supported by inlay fibular graft (Biological nail). (**f**) Clinical appearance of the patient. Walking on '3-point' gait

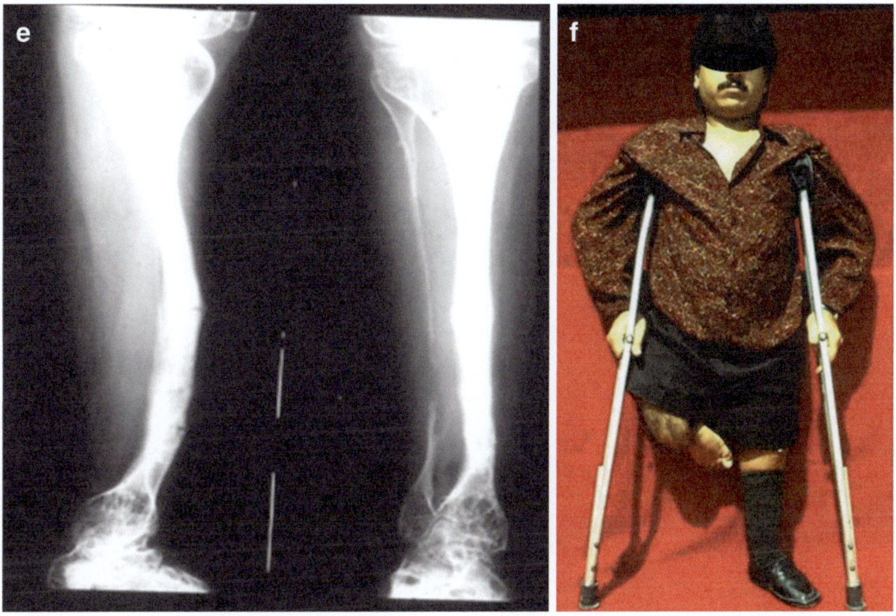

Fig. 9.4 (continued)

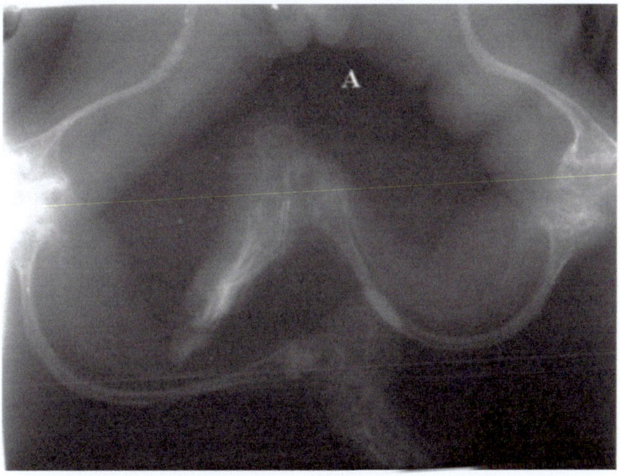

Fig. 9.5 Roentgenogram of a patient with thin lower extremity bones showing severe kyphomelic deformities

calcified at birth. Numerous round cartilage nodules, having a sclerotic margin and a central area of radiolucency are present in the metaphysis and epiphysis of long bones (Fig. 9.6a–f). The 'popcorn calcification' is thought to result from trauma to the growth plate cartilage with consequent abnormal endochondral bone growth. Formation of hyperplastic fracture callus in long bones occasionally results in the formation of a tumour-like mass (Figs. 9.7a–f, 9.8a, b, 9.9a–d).

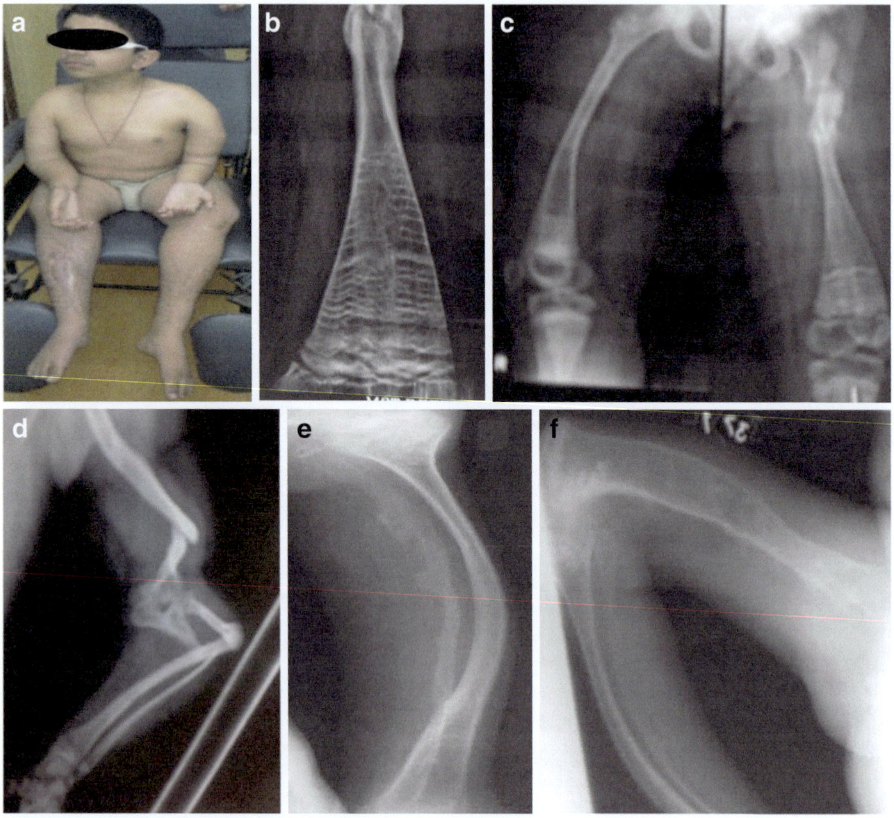

Fig. 9.6 (**a**) A 14-year-old boy revealing multiple upper limb deformities in a case of osteogenesis imperfecta type III. (**b**) The lower end of the femur reveals multiple Park lines. (**c**) Both femora showing evidence of old healed fractures. (**d**) Roentgenogram of a patient with severe deformities of un-united fractures on either side of the elbow. (**e**) Deformities in osteogenesis imperfecta with broadening of the ends of the long bones (trumpet) type of deformity. (**f**) Radiograph revealing flexion contracture of the knee. Note the expansion of the lower end of the femur in a case OI type III

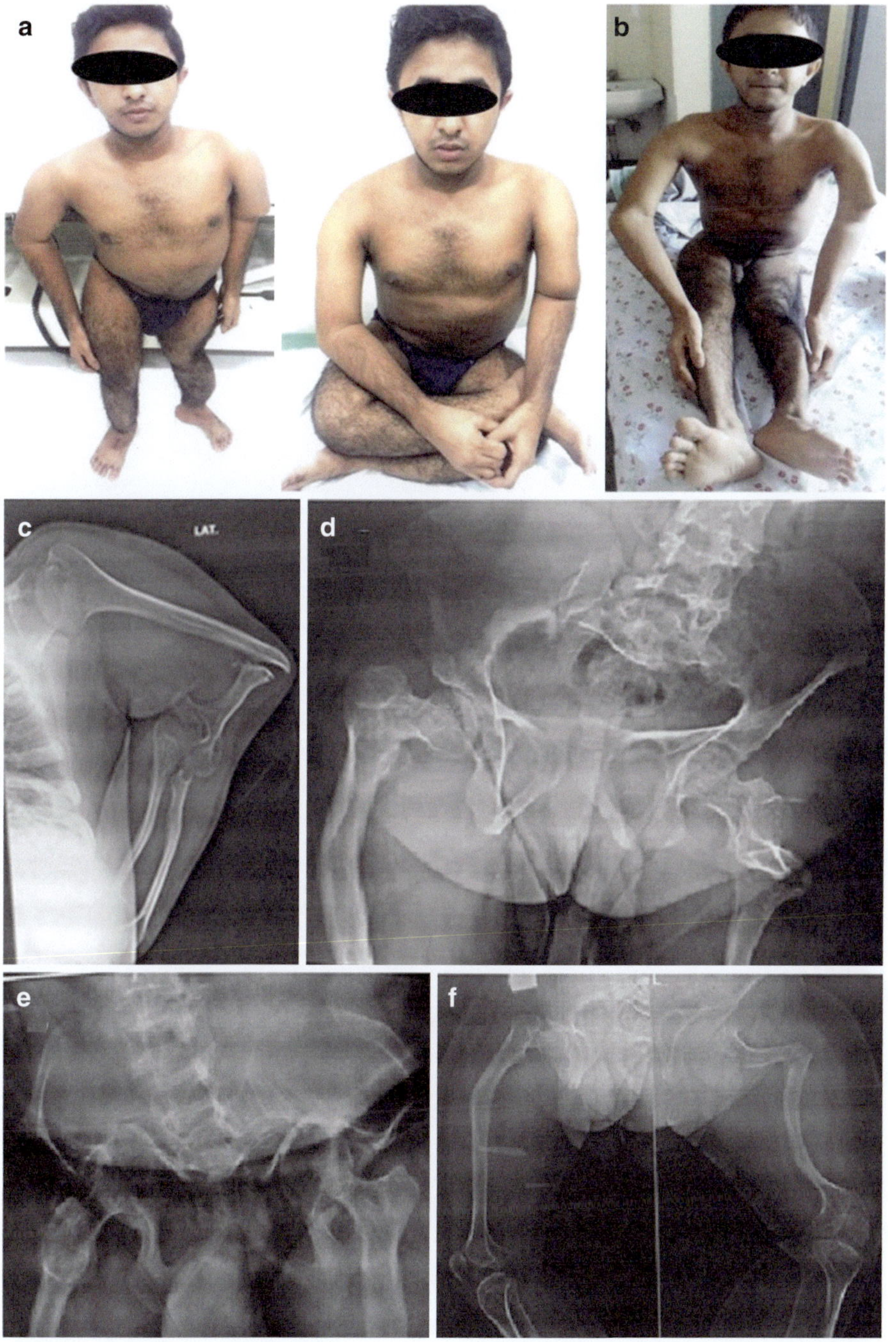

Fig. 9.7 (**a**) A 20-year-old patient of Osteogenesis Imperfecta had multiple fractures of the left upper and lower limbs-OI type III. (**b**) Patient revealing severe deformity of the left arm with non-union of the humeral shaft. (**c**) Roentgenogram of the left humerus reveals non-union with deformity—OI type III. (**d**) The pelvis end of the proximal femora reveal multiple deformities. (**e**). Osteogenesis imperfecta—type III. Acetabular protrusion with 'trefoil' shaped pelvis in a 25 year male. (**f**) Fractured long bones with bowing in an 18-year-old boy

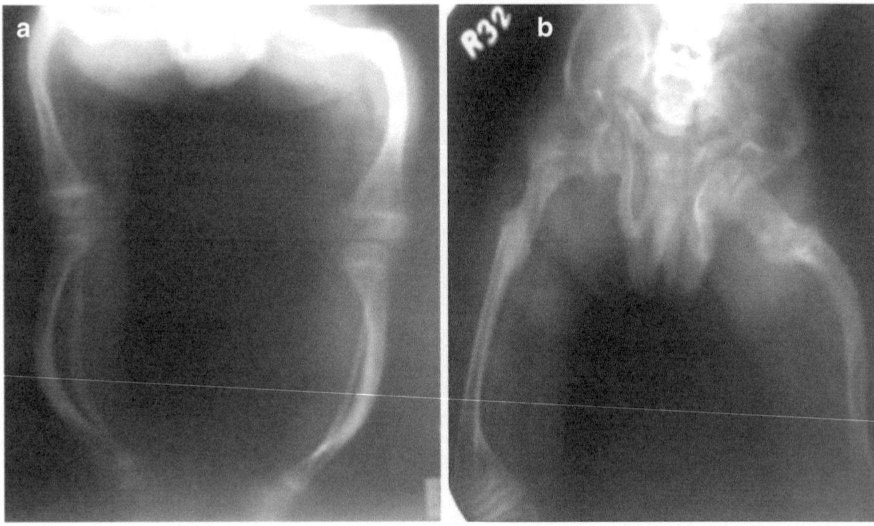

Fig. 9.8 (**a**) X-Ray revealing varus deformity of the lower limb bones with evidence of healed fractures—OI type III. (**b**) X-Ray showing multiple trefoil deformity of the pelvis and healed fractures of both femora

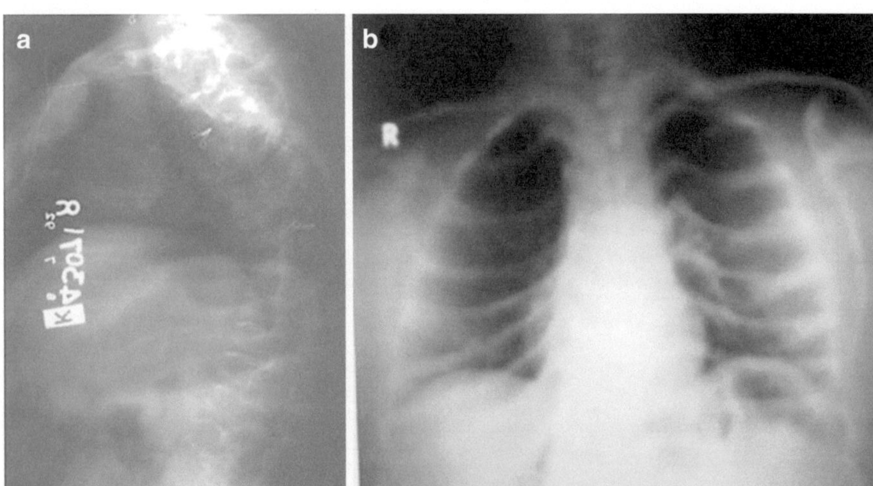

Fig. 9.9 (**a**) Dorso-lumbar spine revealing platyspondyly of the thoracolumbar spine at the age of 6 years. (**b**) Radiograph of the chest of a 19-year-old boy. Note the delicate outline of the ribs with existence of multiple fractures of ribs and beaded appearance on both sides. (**c**) Roentgenogram of the right chest revealing thin and deformed ribs after multiple fractures. (**d**) Roentgenogram shows thin and wide scapula-thoracic joint on the left side

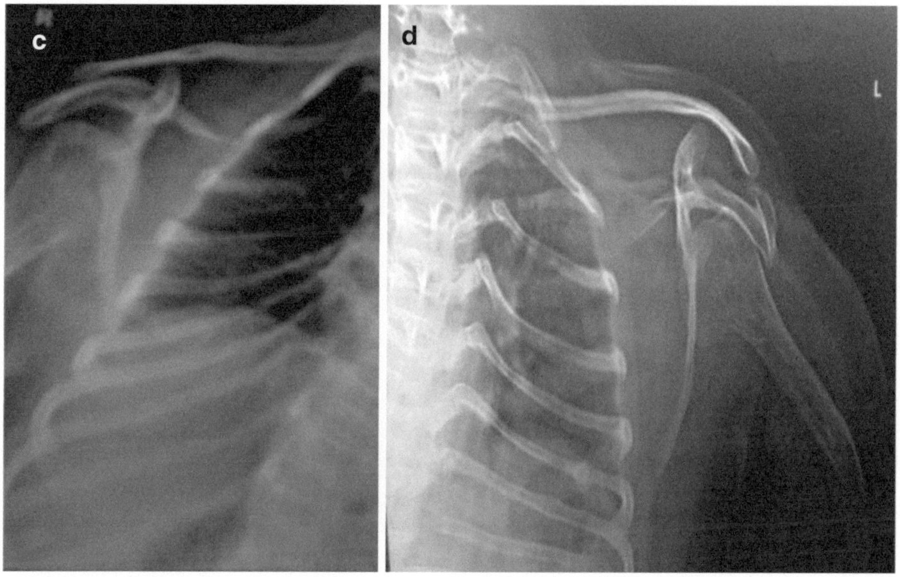

Fig. 9.9 (continued)

9.5 Osteogenesis Imperfecta IV

This form of the disorder is associated with moderately severe deformity and is inherited autosomal dominant (Fig. 9.10a, b). Mild-to-moderate skeletal deformity and disability may occur. The severity of this condition is intermediate between that of type I and type II OI (Fig. 9.11a, b). The incidence of fracture tends to decrease with increasing age. Scoliosis may occur.

There is marked generalized skeletal osteopenia with thinning of cortical bone and medullary bone rarefaction (Fig. 9.12a–c). There may be vertebral compression fractures and cystic change may occur in long bones. There may also be evidence of previous fracture, deformity in long bones and contractures (Fig. 9.13a–d). There is variable short stature (Fig. 9.14.).

The diagnosis at times has to be differentiated with Rickets. In rickets, however, the radiological picture is characteristic where the epiphyseal cartilage is broad and fuzzy, the junction of metaphysis with zone of classified cartilage is irregular and poorly defined. The metaphyses are cup shaped and expanded. On the other hand, the rarefaction of the rest of the bone is considered less than in osteogenesis imperfecta.

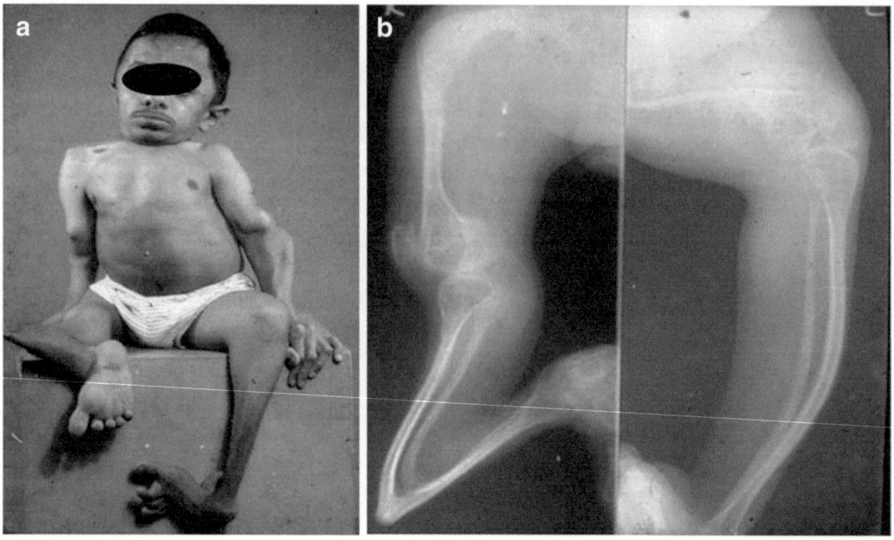

Fig. 9.10 (**a**) A patient with Osteogenesis imperfecta revealing severe angular deformity of the right leg with other deformities in upper and lower extremities. (**b**) Roentegnogram showing acute angulatory deformity of the right tibia and fibula

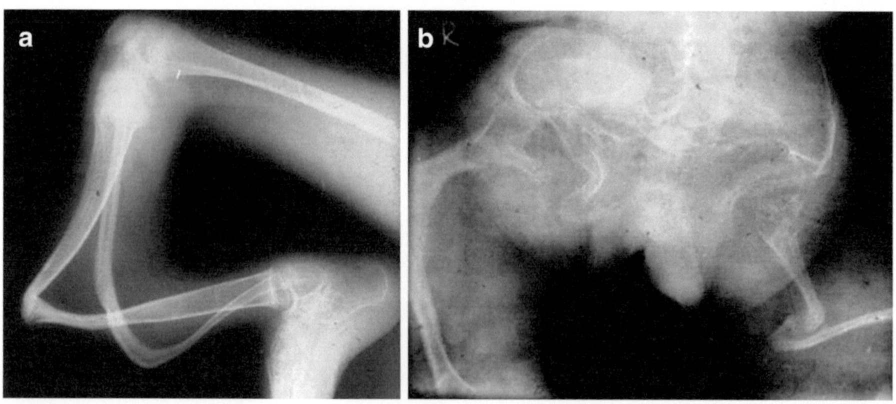

Fig. 9.11 (**a**) Roentgeogram reveals severe angulatory (Kyphotic) deformities of tibial and fibula. Note the trumpet appearance at either ends of the tibia. (**b**) Osteogenesis imperfecta, type IV. 'Trefoil' shaped pelvis and acetabular protrusion of a 26-year-old patient. A malunited fracture of the left femoral shaft with severe angulatory deformity of the femoral shaft

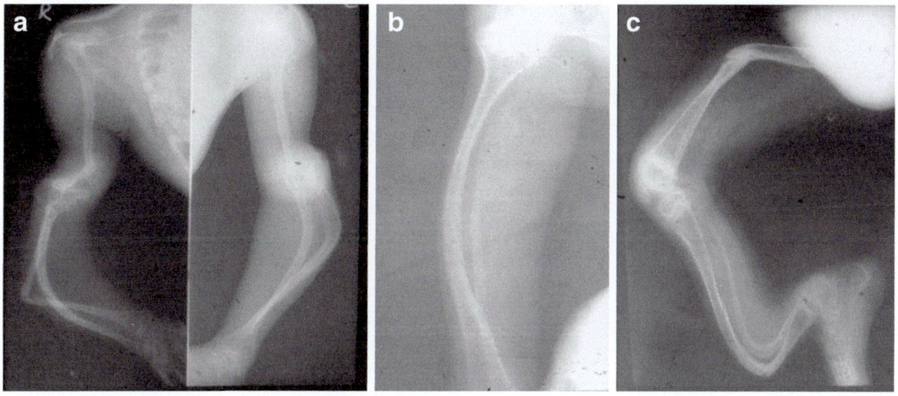

Fig. 9.12 (**a**) Bent Bones, deformities and contracture of various joints in a case of osteogenesis imperfect type IV. (**b**) Radiograph reveals old healed fracture of the femoral shaft. Note congenital pseudoarthrosis of the distal end of the tibia. (**c**) Radiograph reveals old healed fracture of the shaft of femur and anterior angulation of the distal third of tibia and fibula

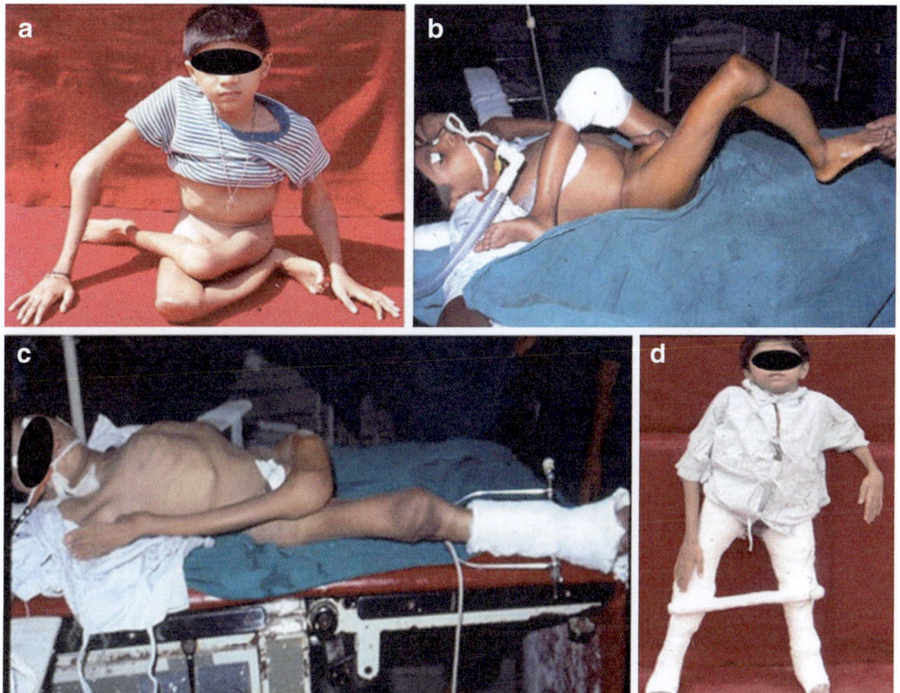

Fig. 9.13 (**a**) A 14-year-old patient with severe deformities of the lower limbs and contracture of hips and knees. (**b**) Pre-operative picture of the patient on the O.T. table for correction of flexion deformity of the right knee. Note the severe flexion deformity of the left hip and the knee. (**c**) The flexion deformity of the right knee is corrected. Surgical correction of the left hip and knee is to be attempted by extensive soft tissue release and multiple osteotomies. (**d**) Patient stands in Double Hip-Spica after complete correction of deformities on both sides

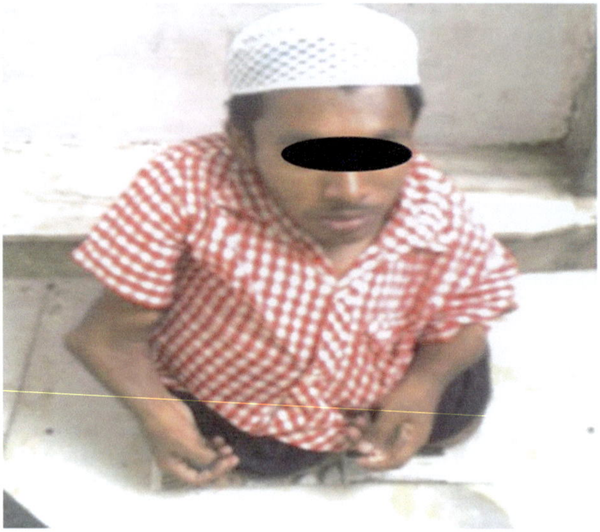

Fig. 9.14 A patient with severe osteogenesis imperfecta revealing the deformity of the right upper extremity. The patient has been sitting on the deformed lower limbs which are too small

9.6 Marfan's Syndrome

Marfan's syndrome is a genetic disorder of connective tissue disorder characterized by diverse skeletal, ocular, cardiovascular and other anomalies. Patients are tall with disproportionately long limbs. There is arachnodactyly (abnormally long and slender digits and long narrow limbs. The ocular findings are myopia and superior displacement of the lens. The cardiovascular findings may be in the form of aortic regurgitation, aortic dilatation, dissecting aneurysm and aortic rupture. The sternum may be depressed (Pectus excavatum) or protruding ((Pectus carinatum). Most of the patients may reveal subluxation of the lens of the eye (Ectopia lentis), myopia and retinal detachment. There may be reduction in the thoracic curve. Scoliosis or kyphoscoliosis may be observed. Such patients may need brace treatment. Kyphotic deformities may require sound fusion. Patients may develop hernia. Pneumothorax and cystic changes in the lungs may also occur.

Radiologically there are various changes in the axial and appendicular skeleton in the form of arachnodactyly, scoliosis and long ribs (Fig. 9.15).

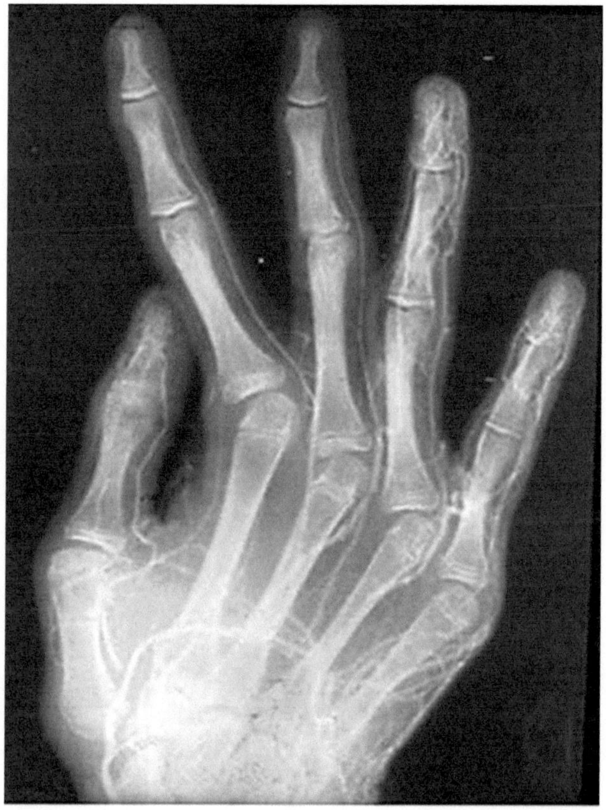

Fig. 9.15 Angiogram of the hand shows arachnodactyly in Marfan's syndrome. Note the symmetrical elongation of the metacarpals and the phalanges with lateral subluxation

9.7 Ehlers-Danlos Syndrome

Ehlers-Danlos syndrome is a group of disorders characterized by skin hyper extensibility, increased joint mobility and tissue fragility. The skin is soft and hyper extensible. Orthopaedically there may be recurrent joint subluxation and dislocation and congenital dislocation of the hip (Fig. 9.16a–j). Scoliosis, clubfoot and spondylolisthesis may also be observed. Radiologically there may be joint dislocation, spinal abnormalities and secondary osteoarthrosis. Hand anomalies include long and cylindrical fingers. Cardiovascular changes may also be observed. Radiologically the skeletal abnormalities may include joint subluxation/dislocation. In late life there may be evidence of osteoarthrosis.

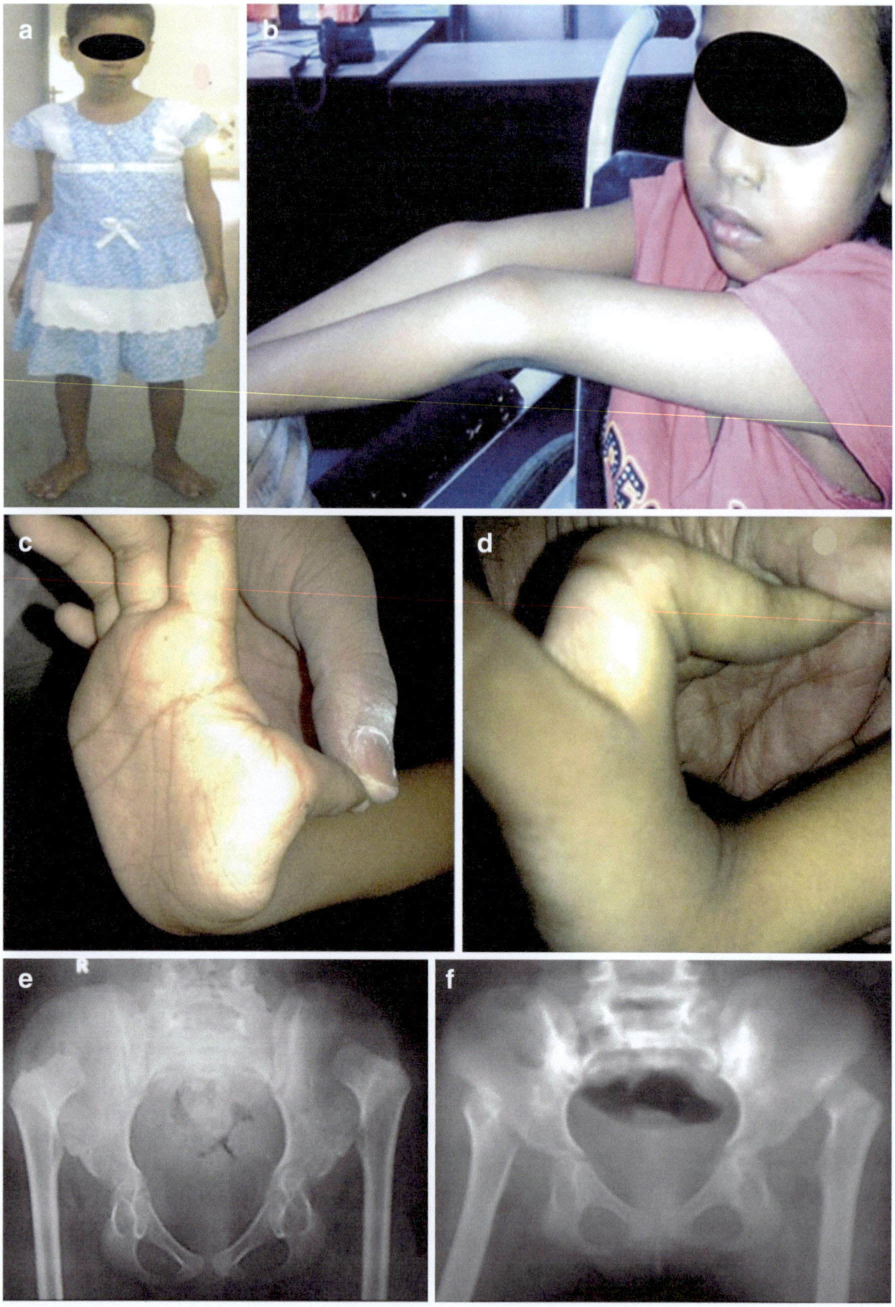

Fig. 9.16 (**a**, **b**) An 8-year-old girl had waddling gate and hyper extension of both the elbows. (**c**, **d**) The hand in a case of Ehlers-Danlos syndrome reveals hypermobility of the fingers and the thumb. (**e**, **f**) The same patient had very high dislocation of both the hips. (**g**) Open reduction was attempted on both sides. The proximal end of the femur could be reduced into the acetabulum easily and without any additional procedure. (**h**) Roentgenogram taken 2 years after the reduction of the hips. (**i**) The same patient standing without any difficulty. (**j**) The patient could squat on the floor

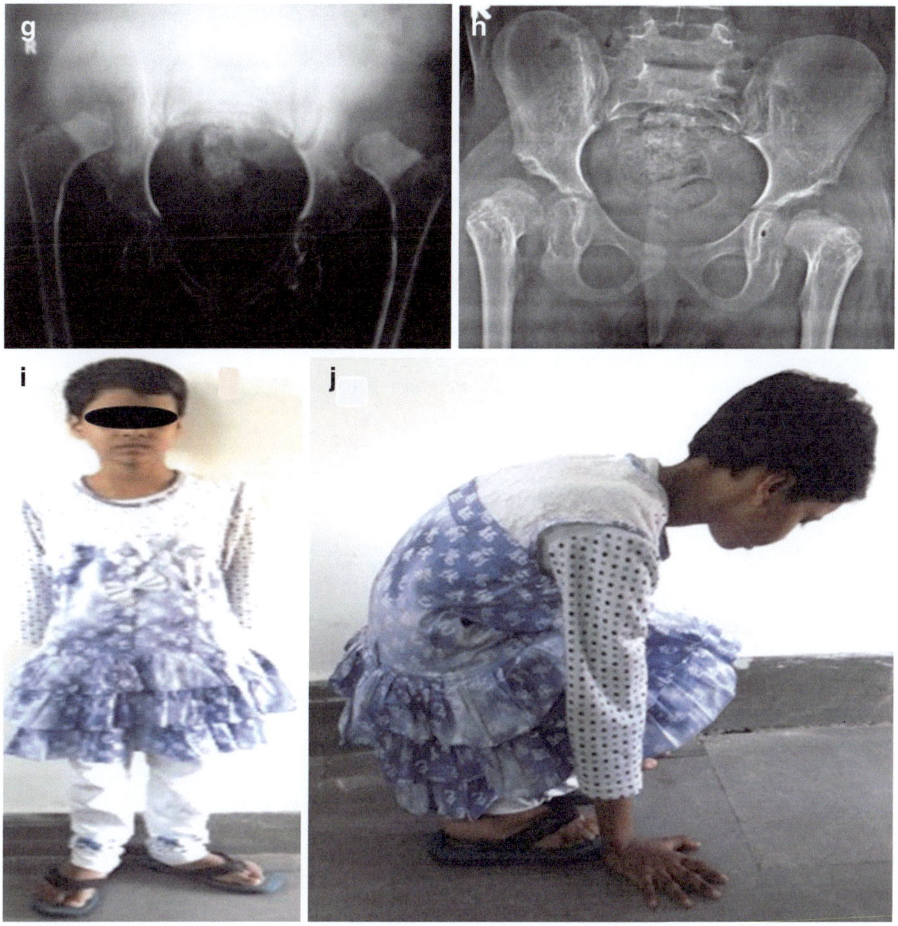

Fig. 9.16 (continued)

Suggested Readings

Cole WG. The molecular pathology of osteogenesis imperfecta. Clin Orthop. 1997;343:235–48.

Cole NL, Goldberg MH, Loftus M, et al. Surgical management of patients with osteogenesis imperfect. J Oral Maxillofac Surg. 1982;40:578.

Rowe DW, Shapiro DW. Osteogenesis imperfect. In: Avioli LV, Kranc S, editors. Metabolic bone disease. W.B. Saundera; 1998. p. 651–95.

Tsipouras P. Osteogenesis imperfect. In: Beighton P, editor. McKusick's heritable disorders of connective tissue. Mosby; 1992. p. 281–314.

Bullough PG, Davidson D, Lorenzo JC. The morbid anatomy of the skeleton in osteogenesis imperfect. Clin Orthop. 1981;159:42–57.

Bauze RJ, Smith R, Francis MJO. A new look at osteogensis imperfecta. J Bone Joint Surg (Br). 1975;57:2.

Francis MO, Bauze RJ, Smith R. Osteogenesis imperfecta: a new classification. Birth Defects. 1985;11:99.

King JD, Bobechko WP. Osteogenesis imperfecta. An orthopaedic description and surgical review. J Bone Joint Surg (Br). 1971;53:72.

Miller EA. Observation on the surgical management osteogenesis imperfecta. Clin Orthop. 1981;159:154.

Pyeritz RE. The Marfan syndrome. In: Royce PM, Steinman B, editors. Connective tissue and its heritable disorders. Molecular, genetic and medical aspects. Wiely-Liss; 1993. p. 437–68.

Pyeritz RE, McKusick VA. The Marfan syndrome: diagnosis and management. N Engl J Med. 1979;300:772.

Robins PR, Moe JH, Winter RB. Scoliosis in Marfan's syndrome. J Bone Joint Surg Am. 1975;57:358.

Godfrey M. The Marfan syndrome. In: Beighton P, editor. McKusick's heritable disorders of connective tissue. Mosby; 1992. p. 51–135.

Beighton P. The Ehlers-Danlos syndrome. In: Beighton P, editor. McKusick's heritable disorders of connective tissue. Mosby; 1995. p. 189–215.

Stanitski DF, Nadjarian R, Stanitski CL, Bawle E, Tsipouras P. Orthopaedic manifestations of Ehlers-Danlos syndrome. Clin Orthop Relat Res. 2000;376:213–21.

Grahame J. Joint hypermobility and genetic collagen disorders. Are they related? Arch Dis Child. 1999;80:188–91.

Disorders of Skeletal Development Due to Lysosomal Storage Disease

10

10.1 Mucopolysaccharidoses: Dysostosis Multiplex

The mucopolysaccharidoses are a recessively inherited group of lysosomal storage diseases due to the deficiency of a specific degradative lysosomal enzyme resulting in accumulation of glycosaminoglycans (mucopolysaccharides) in connective tissue cells. These enzymes are necessary for stepwise degradation of the mucopolysaccharide metabolism. The incomplete degradation products accumulate in lysosomes themselves. These are also known as lysosomal storage diseases and accumulate in the tissues such as brain, viscera and joint capsules. The type of mucopolysaccharidosis depends on the enzyme defect and some of which affect the skeletal system. The lysosomal enzyme that degrades the sulphated glucosamine glycans as heparan sulphate, dermatan sulphate, keratan sulphate and chondroitin sulphate.

The severity of clinical features differs, but as a group these disorders are characterized by variable degrees of skeletal dysplasia and the excretion of excessive amounts of glycosaminoglycans in the urine. Common features include a chronic progressive cause, involvement of many systems, enlargement of organs, widely variable bone disease and an abnormal appearance. Hearing, vision (Corneal blurring) and cardiovascular function.

Although there are many types of mucopolysaccharidoses described, the most common and important of these are as follows:

1. Mucopolysaccharidoses I (Hurler's Disease).
2. Mucopolysaccharidoses II (Hunter's Disease).
3. Mucopolysaccharidoses IV (Morquio's Disease).
4. Goucher's Disease.

So far as the skeleton is concerned, the most characteristic features are found in the Hurler and Morquio syndromes. There are hook shaped vertebrae in lateral

S. S. Yadav, *Skeletal Dysplasias & Congenital Malformations*,
https://doi.org/10.1007/978-981-99-5679-1_10

views of the spine and dysplasia of capital femoral epiphyses, coxa-valga, shortening, metaphyseal widening, epiphyseal dysplasia, osteopenic, coarsely laced trabeculum. Major joints reveal severe contractures. Mental retardation is limited to the Hurler and severe Hunter syndromes.

10.2 Hurler Syndrome-MPS I (Gorgoylism)

Hurler syndrome can be recognized within the first 6 months of life and slowly the symptoms become obvious by the age of 2 years. Abnormal craniofacial features include flattening of the nasal bridge, wide-set eyes, bulging cheeks and thickening of the lips (Fig. 10.1a–d). Skeletal complaints include enlarging scaphocephalic head, restricted joint movements and contractures (Fig. 10.2a–c). Corneal clouding, enlargement of the abdomen and hepatosplenomegaly occurs. There is joint stiffness, increasing dorso-lumbar kyphosis and bowing deformity of the limbs. There has persistent rhinorrhoea. Repeated respiratory infections are common. Independent walking is delayed, speech is rudimentary and hoarse. Later, hepatosplenomegaly, dwarfism, corneal opacity, umbilical hernia and cardiac abnormalities become increasingly evident.

The skeleton is osteoporotic. There is variable generalized rarefaction of tubular bones, delay in the appearance of epiphyseal ossification centres and epiphyseal and metaphyseal dysplasia. The femoral neck is valgoid and the acetabulum is shallow. Dermatan or heparin sulphate is usually detected in the urine. Affected individuals usually die, may be in 10–15 years.

The fingers are shortened. There is widening of the phalanges. The metacarpals (2–5) are pointing at their proximal ends (Fig. 10.3a–c). The first metacarpal is short and avoid. The second metacarpal is long and ovoid and its proximal phalanx is long. The carpal bones are absent. The distal radius and ulna are variously involved. Beaking has been common observation in the lumbar region (Fig. 10.3d, e). There is irregular thickening of the diploe of the skull and delay in ossification of the carpus. The femoral necks are valgoid and the acetabulae are shallow with flared ilia. Joint contractures are common. There is mental retardation.

There is craniosynostosis with early closure of cranial sutures. The thickened skull also has bony projections arising from the floor of the middle and posterior fossa. The term dysostosis multiplex is used to describe the above radiological features which are seen to a lesser extent in other types of MPS and other lysosomal storage diseases affecting the skeleton.

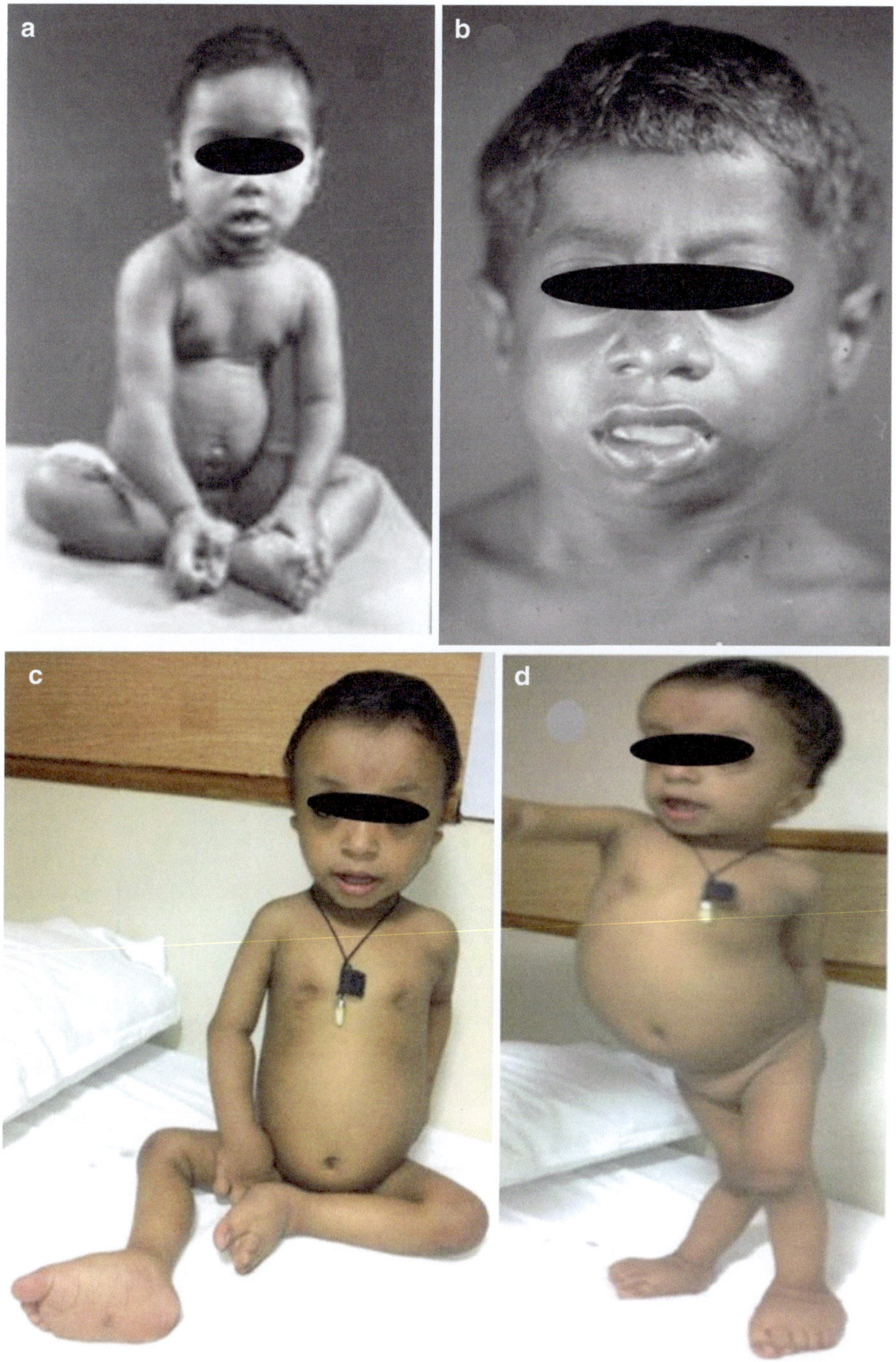

Fig. 10.1 (**a**) A child with mucopolysaccharidosis with open mouth and protuberant tongue and protuberant abdomen. The patient is mentally retarded. (**b**) Note the protruding eyes and tongue, rounded cheeks and thick lips. The patient is mentally retarded. (**c**, **d**) A patient of mucopolysaccharidosis showing large head, bossing of the forehead, open mouth, genu-valgum and recurvatum right lower limb and a protuberant abdomen. Note the broad and short feet

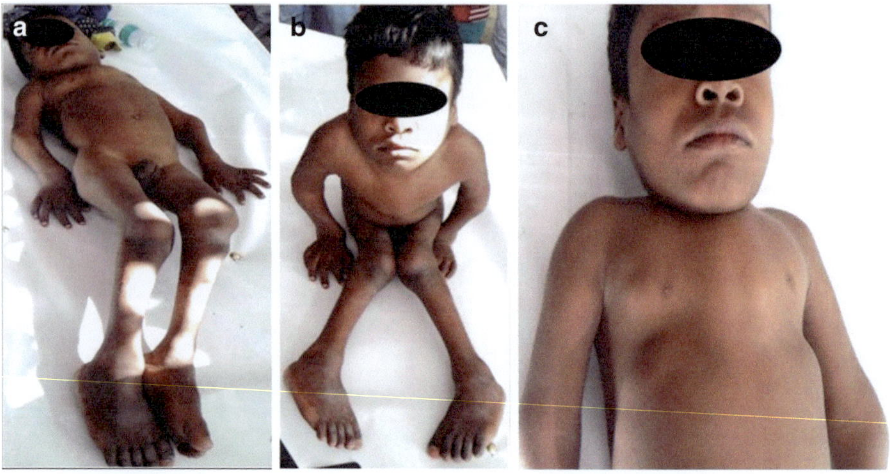

Fig. 10.2 (**a**–**c**) A child with mucopolysaccharidosis showing classical facial appearance along with joint contractures and broad hand and feet

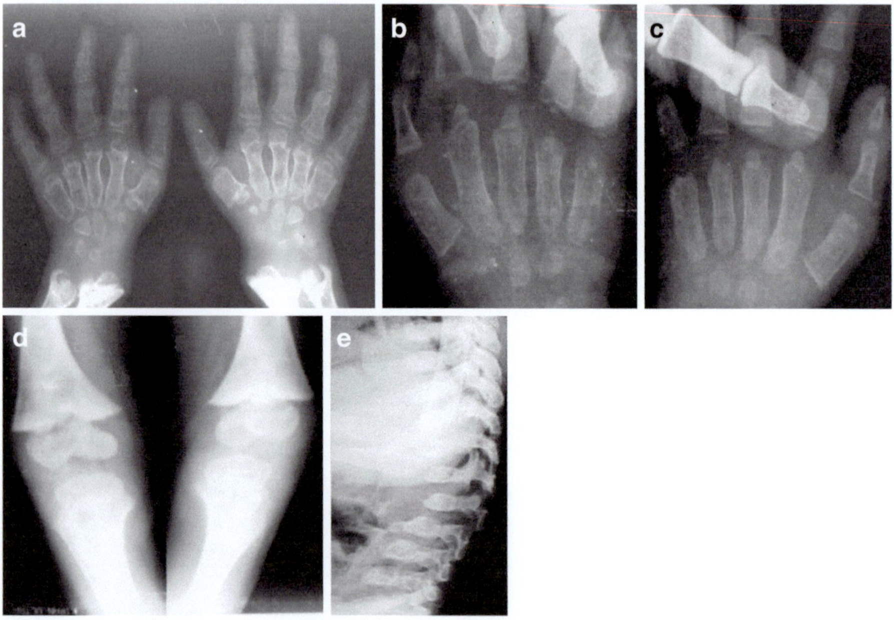

Fig. 10.3 (**a**) Roentgenograms reveal classical tapering of the proximal ends of the metacarpals, (2–5th) along with thickening. Note the epiphysio metaphyseal changes at the lower end of radius and ulna. (**b**, **c**) Another patient showing revealing thickening of metacarpals with tapering of their proximal ends. (**d**) Roentgenogram reveals epiphysio metaphyseal widening along with varus deformity of the knees. (**e**) The lumbar spine reveals hook shaped vertebrae. There is some degree of kyphoscoliosis

10.3 Hunter's Syndrome (MPS II)

The disease occurs in males. In the severe form, the clinical features are essentially similar to those of Hurler's disease but develop more slowly. In the mild form of the disease, patients are of normal intelligence and can survive into adult life. Hepatosplenomegaly and cardiac valve abnormalities occur frequently. The axial and appendicular skeleton show similar changes to those of Hurler's disease.

Although normally milder than the Hurler syndrome, the severe MPS IIA closely resembles Hurler syndrome. It is suspected around the age of 3 with slow speech, recurrent respiratory infections. The head is large, the hands and fingers stiff with joint contractures (Fig. 10.4a–d). The abdomen is enlarged due to hepatosplenomegaly. There may be progressive deafness and corneal blurring. Dermatan or heparin sulphate is usually detected in the urine.

Radiologically there is widening of the phalanges. The metacarpals (2–5) are pointing at their proximal ends (Fig. 10.5a, b). The first metacarpal is usually short and avoid. The carpal bones are absent. The distal radius and ulna are variously involved. The spine reveals kyphoscoliosis and there may be other skeletal deformities (Fig. 10.6a–c).

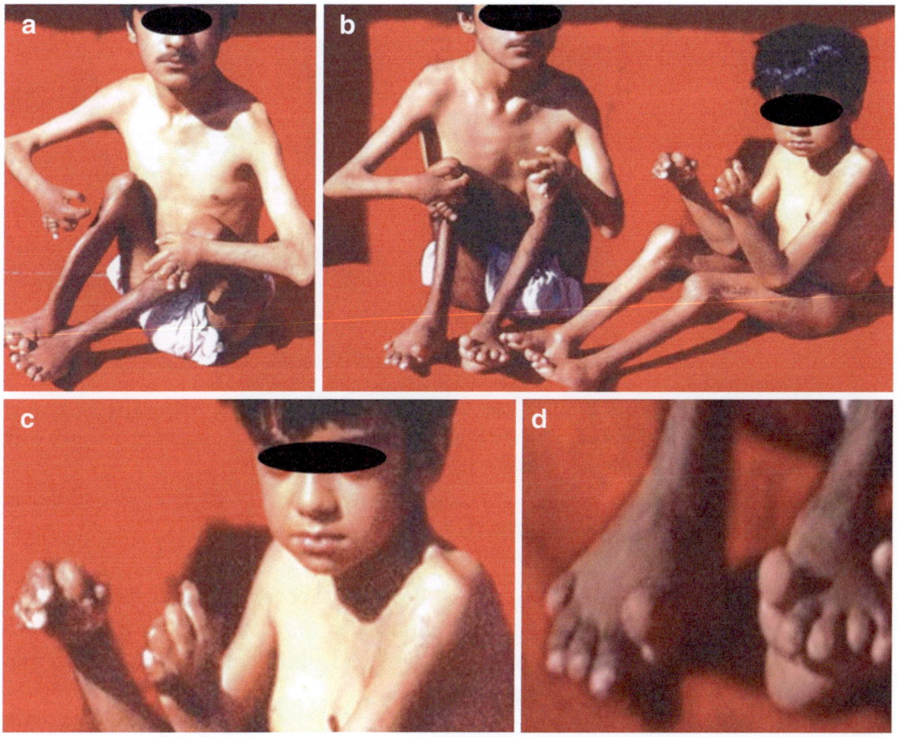

Fig. 10.4 (**a**) An 18-year-old patient of Hunter's disease showing multiple deformities and joint contractures. Note the nodular skin lesions on the hands. The fingers are abnormally shaped. The fingers and toes are short and broad. (**b**) The patient with his younger brother showing similar presentation. (**c**) The same patient revealing corneal blurring. (**d**) The feet are small and broad and reveal deformed big toes

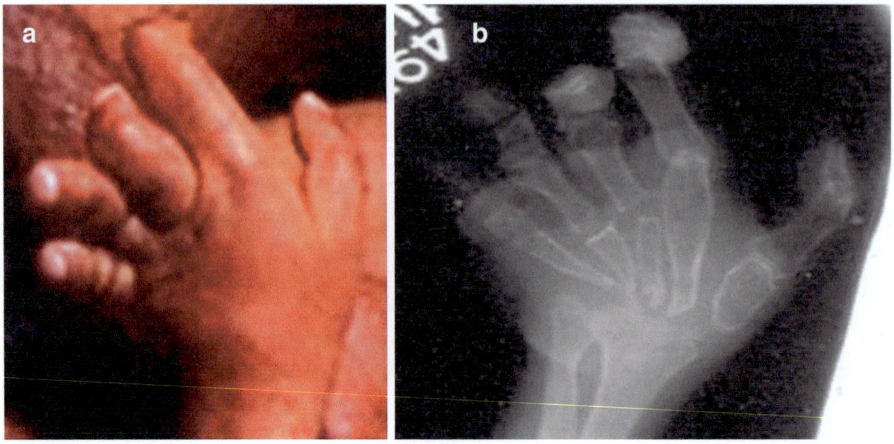

Fig. 10.5 (**a**) The hand is small and wide. The fingers are abnormal in appearance with nodular swelling. There is complete loss of finger flexion. (**b**) The fingers are shortened. There is widening of the phalanges. The metacarpals (2–5) are pointing at their proximal ends. The first metacarpal is short and avoid. The second metacarpal is long and ovoid and its proximal phalanx is long. The carpal bones are absent. The distal radius and ulna are variously involved

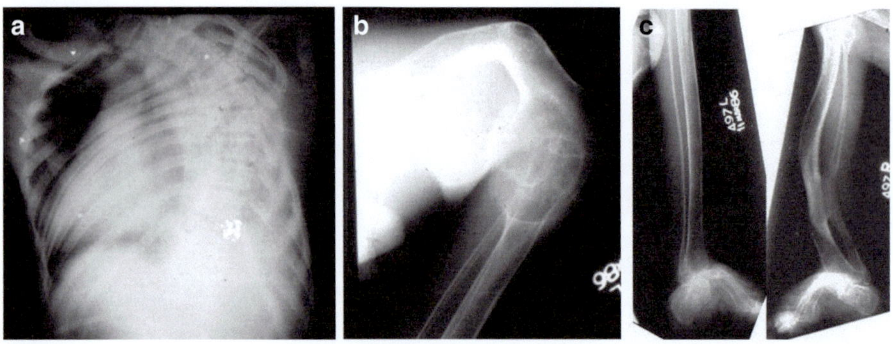

Fig. 10.6 (**a**) Note the kyphoscoliotic spine. (**b**) Roegtenogram revealing malunited fracture of the distal femur with knee joint contracture. (**c**) The tibia and fibula are thinned and attenuated with normal appearance of the epiphysis. Note the cavus deformity of the feet

10.4 Morquio's Disease (MP IV)

It differs from other forms of MPS in that there is joint laxity rather than rigidity. The characteristic short trunk and short stature exist. Facial features are not coarse and intelligence is normal. There may be large head, short neck and pectus carinatum (Fig. 10.7a). Two distinct forms of this syndrome are recognized.

Affected individuals appear normal at birth but develop progressive severe growth retardation. There may be dorso-lumbar Kyphoscoliosis and genu-valgum in early childhood. There is enlargement and development of flexion contractures. Organomegaly is not prominent. Deafness and cardiac abnormalities such as aortic incompetence may occur.

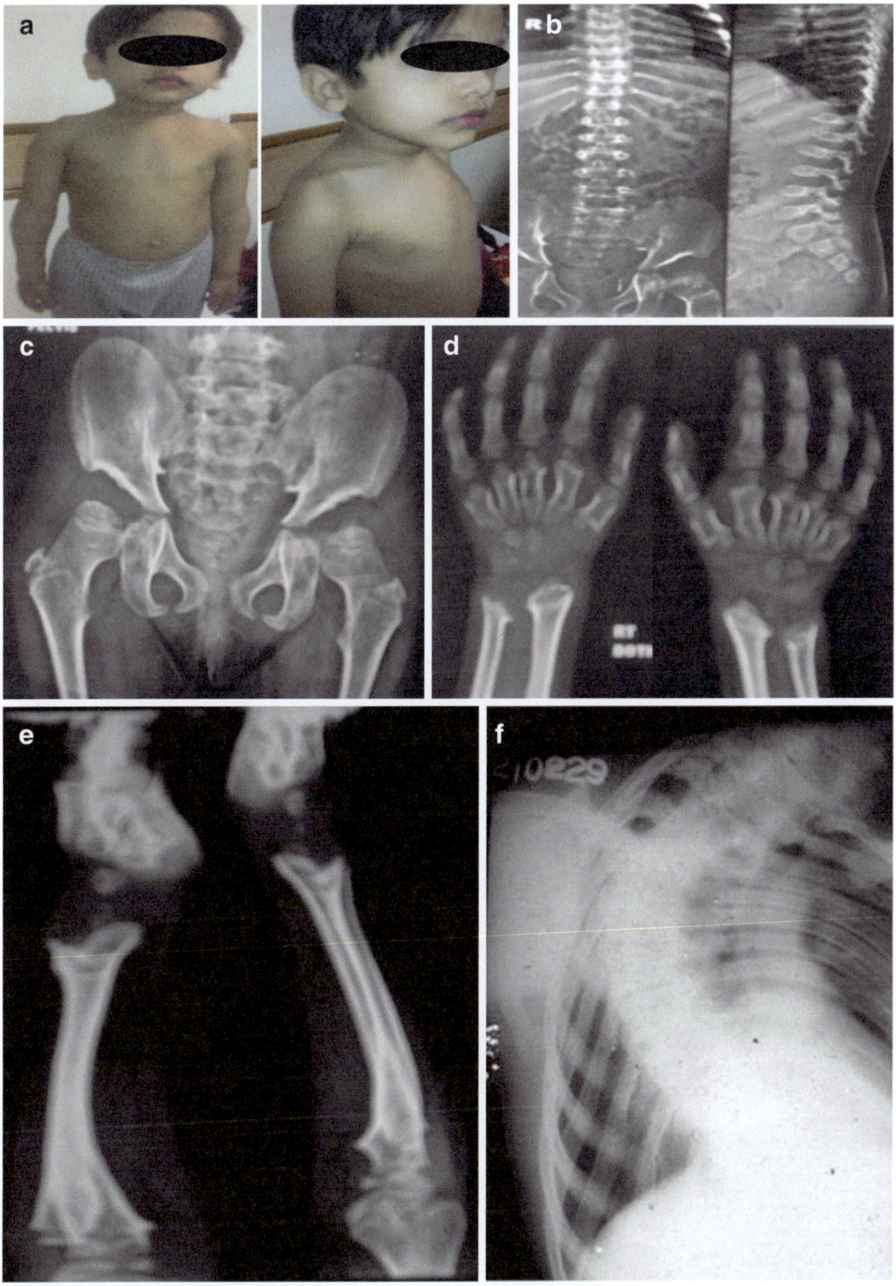

Fig. 10.7 (**a**) A case of Morquio's disease (MPH-IV) has pectus carinatum. (**b**) Roentgenogram of the DL spine reveals abnormal vertebrae in the form of classical anterior beaking (tongued) vertebral bodies. (**c**) Roentgenogram reveals large and flattened acetabulae. Note the coxa-valga with flattening of the heads of the femora. (**d**) The metacarpals reveal diagnostic signs of MPH with pointing proximal ends. The lower ends of the radius in ulna also reveal cupping and other developmental abnormalities. (**e**) Roentgenogram reveals signs of minimally developed carpal bones. The lower end of the radius and ulna reveal classical cupping signs. (**f**) Roentgenogram reveals kyphoscoliosis in a case of Morquio's Disease

Radiologically the vertebral bodies are flattened and there is anterior beaking (Fig. 10.7b). The odontoid process is hypoplastic and atlanto-axial instability may be observed. The tubular bones are short with small and poorly defined ossification centres. There may be epiphyseal irregularity at the femoral heads which may mimic Perthes' disease (Fig. 10.7c, d). Kyphoscoliosis is evident and the disc spaces are widened (Fig. 10.7e, f). Bones of the extremities are small and stubby. Iliac bones are flared. Keratan sulphate is present in the urine of such patients.

10.5 Gaucher's Disease

Gaucher's disease is characterized by Hypersplenism associated with anaemia, leucopoenia and thrombocytopaenia. The skeletal manifestations include bone pain osteopenia, pathological fractures and medullary and subchondral osteonecrosis. Degenerative arthritis and Kyphoscoliosis is common.

Radiologically osteonecrosis may present as an acute osteomyelitis like changes usually in the medial femoral condyle (Fig. 10.8). There is localized tenderness and with all the symptoms of acute infection. There may be failure of remodelling. There is flaring ends of the long bones particularly the distal end of the femur and proximal end of the tibia. There is evidence of medullary infarction of subchondral bone narcosis with collapse of the articular surface. Bone scintigraphy shows increased isotope uptake in affected bones.

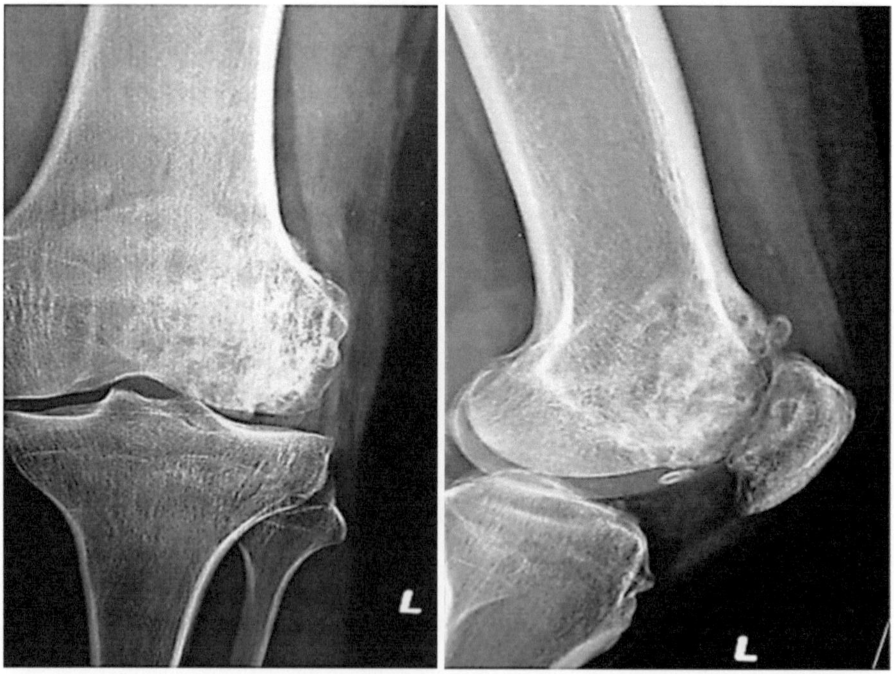

Fig. 10.8 Radiographic appearance of the knee in Goucher's disease showing osteonecrosis of the lateral condyle of the femur

Suggested Readings

Hopwood JJ, Phillip Morris C. The mucopolysaccharidoses. Diagnosis, molecular genetics and treatment. Mol Biol Med. 1990;7:381–404.

Kelly TE. The mucopolysaccharidosis and mucolipidoses. Clin Orthop. 1976;114:116–36.

Schmidt H, Ullrich K, Von Lengerke HJ. Radiological findings in patients with mucopolysaccharidosis I. Pediatr Radiol. 1987;17:409–14.

Beutler E. Gaucher's disease. New Engl J Med. 1991;325:1354–60.

Hermann G, Pastores GM, Abdelwahab IF, Lorberboym AM. Gaucher disease. Assessment of skeletal involvement and therapeutic responses to enzyme replacement. Skelet Radiol. 1997;26:687–96.

Whitley CB. The mucopolysaccharidoses. In: Beighton P, editor. McKusick's heritable disorders of connective tissue. Mosby; 1992. p. 367–499.

Disorders of Skeletal Development Associated with Osteoscalerosis

11

11.1 Osteopetrosis

Osteopetrosis (Marble bones, Albers Schonberg's disease) is a rare familial disease, characterized roentgenographically by a marked increase in the radio density of bones and by abnormalities in the shapes of the long bones, especially the metaphyses (Fig. 11.1a–c). There is a genetic damage resulting impaired activity of the osteoclasts and failure of reabsorption of bone causing abnormal density and thickening in a way, osteopetrosis may be considered as counter part of osteogenesis imperfecta. Regional osteopetrosis is also known to occur in a few bones only. The ends of the long bone may resemble healing rickets.

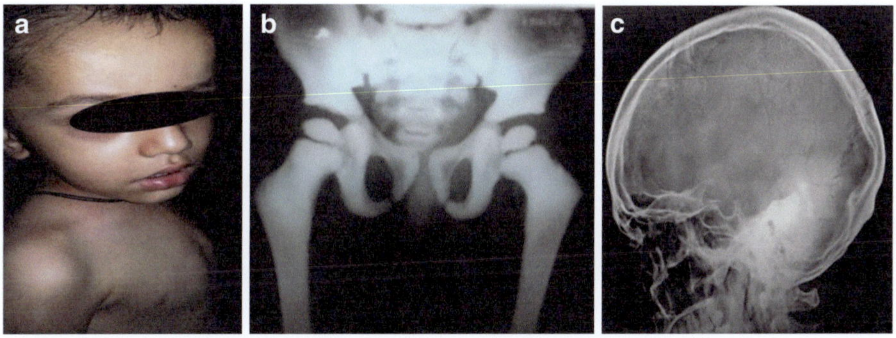

Fig. 11.1 (**a**, **b**) Clinical presentation of a child with marble bone disease along with his radiograph of the pelvis and proximal femora. (**c**) Roentgenogram reveals thickening and sclerosis of the skull bones. Note the evidence of vermian bones in the posterior fossa

S. S. Yadav, *Skeletal Dysplasias & Congenital Malformations*,
https://doi.org/10.1007/978-981-99-5679-1_11

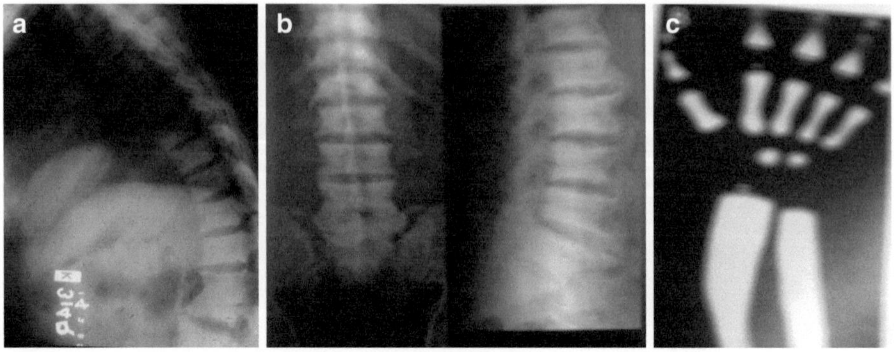

Fig. 11.2 (**a**, **b**) The spine also reveals sclerosis with patchy classification of the long anterior ligament (**c**) Roentgenogram of the forearm and hand showing massive sclerosis with no medullary canal

The condition has not infrequently affected several members of a family. Any or all of the bones may be affected and the condition is usually symmetrical (Fig. 11.2a–c). In the case of the long bones, the usual differentiation into cortex and medulla is lost and the change may affect any part of the bones but it is most common at the metaphyses. The condition affects not only the bone developed from the epiphyseal cartilage, but also that growing from the subperiosteal osteogenic layer. It is well-known that the so-called marble bones are liable to fracture, especially in the adolescent and is not an Index of increased strength (Fig. 11.3a–f). Chalky bones, a condition closely resembling 'marble bones' was found in those working with cryolite, a compound of fluorine, sodium and aluminium.

Fig. 11.3 (**a**) Roentgenogram reveals a case of osteopetrosis with dense bones. There is loss of corticomedullary differentiation. (**b**) The femoral neck on the left side shows fracture. The fracture has been fixed with twin-free fibular struts with Kirschner wires. (**c**) Roentgenogram reveals thick skull vault and amorphous sclerosis of the base with narrowing of the sella turcica. (**d**) Osteopetrosis involving the tibia and fibula. Note the uniform scoliosis of both the bones. (**e**) Shows evidence of healed fracture of tibial diaphysis. (**f**) Roentgenogram reveals old healed fracture of the neck on the left side with varus deformity

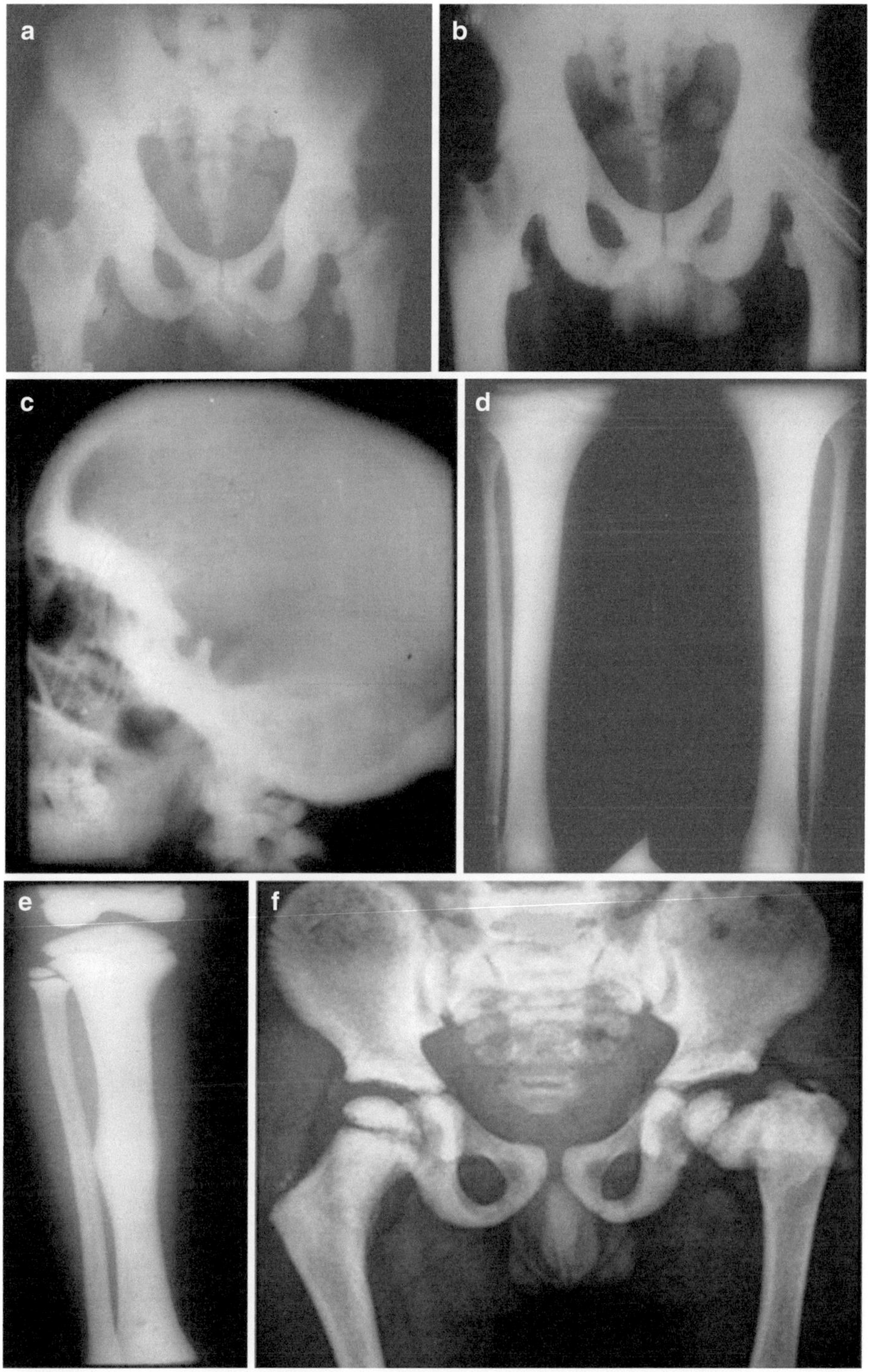
a
b
c
d
e
f

11.2 Pyknodysostosis

Pyknodysostosis is characterized by generalized osteosclerosis, short stature and abnormal skull development. The cranium is large with small faces and receding chin (Fig. 11.4a, b). The anterior fontanelle remains patent and fronto-occipital bulging. There are dental anomalies with high arch palate. Patient may also have proptosis blue sclera and shortening of the terminal phalanges. Dysplasia of the finger nails may also be observed.

The bones are fragile with increased risk of pathological fracture. The hands are short and stubby with spoon-nail deformity with distal osteolysis. The deciduous teeth are malformed and the clavicles are incompletely formed, with hypoplasia of the acromial ends. The entire skeleton displays an increased bone density. The thorax is narrow and there may be pectus excavatum.

Radiologically there is dense and uniform osteosclerosis of long bones due to marked thickening. There is hypoplasia of the terminal phalanges of the hands and the feet (Fig. 11.5a, b). There is delayed closure of cranial sutures with obtuse mandibular angle.

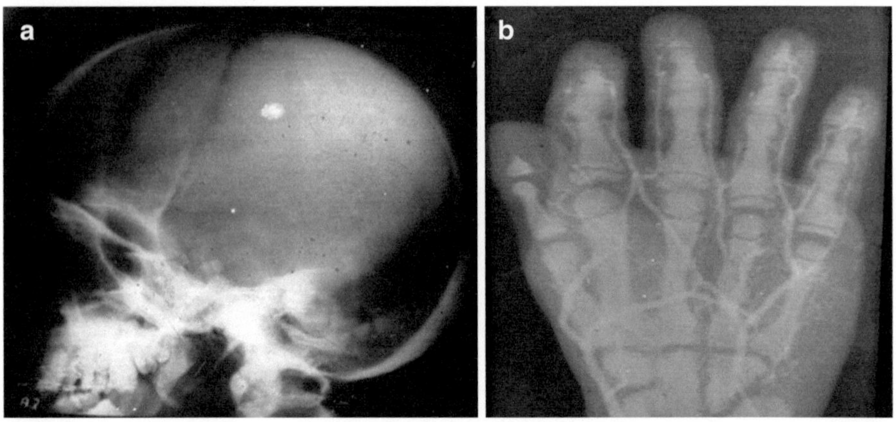

Fig. 11.4 (**a**) The occipital bone reveals Wormian bones. (**b**) An angiogram in a case of c revealing diminished vasculature at the tips of the fingers

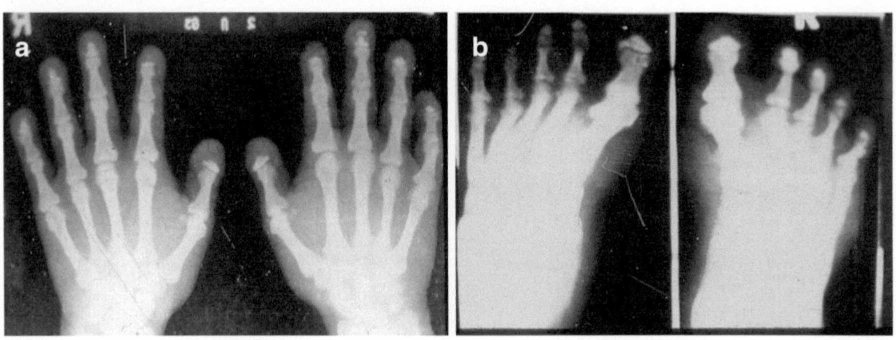

Fig. 11.5 (**a, b**) A classical case of pyknodysostosis with short and conical distal phalanges in the hands and feet

11.3 Melorheostosis

Melorheostosis, described by Leri and Joanny (1922), is a rare, non-familial condition marked by deformity of the characteristic, extra cortical hyperostosis. It is characterized by dense condensation on the one side of several bones of a limb. Men and women are equally affected. There is an endosteal pattern of hyperostosis marked by streakiness of the long bones and spotting of the small bones and the usual subperiosteal or extra cortical pattern of hyperostosis seen in adults which increase with age. Although there may be bone pain and limitation of joint movements with deformities, the bones are not fragile and there is no special susceptibility to fracture. Sometimes the lesion is associated with deformity of the affected bone. Thus in the lower limb, bowing of the femur and of the tibia is common. There may be pain, swelling and stiffness in the neighbouring joints. The disease may be confined to one limb or may be bilateral. Several bones may be involved in the same extremity (Fig. 11.6a–c).

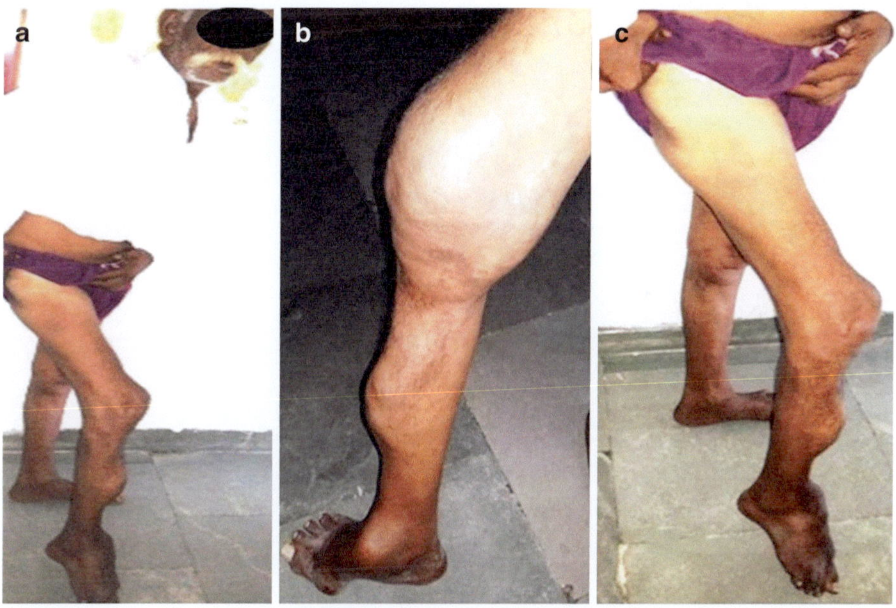

Fig. 11.6 (**a, b, c**) A 60-year-old patient with advanced Melorheostosis of the right lower extremitiy. The total extremity is involved. (**d**) Roentgenogram reveals a dense cortical sclerotic lesion on the medial femoral shaft on the right side and one such lesion under the neck of the femur. (**e**) Radiograph shows continuous involvement from the hip to the knee. (**f, g**) Roentgenograms revealing classical presentation of melorheostosis in advance stages at various sites at femur and tibia. (**h**) Roentgenograms of the feet show classical 'dripping candle wax' changes in the first meta-tarsal and the big toe. (**i, j**) Roentgenograms revealing scalerotic course lines in the body of the talus

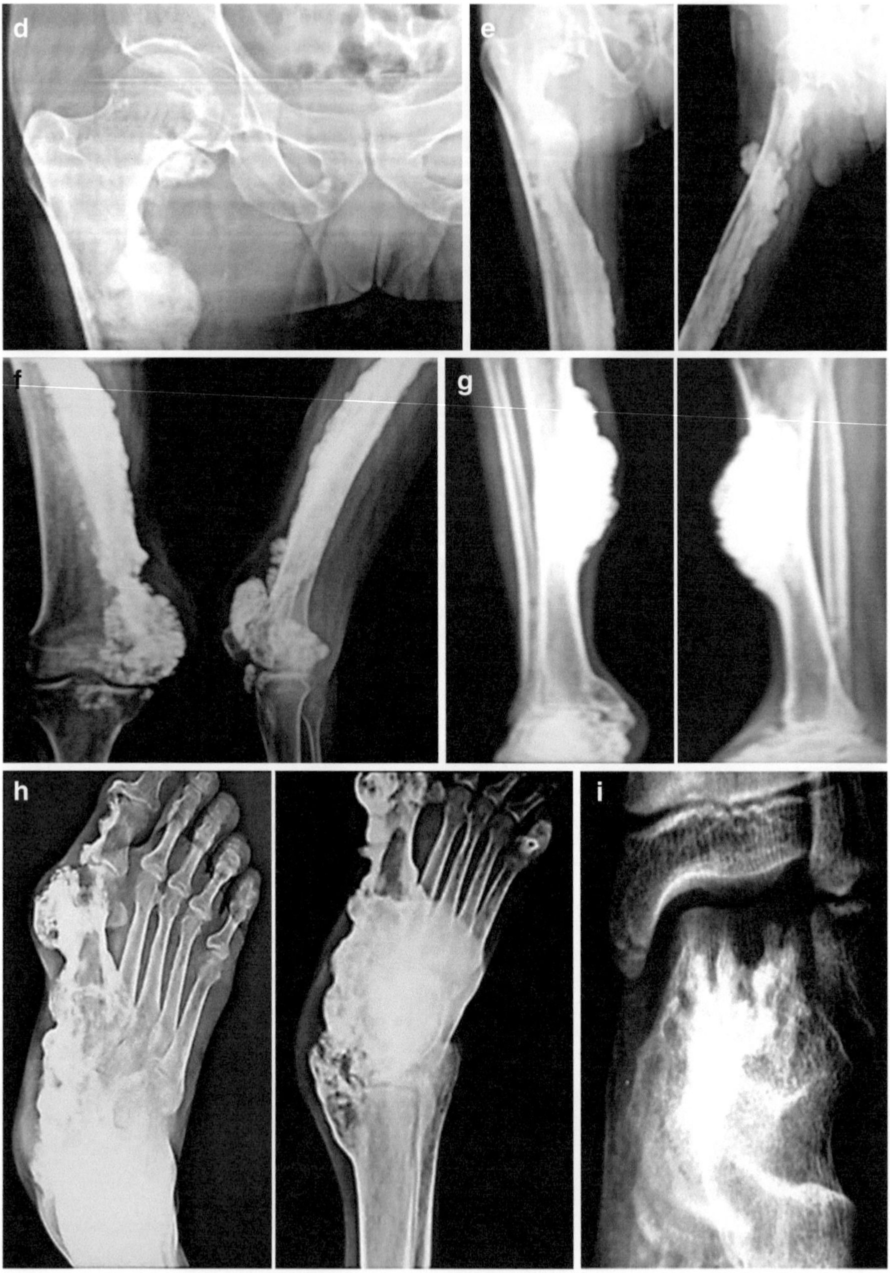

Fig. 11.6 (continued)

Fig. 11.6 (continued)

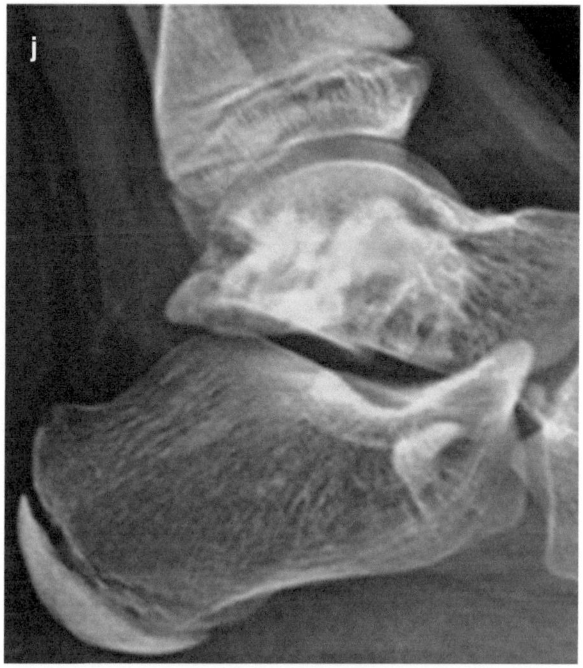

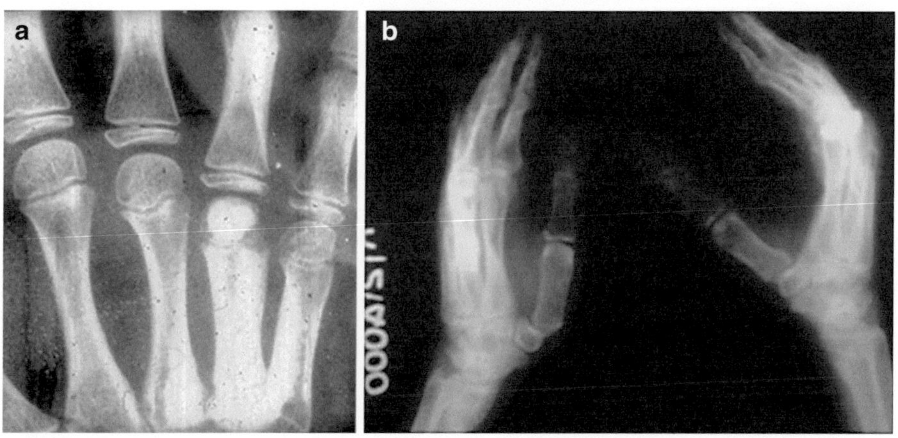

Fig. 11.7 (**a**) Roentgenogram showing scalerosis of the fourth and fifth metacarpals and their respective proximal phalanges. (**b**) There is evidence of dripping candle wax in the second and third metacarpals

Radiologically there is irregular patchy osteosclerosis with thickening of the endostieal and periosteal surfaces of long bone. The appearance resembles those of 'dripping candle wax' (Fig. 11.6d–j). There may also be shortening and deformity of the affected bones (Figs. 11.7a, b and 11.8a, b). The differential diagnoses for melorheostosis include osteopoikilosis, osteopathic striata, myositis ossificans, parosteal osteosarcoma and osteoma. The treatment in most instances has been symptomatic.

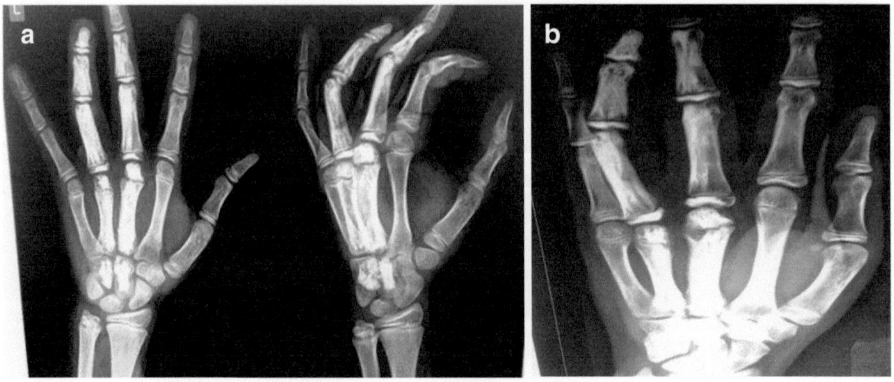

Fig. 11.8 (**a**) Roentgenogram reveals classical changes of melorheostosis in third and fourth ray. (**b**) Radiograph shows scalerotic changes in second, third and fourth rays

11.4 Osteopoikilosis

In osteopoikilosis (spotted bones) there are rounded sclerotic foci of various sizes and increased density located primarily in epiphyses and metaphyses of the involved bones (Fig. 11.9a, b). These may also be scattered throughout the entire bone and are often arranged in the long axis of the bone (Fig. 11.10a, b). The skull, vertebrae and ribs seems to be exempt. The spots are grouped towards the end of the bone in the epiphyses. They give rise to no symptoms and are usually discovered by chance.

Various skin lesions like scleroderma have been found in the number of cases osteopoikilosis overline the bony lesion.

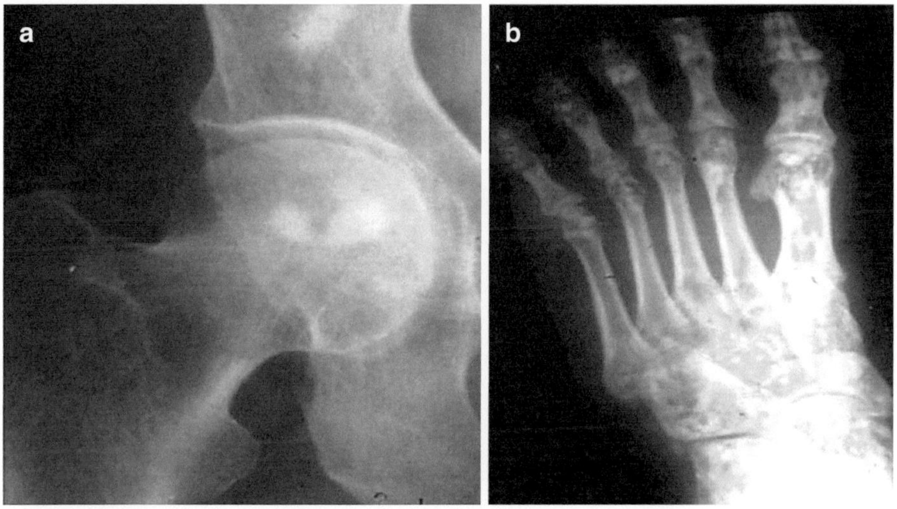

Fig. 11.9 (**a**) Note the radio-opaque islands in the head of the femur and supra-acetabular area. (**b**) Roentgenogram reveals multiple well defined sclerotic lesions (nodules) centered on the joints

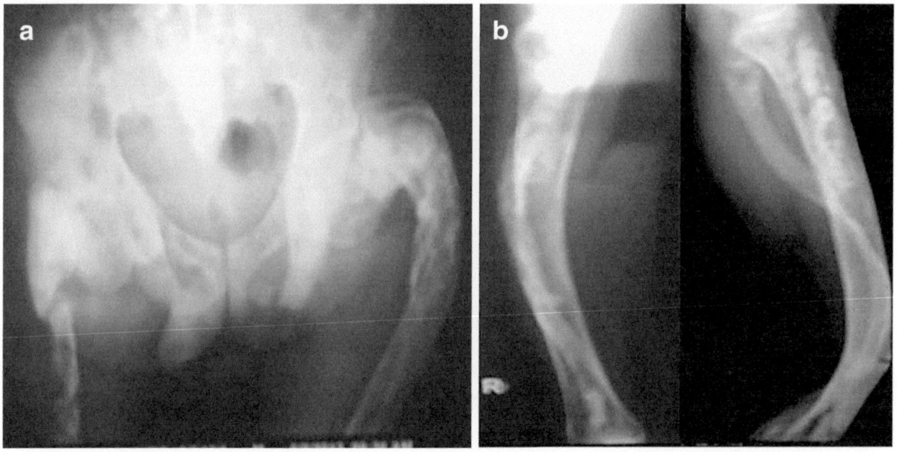

Fig. 11.10 (**a**) Roentgenogram of the pelvis and proximal femora revealing multiple deformities. Calcific bodies can be seen with fracture at the proximal end of the right femur. (**b**) Roentgenogram of the tibiae showing irregular sclerosis in the medullary canal with many island of calcification in a case of osteopoikilosis. These are round/oval with their long axes parallel to the long axis of the involved bones. Note the bowing deformities

11.5 Osteopathia Striata

Osteopathia Striata is characterized by longitudinal striations of increased bone density, usually towards the ends of the long bones and the pelvis (Fig. 11.11a–e). There are lines of increased density are seen in the metaphysis (Fig. 11.12a–d). The

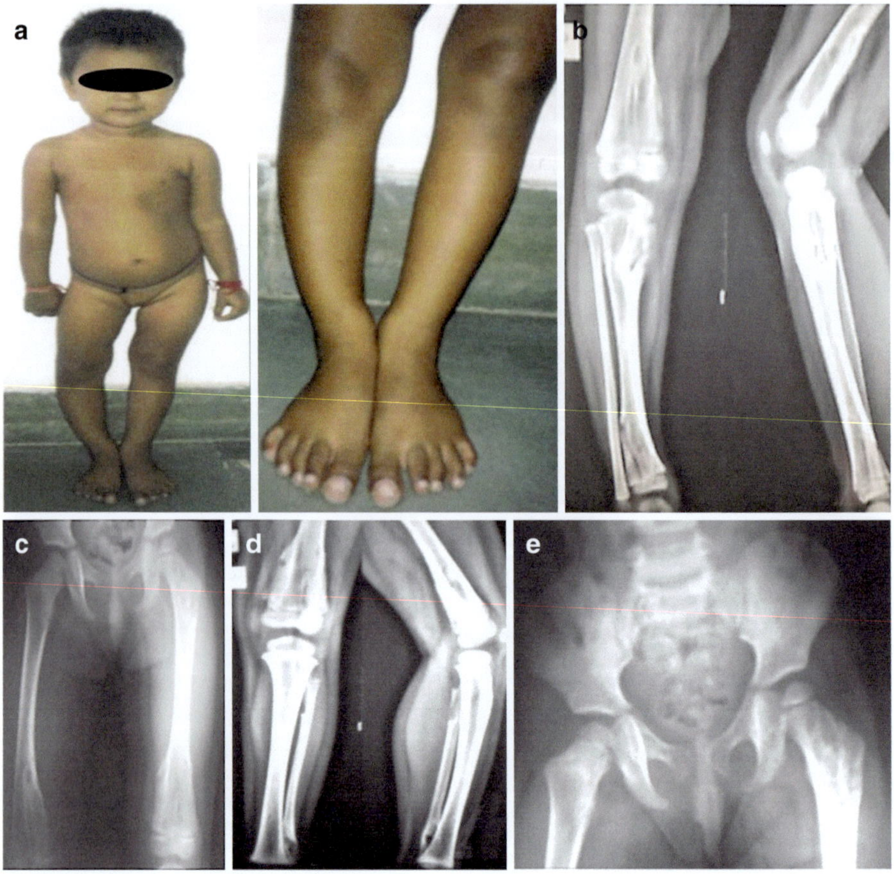

Fig. 11.11 (**a**) A child with bilateral osteopathia striata with shortening on the right side. (**b**, **c**) Roentgenogram showing the changes in the lower end of the femur and both ends of the tibiae. (**d**) Roentgenogram reveal changes in the lower end of the femur and both ends of fibula. (**e**) Similar changes are also seen in the proximal end of the left femur

less dense areas, however, are not due to masses of cartilage but because of the bones of normal texture. It is characterized by multiple condensations of cancellous bone tissues which begin at epiphyseal line and extend into the diaphysis (Fig. 11.13a–d). The iliac bones may be 'Fan-like' and patient is usually asymptomatic. The base of the skull is thickened.

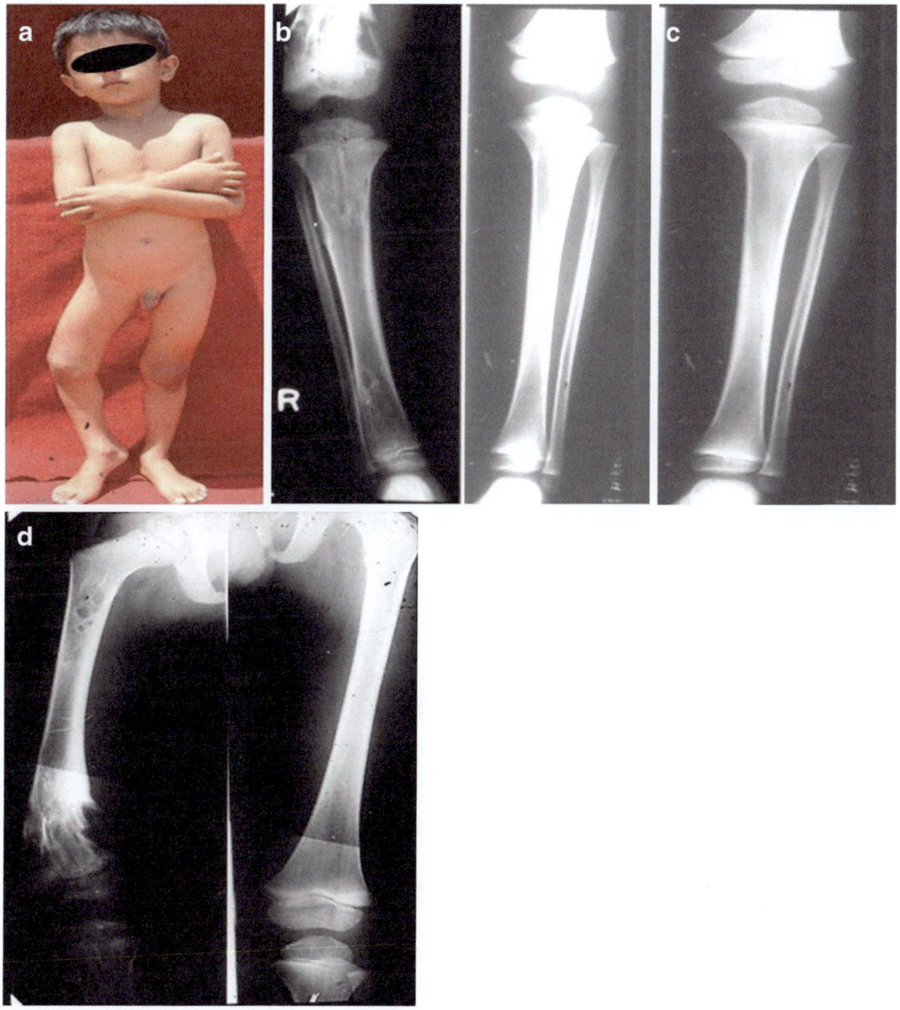

Fig. 11.12 (**a**) A case of osteopathia striata with shortening in the right lower limb. (**b**, **c**) Roentgenogram reveals linear chondromas at the either end of the tibia from epiphysis to the diaphysis. Note the late changes at the lower end of the right femur. (**d**) Note the striations in the distal end of the femur. Early changes are also seen in the proximal femur

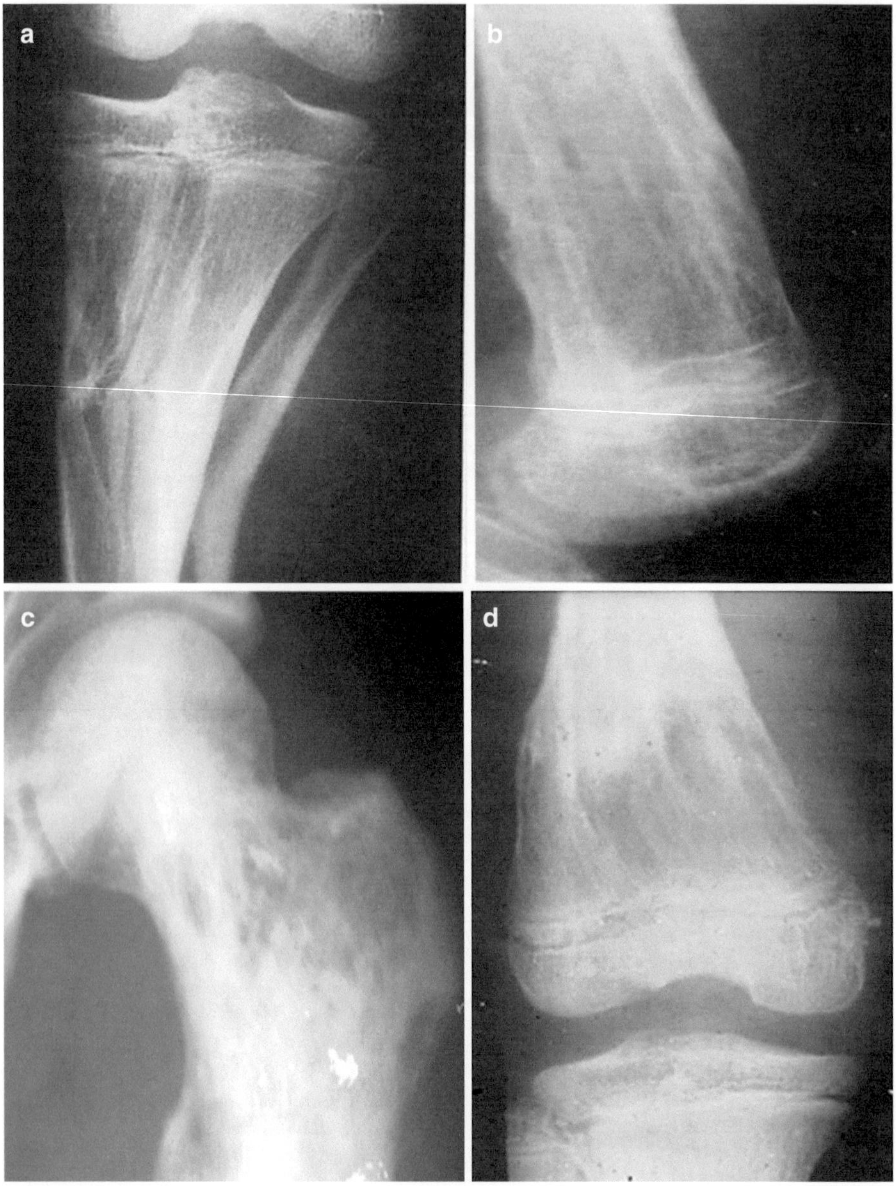

Fig. 11.13 (**a**, **b**) Roentgeogram reveals multiple condensation of the cancellous tissue which begins from the epiphyseal line towards the diaphysis in the form of multiple alternating sacroiliac bands in a case of osteopathia striata. (**c**, **d**) Roentgeogram reveals presence of multiple condensation of cancellous bone tissue. The striations start from the epiphyseal line and extend towards the diaphysis

11.6 Metaphyseal Dysplasia (Pyle's Disease)

Metaphyseal dysplasia (Pyle's disease), sometimes known as craniometaphyseal dysplasia, is a disturbance of endochondral bone growth in which there is a failure of modelling of cylindrical bones and the ends of the shafts remain larger than normal in circumference. Thus it appears that there is a lack of osteoblastic activity in the metaphyses and diaphyses, affecting the periosteum as in osteopetrosis, but there is osteoblastic activity for endochondral bone formation which is unlike osteopetrosis. The metaphyses appear wide resulting in an 'Erlenmeyer Flask' shaped appearance (Fig. 11.14a–c). The patients are tall and have broadening of the metaphyseal areas (Fig. 11.15a, b). All long bones are affected but changes are most dramatic in distal femur and proximal tibia. The involved bone ends shows flaring with thin cortex, osteoporosis and fracture (Fig. 11.16a, b). There is mild thickening of the calvaria and slight frontal bulging.

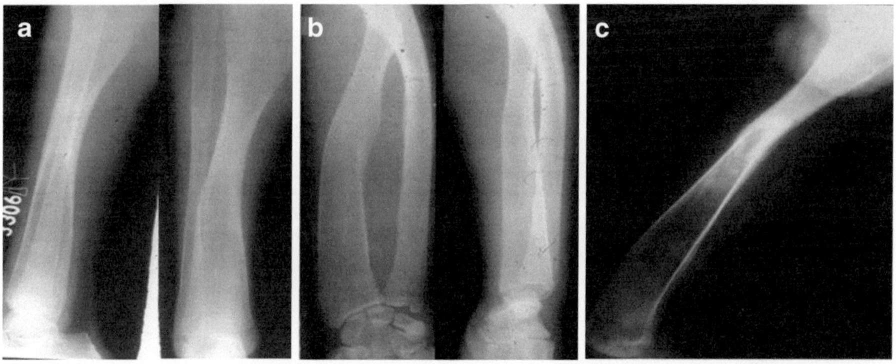

Fig. 11.14 (**a**) Roentgenogram of the tibia is quite characteristic of 'Erlenmeyer-Flask' deformity in piles disease. (**b**) There is clear splaying of the metaphysio-diaphyseal area of the lower end of the radius which is quite characteristic of the disease. (**c**) Roentgenogram revealing is clear splaying of the ends of the femur

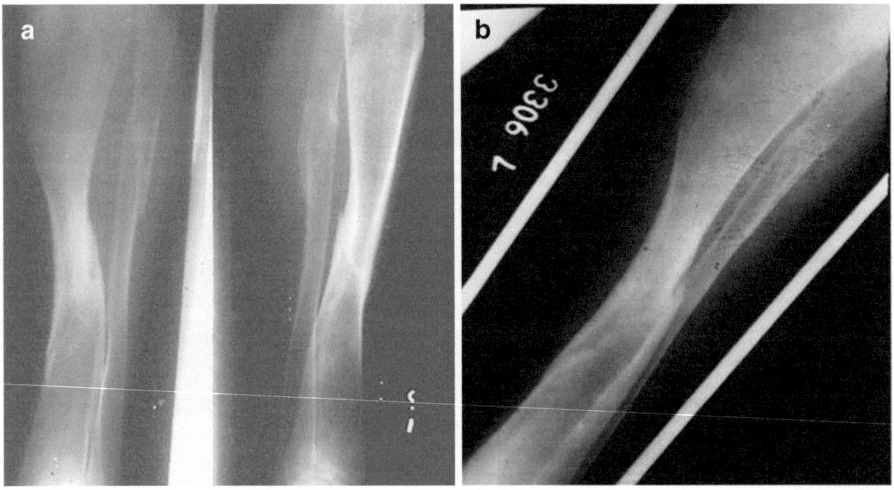

Fig. 11.15 (**a**) Roentgenogram shows metaphyseal dysplasia. Note the 'Erlenmeyer-Flask' deformity and normal tibial diaphysis. (**b**) Roentgenogram reveals widening of the metaphyses on either ends of the tibial shaft

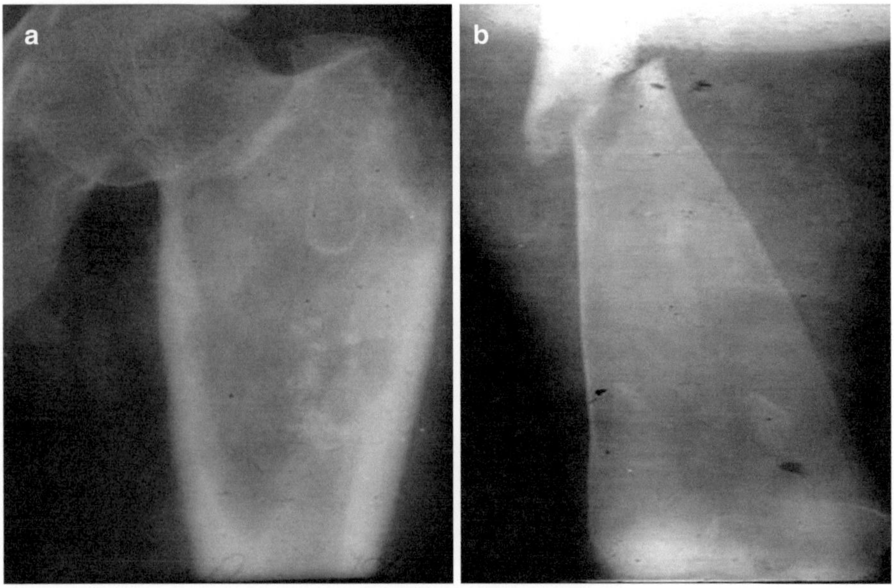

Fig. 11.16 (**a**) Roentgenogram revealing wide metaphysis which is quite characteristic of Pyles disease. (**b**) The lower end of the femur is quite broad. The lower femoral shaft has resulted in a fracture at its proximal end

11.7 Progressive Diaphyseal Sclerosis

Diaphyseal Sclerosis (Progressive diaphyseal dysplasia, Camurati-Engelmann-ribbing disease) is a group of rare conditions with symmetrical fusiform enlargement and sclerosis of the shaft of the long bones limited to diaphysis and the metaphysic and epiphysis are not involved. Commonly seen in males, there is weakness. There is thickening of the mid-diaphyseal cortex which affects mainly the tibia and the femur. There may be eventually being inadequate space for haematopoiesis, with development of anaemia and enlargement of spleen and liver. The condition is frequently familial and probably hereditary. The gait is described as shuffling or waddling and movement in some cases may be painful and restricted due to tenderness of the bones. The bones are strong and pathological fractures are not reported. The head is enlarged and forehead is prominent. The muscles are small, flabby and weak. There may be proptosis and cranial nerve palsies. The serum alkaline phosphatase may be raised.

Radiologically there is variable amount of periosteal and endosteal cortical thickening of the diaphysis of long bones. The epiphyses and the metaphyses are normal (Fig. 11.17a, b). The base of the skull reveals thickening. There is increased uptake on bone scintigraphy.

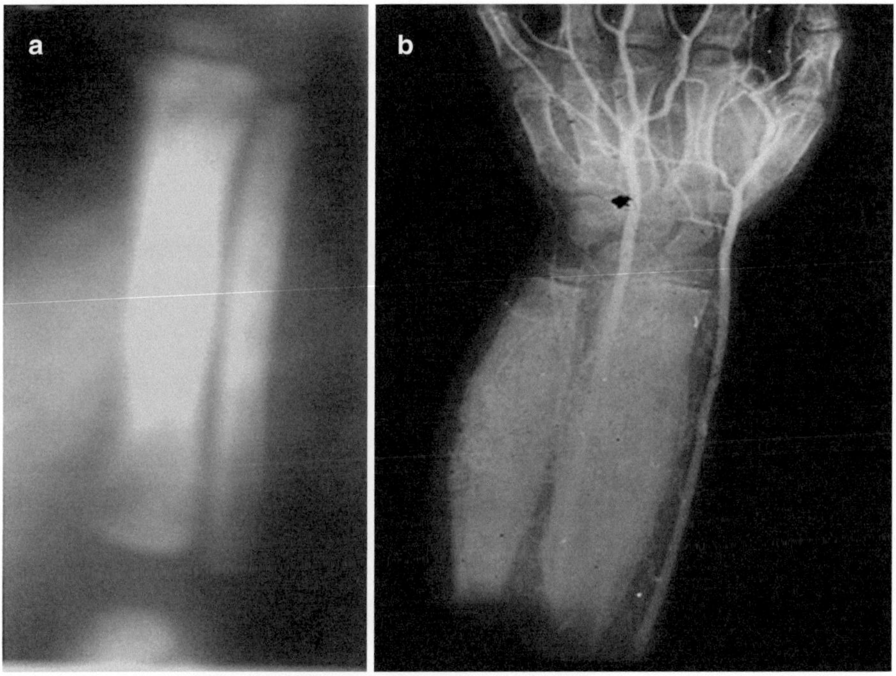

Fig. 11.17 (**a**) Roentgenogram revealing diaphyseal sclerosis of the tibial and fibular shafts. The metaphyseal and epiphyseal areas are unaffected. (**b**) Angiogram of the patient showing radius and ulna with diaphyseal sclerosis. The ends of the bones are spared

11.8 Infantile Cortical Hyperostosis

Infantile cortical hyperostosis (Caffey's Disease) is characterized by the formation of subperiosteal bone on the shaft of long bones and on the mandible. The infant usually become feverish and hyperirritable. In a few cases swellings have first appeared in the legs, face and arms and there is limitation of movements of the limbs and tenderness. The swellings are wood-hard, deeply situated and fixed to bone. The E.S.R is increased and serum alkaline phosphatase is elevated. It is mostly a self-limiting condition affecting young infants.

Radiologically the cortical walls of bones which underlie the swellings are thickened externally. The subperiosteal shadow encloses a varying length of the shaft or the whole of it. The mandible is thickened. Gradually the enlarge porotic bone shrinks away and the shaft resumes its normal appearance.

The diagnosis may be delayed as the disorder mimics a wide range of diseases including osteomyelitis, scurvy, hyper vitaminosis-A, bone tumours and child abuse. The emphasis here is to be reminding the clinicians about the existence of the disease.

11.9 Mixed Sclerotic Dysplasia

Mixed sclerotic dysplasia (mixed sclerotic bone dystrophy) is a rare presence of osteopoikilosis, melorheostosis and osteopathia striata in various combinations, with or without other defects in same individuals (Fig. 11.18). Pathologically all these conditions can be distinguished from osteopetrosis by the absence of cartilage in these sclerotic bones.

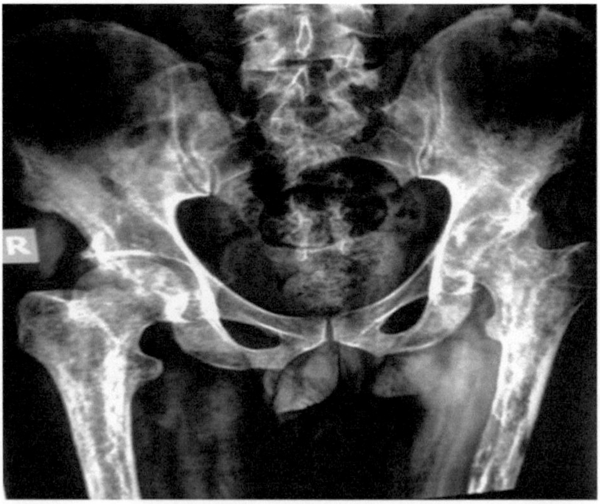

Fig. 11.18 Radiograph of mixed sclerotic dysplasia revealing signs of melorheostosis, osteopoikilosis and osteopathia striata

Suggested Readings

Milgram JW, Jasty M. Osteopetrosis. J Bone Joint Surg Am. 1982;64:912.

Hinkel CL, Beiler DD. Osteopetrosis in adults. AJR Am J Roentgenol. 1955;74:46.

Dent CE, Smellie JM, Watson L. Studies in osteopetrosis. Arch Dis Child. 1965;40:7.

Popoff SN, Marks SCJ. The heterogeneity of osteopetrosis reflects the diversity of cellular influences during skeletal development. Bone. 1995;17:437–45.

Shapiro F, Glimcher MJ, Holtrop ME, et al. Human osteopetrosis. A histological, ultrastructural and biochemical study. J Bone Joint Surg. 1980;62A:384–99.

Shapiro F. Osteopetrosis: current clinical considerations. Clin Orthop. 1993;294:34.

Milgram JW, Jasty M. Osteopetrosis. A morphological study of 21 cases. J Bone Joint Surg. 1982;64A:912–29.

Gelb BD, Shi GP, Chapman HA, Desnick RJ. Pychodysostosis, a lysosomal disease caused by cathepsin K deficiency. Science. 1996;273:1236–8.

Edelson JG, Obad S, Geiger R, et al. Pycnodysostosis, orthopaedic aspects with a description of 14 new cases. Clin Orthop. 1992;280:263–76.

Fryns JP, et al. Melorheostosis in a 3 year old girl. Acta Pediatr Belg. 1980;33:185.

Young D, Drummond D, Herring J, Cruess RL. Melorheostosis in children. Clinical features and natural history. J Bone Joint Surg. 1979;61B:415–8.

Campbell CJ, Papademetriou T, Bonfiglio M. Melorheostosis. A report of the clinical, roentgenographic and pathological findings in 14 cases. J Bone Joint Surg. 1968;50A:1281–303.

Hundley JD, Wilson FC. Progressive diaphyseal dysplasia. Review of the literature and report of 7cases in one family. J Bone Joint Surg. 1987;55:461–74.

Johnston CE. Progressive diaphyseal dysplasia. Orthopaedics. 1984;7:133.

Aggarwal P, Wali JP, Sharma SK. Progressive diaphyseal dysplasia. Case report and literature review. Orthopaedics. 1990;13:901–4.

Lagier R, Mbakop A, Bigler A. Osteopoikilosis. A radiological and pathological study. Skelet Radiol. 1984;11:161–8.

Horan FT, Beighton PH. Osteopathiastriata with cranial sclerosis. An autosomal dominant entity. Clin Genet. 1978;13:201–6.

Greenspan A. Sclerosing bone dysplasia. A target-site approach. Skeletal Radiol. 1991;20:561.

Whyte MP, Murphy WA, Fallon MD, et al. Mixed sclerosing bone dystrophy. Report of a case and review of the literature. Skelet Radiol. 1981;6:95–102.

Beighton P. Sclerosteosis. J Med Genet. 1988;25:200–3.

Gorlin RJ, Koszalka MF, Spranger J. Pyle's disease (familial metaphyseal dysplasia). J Bone Joint Surg. 1970;52A:347–54.

Phatak SV, Kolwadkar PK, Phatak MS. Pictorial essay: infantile cortical hyperosteosis (Caffey's disease). Indian J Radiol Imaging. 2004;14:185–6.

Berstein RM, Zaleske DJ. Familial aspects of Caffey's disease: a report of a family and review of the English literature. Am J Orthop. 1995;24

Pazzaglia UE, Byers P, Beluffi G, et al. Pathology of infantile cortical hyperosteosis (Caffey's disease). J Bone Joint Surg Am. 1985;67:1417.

Caffey's J. Pediatric x-ray diagnosis. 5th ed. Chicago: Year Book Medical Publishers; 1967.

Staheli LT, Church CC, Ward BH. Infantile cortical hyperosteosis (Caffey's disease). JAMA. 1968;203:96.

Navarre P, Pehivanov I, Morin B. Recurrcnce of infantile cortical hyperosteosis: a case report and review of the literature. J Peciatr Orthop. 2013;33:e10–7.

Dysplasia Associated with Heterotopic Bone Formation 12

12.1 Fibrodysplasia Ossificans Progressiva

Fibrodysplasia ossificans progressiva (Myositis ossification circumscripta) is a rare condition which is characterized by progressive heterotopic ossification in soft tissue of the trunk and extremities (Fig. 12.1a, b). The disease usually begins in childhood with episodes of painful soft tissue swelling. It usually develops in the paraspinal muscles, the muscles of the neck and proximal limb girdles (Fig. 12.2a–c). The swelling become indurate and eventually ossify. Episodes of ossification often increase prior to puberty (Fig. 12.3a–f). Early lesions show no radiographic evidence of calcification or ossification in soft tissue but may appear as a poorly-defined soft tissue swelling. Within a couple of weeks, focal poorly-defined radio-opaque areas begin to appear. After 4 weeks, the edge of the lesion becomes increasingly well-defined from surrounding soft tissue (Fig. 12.4a–g). A history of trauma or heavy activity is often but not always elicited. Biopsy is usually contraindicated in this condition. Muscles and joint contractures and fixed deformities result from progressive soft tissue ossification (Fig. 12.5a, b). Progressive kyphoscoliosis and increasingly limited thoracic movements compromise respiratory function. Plasma calcium and phosphate are normal but the alkaline phosphate may be elevated.

This category of lesion comprises a wide variety of conditions which are thought not to be neoplastic but rather development or reactive in origin. Although these lesions are designed tumour-like they present with clinical and radiological features that resemble those of a bone neoplasm (Fig. 12.6a, b).

Congenital malformations of the great toes, most often shortening with a single phalanx are characteristically present in these patients. There may be a hallux valgus deformity. The thumbs may also be short.

S. S. Yadav, *Skeletal Dysplasias & Congenital Malformations*,
https://doi.org/10.1007/978-981-99-5679-1_12

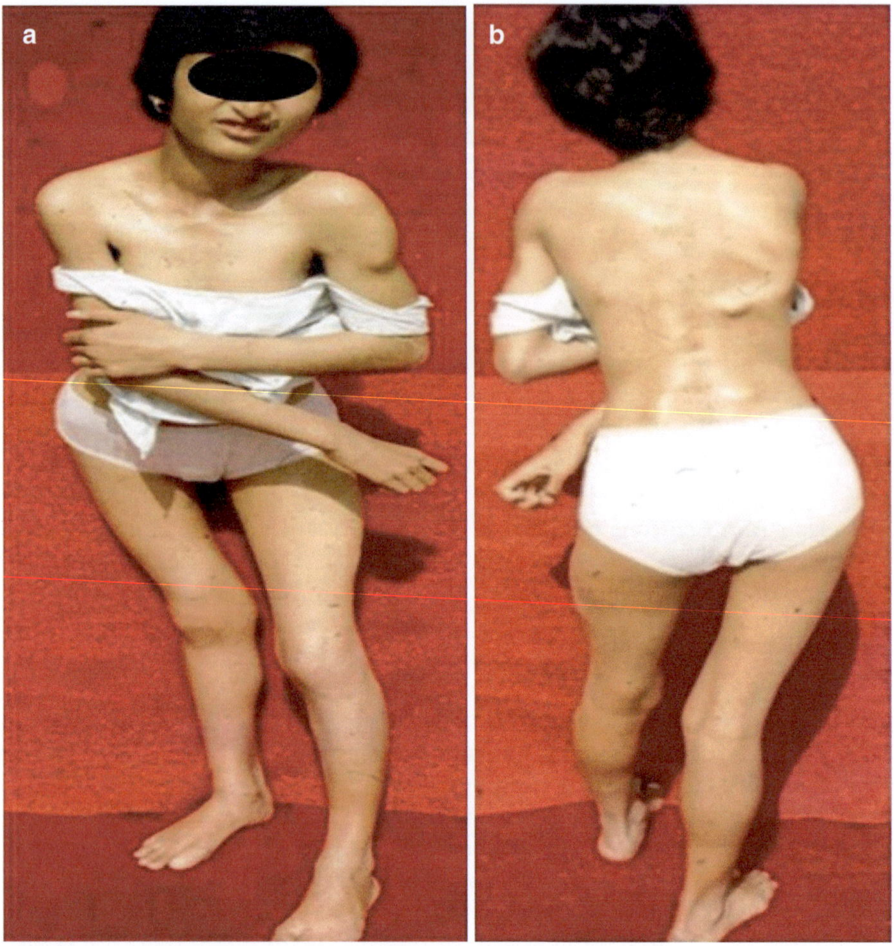

Fig. 12.1 (**a**) A 20-year-old girl showing deformities and bony projections on the left arm. Note the shortening of the right big toes. (**b**) The same patient reveals bony swelling at the back also

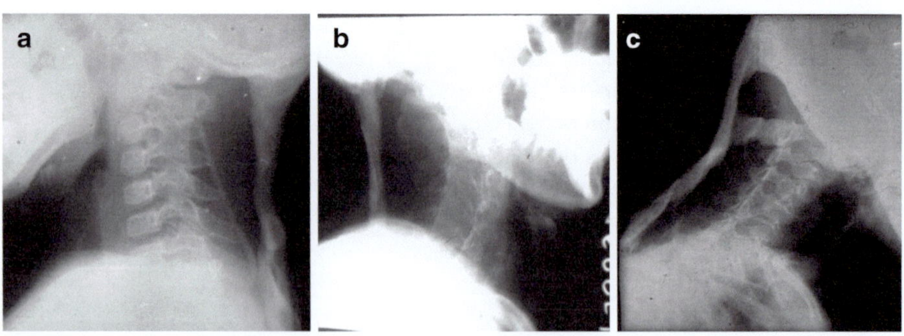

Fig. 12.2 (**a**) Roentgenogram reveals bridging ossification in the cervical extensors. (**b**) This case shows similar changes but the cervical spine is completely fused. (**c**) Note the bridging ossification of the cervical extensors. The spine of C2 is fused with the bridge. The cervical vertebrae reveal partial fusion

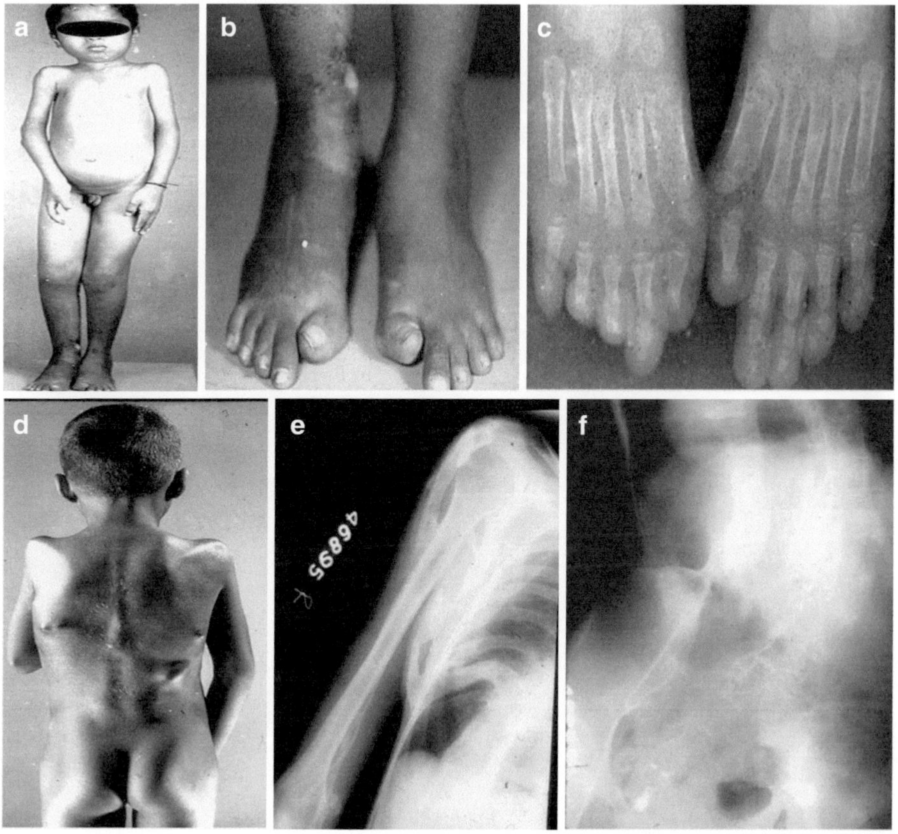

Fig. 12.3 (**a**) A patient with Fibrodysplasia ossificans progressiva revealing stiff body. (**b**) The big toes are short-a classical presentation in fibrodysplasia ossificans progressiva. (**c**) The X-ray reveals shortening of the big toes with a single phalanx—a classical diagnostic sign. (**d**) A 10-year-old boy revealing bony projections under the skin of the back. (**e**) Gross new bone formation around the shoulder and in the axilla with severe adduction deformity of the shoulder. (**f**) Radiograph reveals new bone formation between the ilium and the spine. Note the elongation of the iliac crest

12.1.1 Progressive Osseous Heteroplasia

Progressive osseous heteroplasia is a disorder in which bone forms within skin and muscle tissue. Bone that forms outside the skeleton is called heterotopic or ectopic bone. In progressive osseous heteroplasia, ectopic bone formation beings in the deep layers of the skin (dermis and subcutaneous fat) and gradually moves into other tissues such as skeletal muscle and tendons. Signs and symptoms of progressive osseous heteroplasia usually become noticeable during infancy. In some affected individuals, however, the disorder may not become evident until later in childhood or in early adulthood. The bony lesions within the skin may be painful and may develop into open sores (ulcers). Over time, joints can become involved, resulting in impaired mobility.

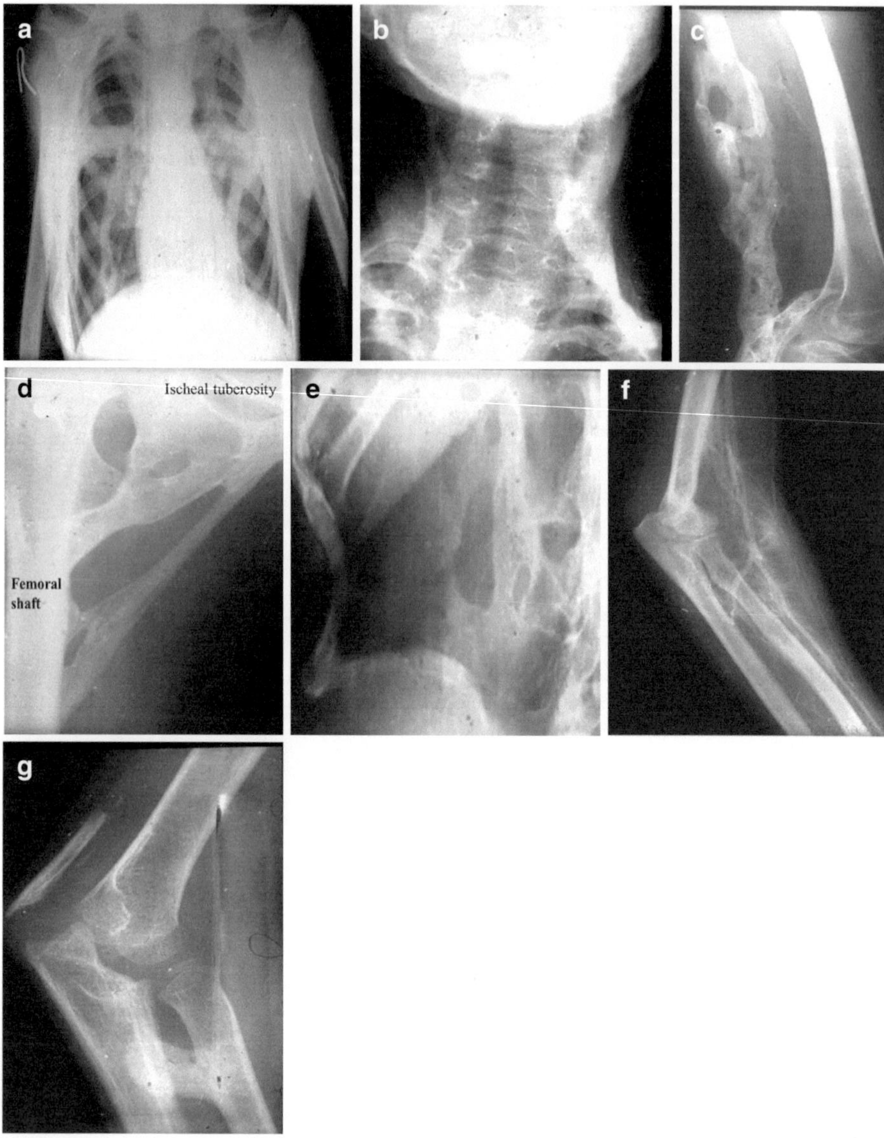

Fig. 12.4 (**a**) Roentgenogram reveals extensive heterotopic ossification around the ribs and the proximal humerus. (**b**) Roentgenogram reveals new bone formation in the left sternomastoid. (**c**) New bone formation in the back muscles of the thigh. (**d**) Roentgenogram showing struts of new bone formation between the right ischium and the proximal femur. (**e**) Roentgenogram reveals ectopic ossification between the ribs and the ilium. (**f**) An angiogram revealing increased vascularity in early stages of the disease. Note the early signs of heterotopic ossificans anterior to the elbow. (**g**) Radiograph showing bony strut in the triceps and cross union between radius and ulna

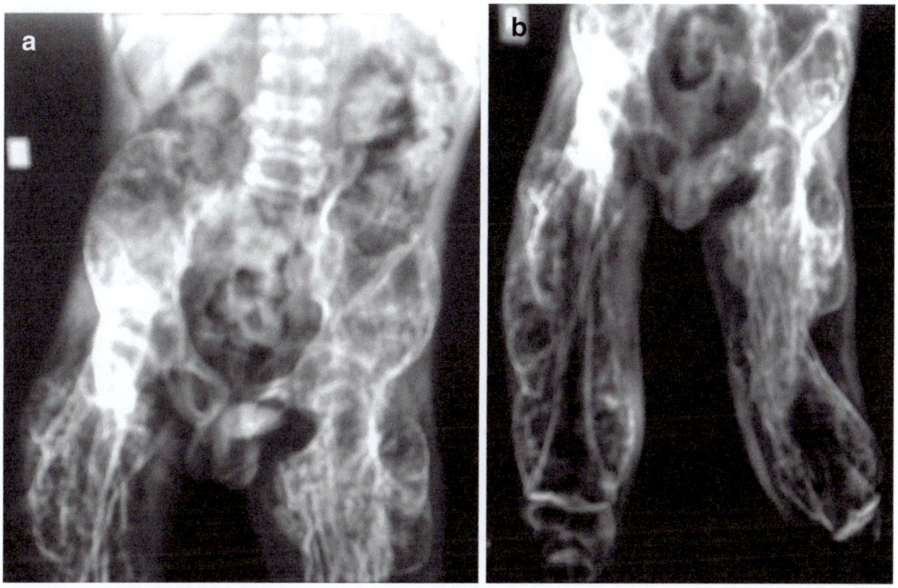

Fig. 12.5 (**a**) Roentgenogram of a 10-year-old boy revealing extensive new bone formation over both the hip joints and proximal thighs. (**b**) Roentgenogram of the same patient showing fracture of the distal third of the left femur with areas of patchy sclerosis and lyses in both the thighs

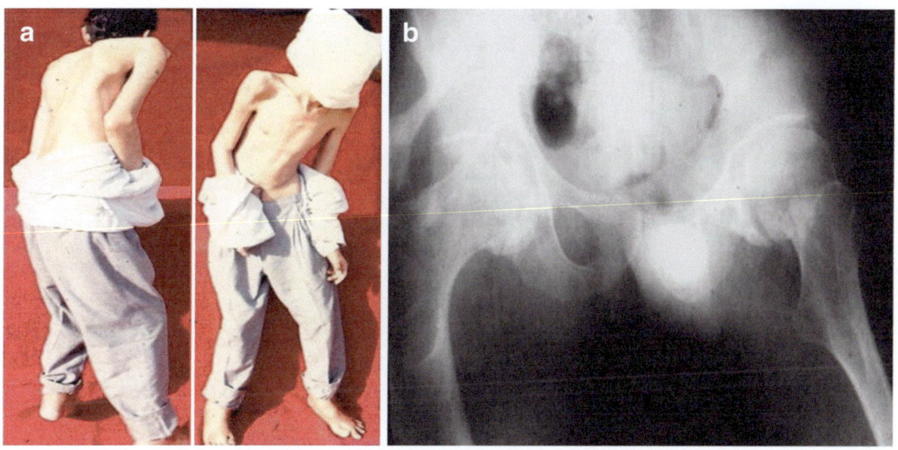

Fig. 12.6 (**a**) A 22-year-old boy revealing various deformities and complete stiffness of the body. Pectum carinatum is also observed along with short big toes. (**b**) Roentgenogram of the same patient with extensive heterotopic ossification in and around the pelvis with fusion at the hip joints

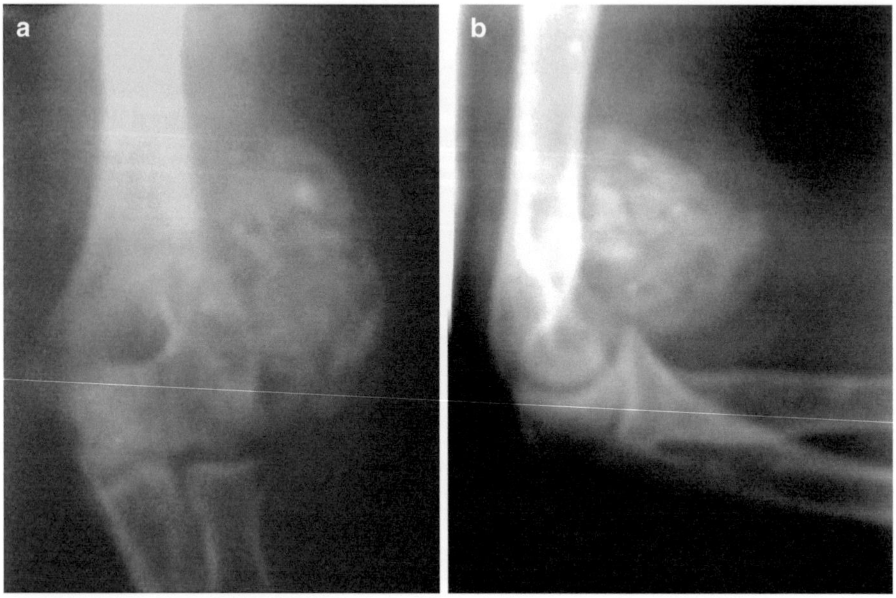

Fig. 12.7 (**a**) Roentgenogram revealing myositis ossificans traumatica. The patient did give history of local massage. (**b**) Roentgenogram of the same patient showing the mass in the cubital fossa limiting flexion at the elbow

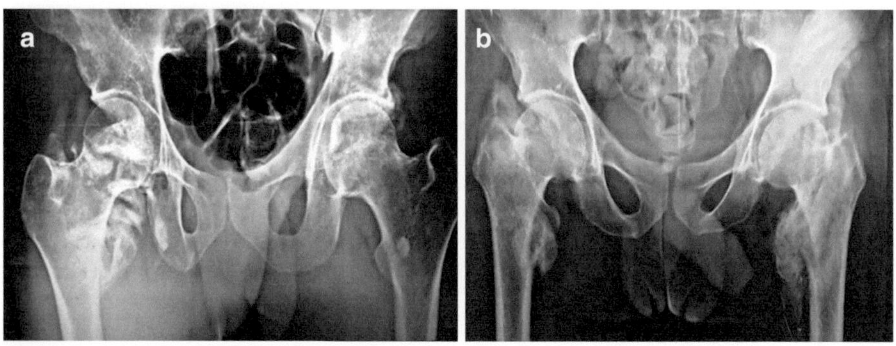

Fig. 12.8 (**a**) Roentgenograms reveal early formation of new bone around the right hip. The movements are greatly restricted. (**b**) Roentgenogram revealing early myositis ossificans traumatica. The joints are not yet involved

12.1.2 Myositis Ossificans Traumatica

Myositis ossificans refers to be benign, mostly a solitary mass in soft tissue. There may or may not be history of trauma (Fig. 12.7a, b). It is mainly thought of as a disorder that may be diagnosed before skeletal maturity, usually in teens and pre-teens.

Symptoms include a tender, progressively expending mass which is often observed 2–3 weeks after injury, mainly around the elbow and the hip (Fig. 12.8a, b). The lesion becomes harder and gradually non tender. Usually a history of local massage can be option (Fig. 12.9a, b). Glutei, Quadriceps and Brachialis are the most commonly involved muscles.

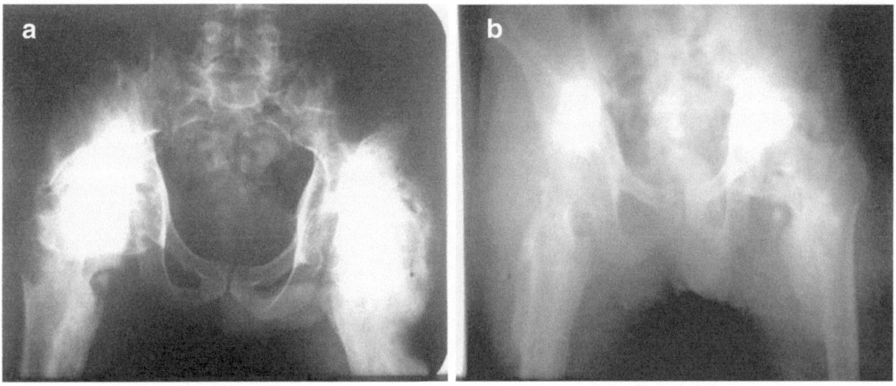

Fig. 12.9 (**a**) Roentgenogram revealing massive new bone formation around both the hips in a head injury patient. (**b**) Roentgenogram showing excision of the bone clearing the hip joints. The patient after extensive physiotherapy had regained functional range of moments

Suggested Readings

Kaplan FS, Hahn GU, Zasloff MA. Heterotopic ossification. Two rare forms and what they can tech us. J Am Acad Orthop Surg. 1994;2:228–96.

Kaplan FS, Craver R, MacEwen GD, et al. Progressive osseous heteroplasia. A distinct developmental disorder of heteroplasia ossification. J Bone Joint Surg. 1994;76A:425–36.

Connor JM, Evans DAP. Fibrodysplasia ossificans progressive. The clinical features and natural history of 34 patients. J Bone Joint Surg (Br). 1982;64:76.

Heifetz SA, Galliana CA, DeRosa GP. Myositis (fasciitis) ossificans in an infant. Pediatr Pathol. 1992;12:223.

Smith R, Russell RGG, Woods CG. Myositis ossificans progressiva. J Bone Joint Surg. 1976;57B:41–52.

Ogilvie-Harris DJ, Fornasier VL. Pseudomalignant myositis ossificans. Heterotopic new-bone formation without a history of trauma. J Bone Joint Surg Am. 1980;62:1274.

Pazzaglia UE, Beluffi G, Colombo A, et al. Myositis ossificans in the newborn. J Bone Joint Surg Am. 1986;68:456.

Athanasou NA, Benson MK, Brenton DP, Smith R. Progressive osseous heteroplasia. Bone. 1994;15:471–5.

Other Abnormalities of Muscolo-Skeletal Development 13

13.1 Neurofibromatosis

Neurofibromatosis is a single gene disorder and autosomal dominant. Neurofibromatosis is characterized by the development of multiple nerve sheath tumours and skin lesions (Fig. 13.1a–d). It occurs in two main forms:

Type I neurofibromatosis (peripheral neurofibromatosis; Von Recklinghausen's disease), which is a relatively common inherited disease. There are:-

- Café-au-lait spots.
- Neurofibromas.
- Dysostosis.
- Congenital pseudoarthrosis of tibia.
- Scoliosis.

Type II neurofibromatosis is a rare condition with central neurofibromatosis; bilateral acoustic neurofibromatosis.

Only 50% of patients with neurofibromatosis have a family history of the disease and is more common in males than females. There may be scoliosis or kyphoscoliosis (Fig. 13.2a–d). These are most commonly solitary neurofibromas but Plexiform and diffuse neurofibromas, as well as other types of neural tumours may be present at birth. They are most commonly found in skin and superficial soft tissue but may occur in any location (Fig. 13.3). Congenital bowing and pseudoartrosis of long bones occurs in an approximately 10% of cases (Fig. 13.4a–c). Intra osseous lesions have been observed which need suitable excision and bone grafting (Fig. 13.5a–c). Biopsy is important in such cases (Fig. 13.6a–e).

S. S. Yadav, *Skeletal Dysplasias & Congenital Malformations*,
https://doi.org/10.1007/978-981-99-5679-1_13

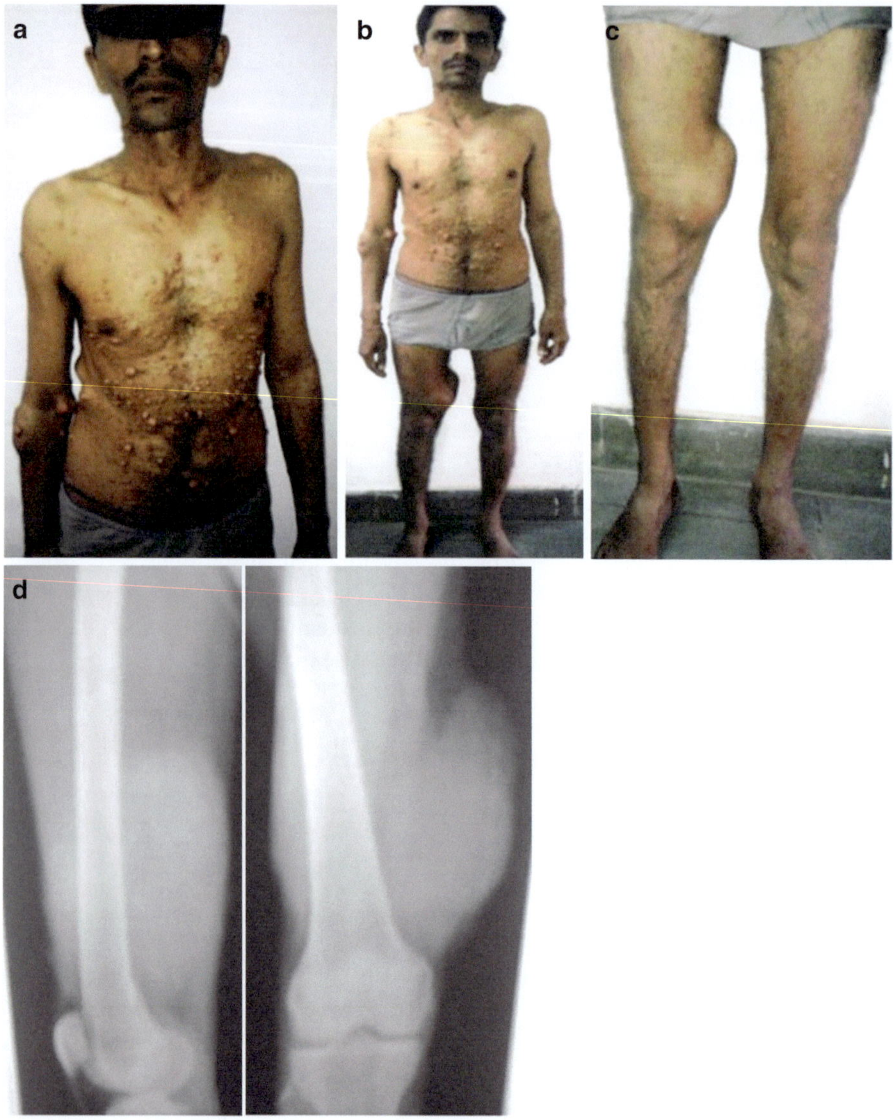

Fig. 13.1 (**a**) A 35-year-old male with extensive neurofibromatosis all over the body. (**b**) The lesions on the right elbow and medial aspect on the right thigh are large. (**c**) The biopsy revealed neurofibrosarcoma. (**d**) The roentgenogram reveals the soft tissue lesion

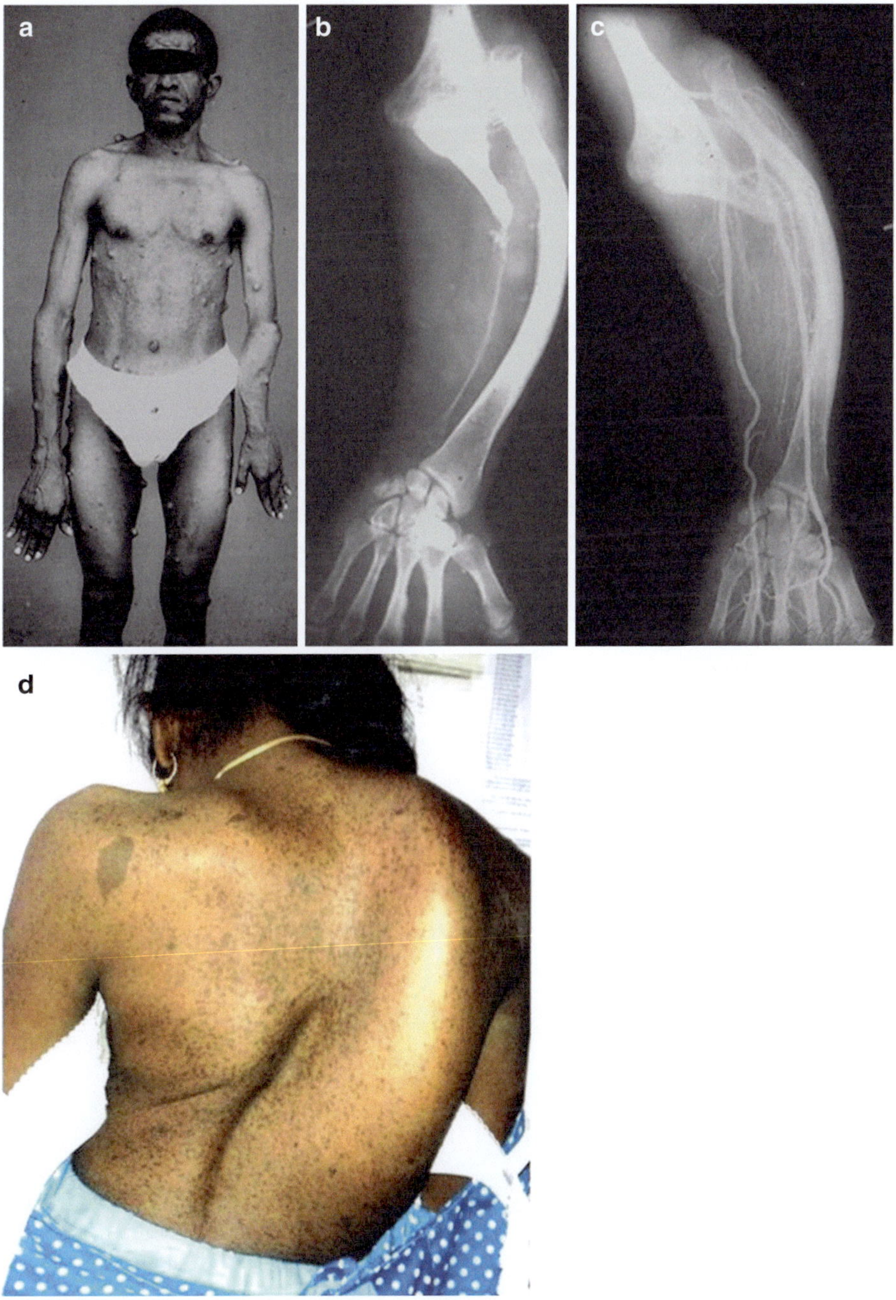

Fig. 13.2 (**a**) A 40-year-old male revealing multiple nodules (peripheral type) all over the body and deformity of the left forearm. (**b**) Radiograph of the left forearm of the same patient reveals bowing of the radius with dislocation of its head. The distal three-fourth the ulna reveals a streak of bone. (**c**) An angiogram of the forearm. (**d**) An 18-year-old female revealing kyphoscoliosis, multiple nodules and Café-au-lait spots at the back

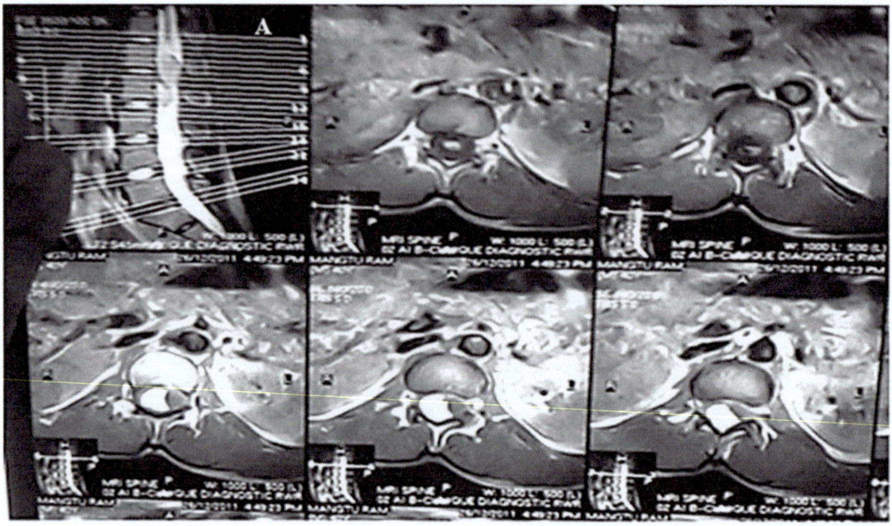

Fig. 13.3 MRI of the lumbar spine in a case of neurofibromatosis reveals a dumbbell-shaped extra dural lesion

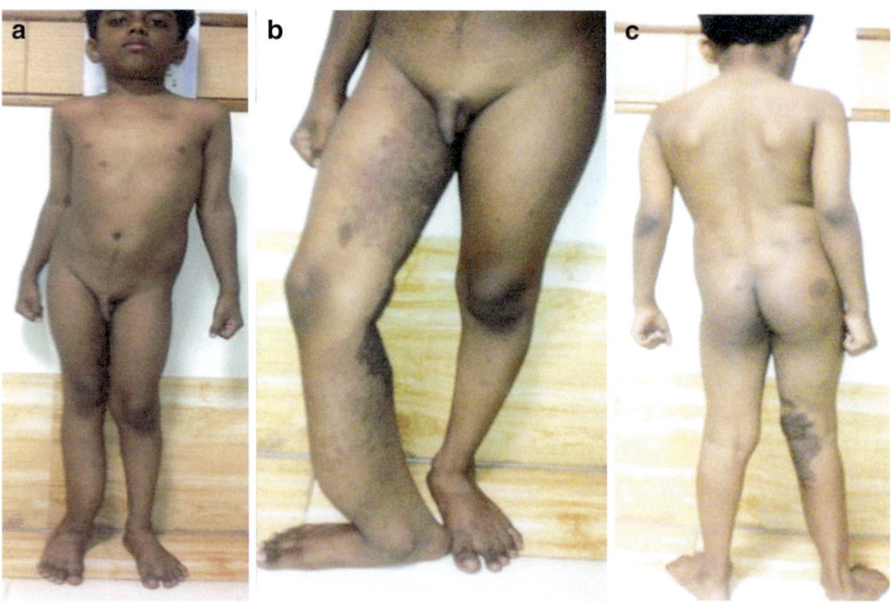

Fig. 13.4 (**a**, **b**, **c**) A boy showing hemi hypertrophy of the right lower limb with anterior bowing of the distal leg (Congenital pseudoarthrosis of tibia). Note the pigmentation over the extremity

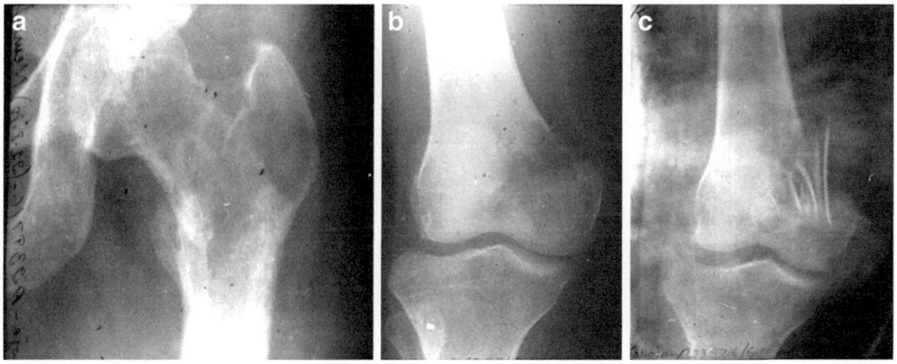

Fig. 13.5 (**a**) Roentgenogram reveals neurofibromatosis of the proximal end of the right femur. (**b**) Roentgenogram reveals lytic lesion of the medial condyle of the right femur. (**c**) The lesion was locally excised and filled with cancellous bone chips

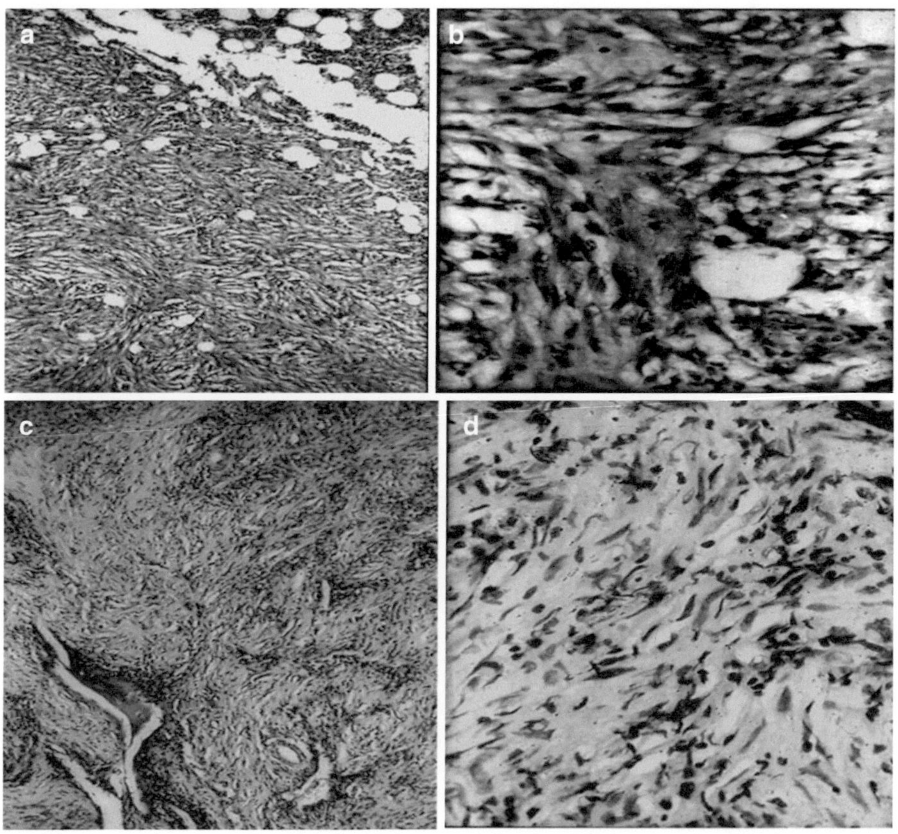

Fig. 13.6 (**a**, **b**) Histopathological changes in a case of neurofibromatosis of the bone. (**c**, **d**) Histopathology revealed evidence of neurofibromatosis. (**e**) Biopsy of the lesion confirmed the diagnosis

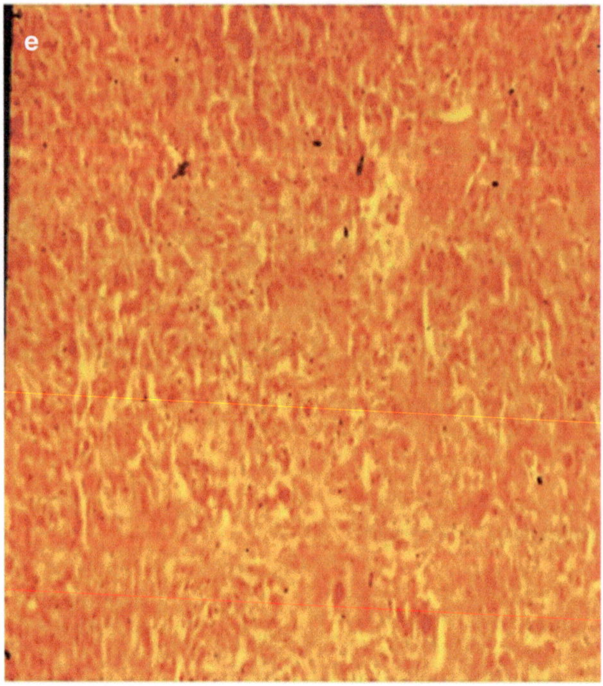

Fig. 13.6 (continued)

13.2 Orthopaedic Manifestations of Neurofibromatosis

1. Unusual scoliosis.
2. Non-Union of long bones.
3. Overgrowth of a part.
4. Radiologically a curious bone lesion.
5. Can mimic benign or malignant bone lesion.
6. Coxa-valga (20% cases) and may be protrusio acetabuli.
7. 5% may develop malignancy—Neurofibrosarcoma.

13.3 Fibrous Dysplasia

Fibrous Dysplasia is probably not a true neoplasia but a development abnormality. It is a common disorder that produces a variety of complaints and physical findings. It can be divided into Monostotic and Polyostotic. The majority of patients (approximately 85%) have a single skeletal lesion, monostotic fibrous dysplasia (Fig. 13.7). A rib is the most common location of monostotic fibrous dysplasia, but any bone can be involved.

Most patients present with swelling deformity and bone pain which may be of long duration and may be associated with pathological fracture. Any bone may be affected but those most often involved are the skull, jaw, rib, pelvis and the long tubular bones, especially the femur and tibia. There may be associated

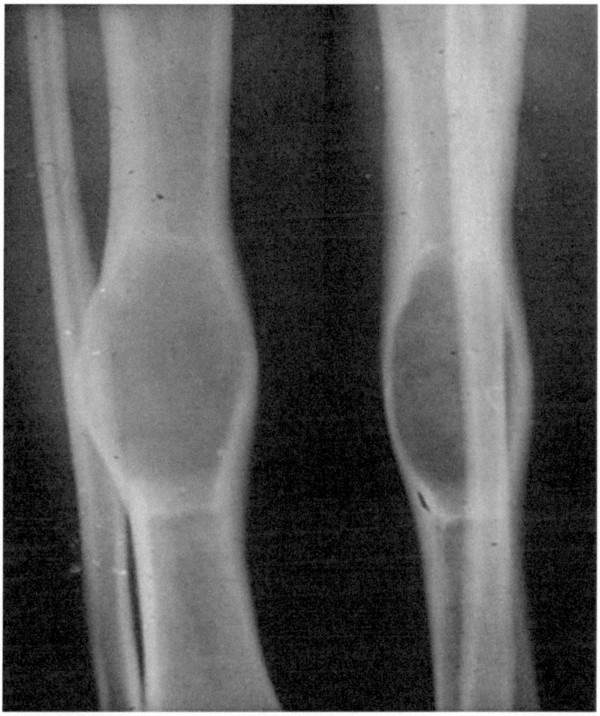

Fig. 13.7 Roentgenograms reveal mono-osteotic involvement of the tibiae

bony deformity (i.e., shepherd's crook deformity of the proximal femur) (Fig. 13.8a–f). There is often a sclerotic rim around the lesion in long bones the metaphysis and diaphyses are most often involved. The treatment of fibrous dysplasia is mainly surgical in the form of curettage and bone grafting (Fig. 13.9).

Polyostotic fibrous dysplasia has numerous lesions. The patients with polyostotic fibrous dysplasia may have only two or three small areas of involvement or may have extensive skeletal abnormalities with grossly deformed bones (Fig. 13.10a–g). The lesion is usually well-defined, radiolucent area which has a ground glass appearance (Fig. 13.11a–c). Occasionally the child presents with a pathologic fracture or angular deformity. Skin abnormalities are frequently seen in more florid types of Polyostotic fibrous dysplasia. These lesions are macular and light brown (Café-au-lait-spots), unlike similar lesions with a smooth border seen in neurofibromatosis. In some patients a very severe form of Vitamin D—resistant rickets may be present.

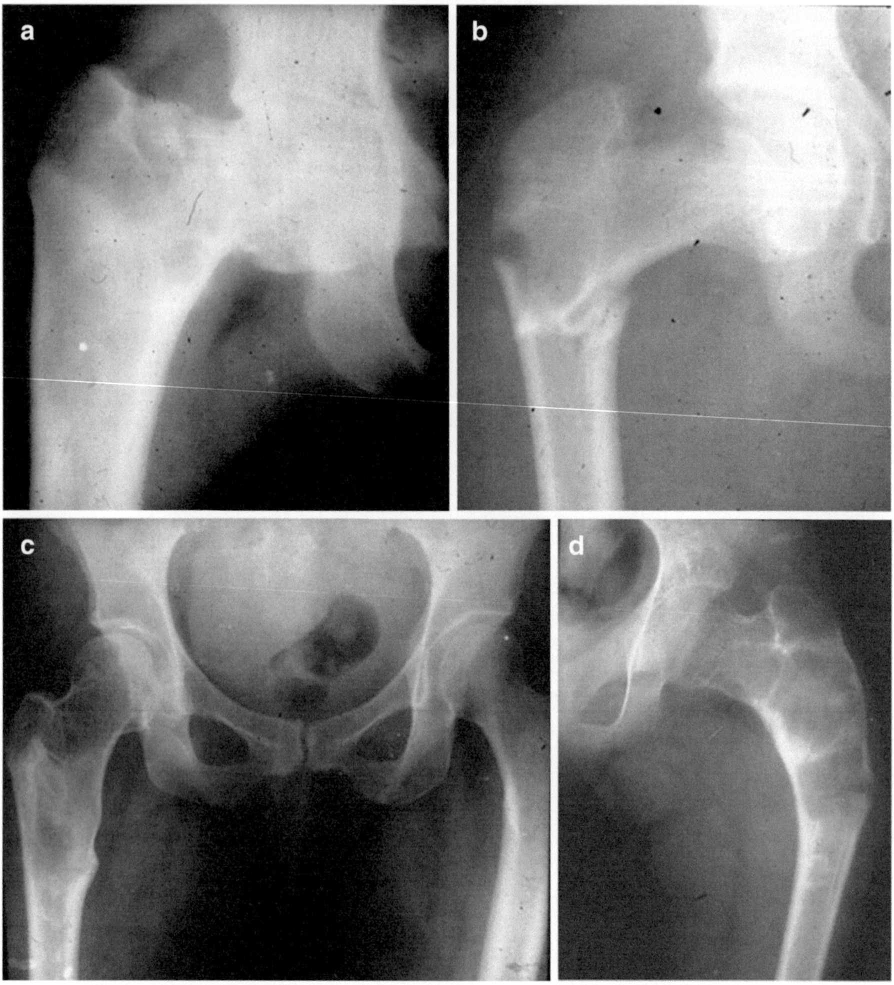

Fig. 13.8 (**a**) Roentgenogram reveals fibrous dysplasia of the proximal femur with undisplaced pathological fracture resulting in coxa-vara. (**b**) Roentgenogram showing fibrous dysplasia with pathological fracture of proximal femoral shaft. (**c**) Roentgenogram reveals classical signs of fibrous dysplasia of the proximal femur. (**d**) Roentgenogram shows involvement of the proximal femur with shepherd's crook deformity. (**e**, **f**) Roentgenograms reveal shepherd's crook deformity in proximal femora involved in fibrous dysplasia

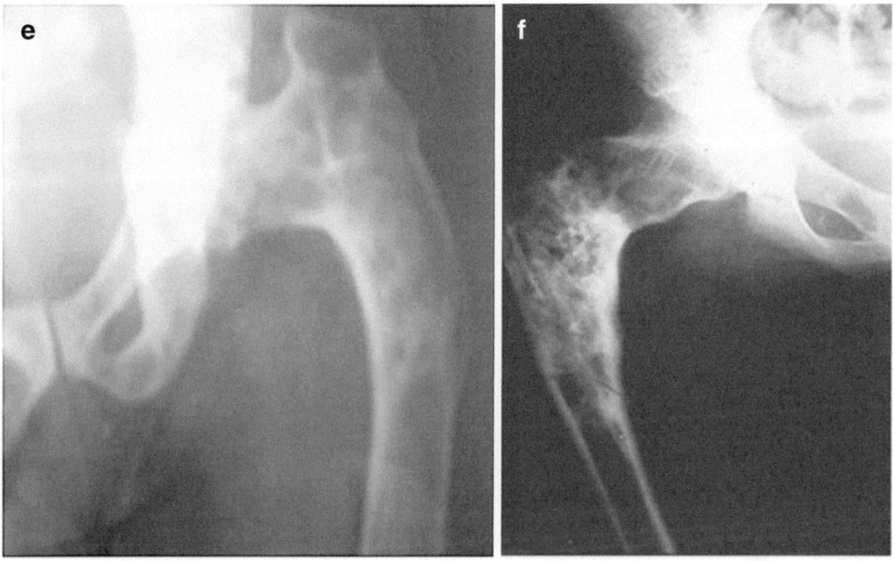

Fig. 13.8 (continued)

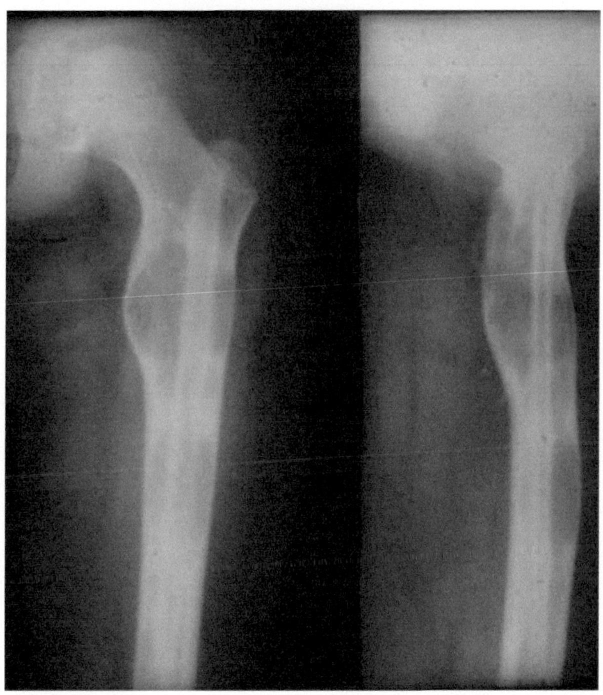

Fig. 13.9 Post-operative roentgenogram in a patient with fibrous dysplasia of proximal femur treated by curettage and inlay fibular strut graft

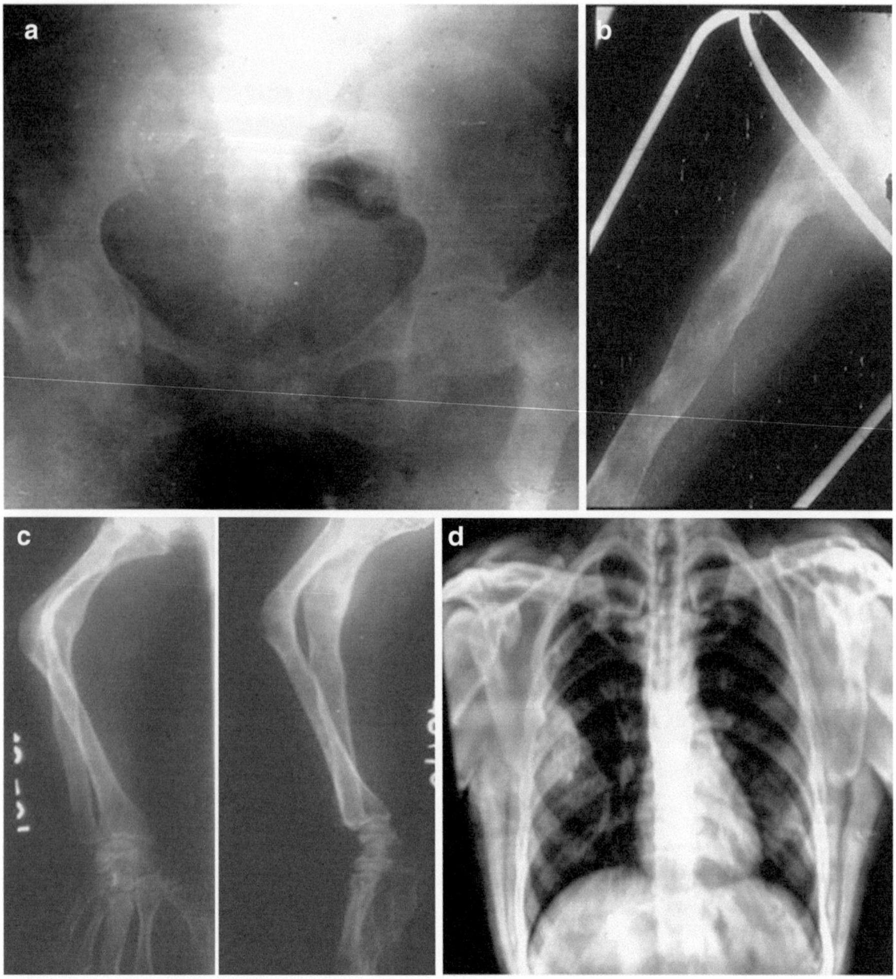

Fig. 13.10 (**a**) Roentgenograms showing fractures of the femoral neck on either side in a case of polyostotic fibrous dysplasia. (**b**) Note the extensive involvement of the femoral diaphysis. (**c**) Roentgenogram reveals a mal-united forearm fracture in a case of polyostotic fibrous dysplasia. (**d**) Roentgenogram reveals extensive involvement of the scapulae and clavicles on both sides in a case of polyostotic fibrous dysplasia. (**e**) Extensive involvement of the proximal tibia. (**f**) Roentgenograms reveal extensive involvement of the tibia in a case of polyostotic fibrous dysplasia. (**g**) The tibia reveals a pathological fracture

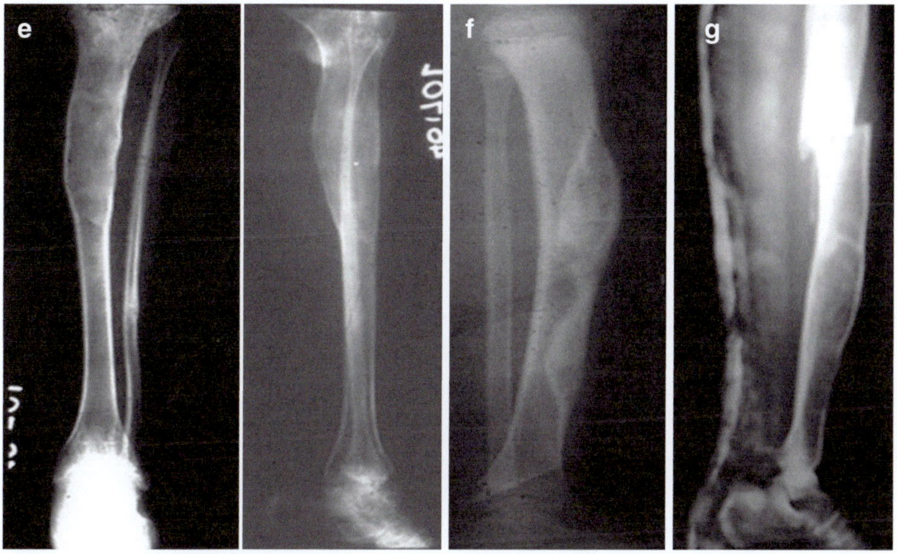

Fig. 13.10 (continued)

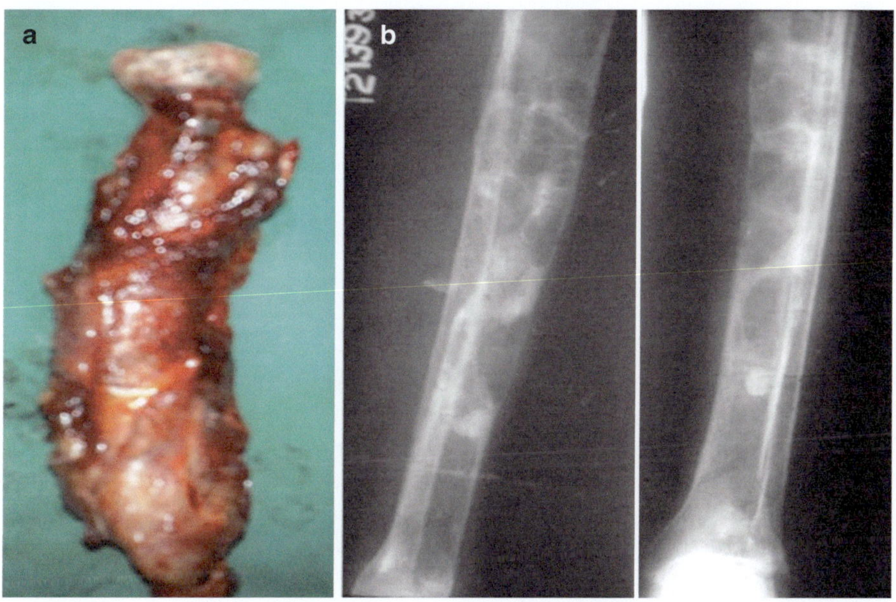

Fig. 13.11 (**a**) An excised specimen of proximal radius grossly involved with fibrous dysplasia. (**b**) Extensive involvement of the tibial shaft by fibrous dysplasia. (**c**) Roentgenograms reveal mono-osteotic involvement of the tibiae

Fig. 13.11 (continued)

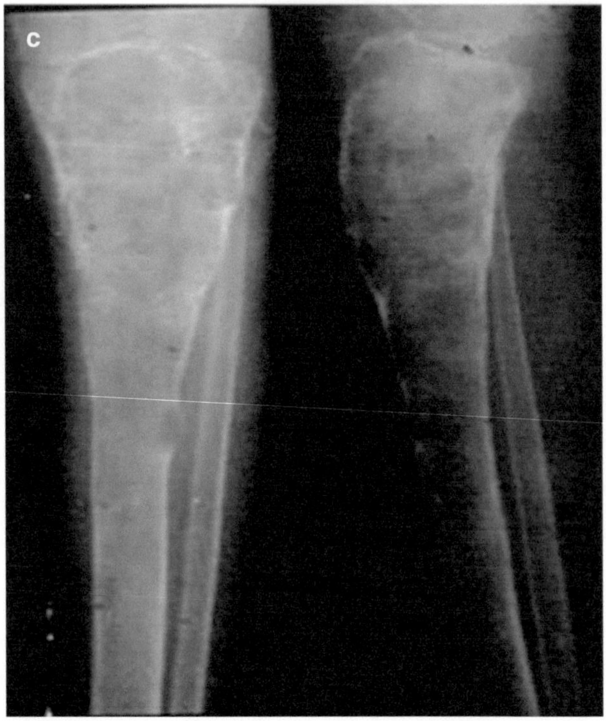

13.4 McCune-Albright Syndrome

- Polystotic fibrous dysplasia
- Sexual precocity
- Hyperplastic endocrine disorder
- Café-au-lait spots.

It is an autosomal dominant disorder and the pathogenesis is obscure. Lower limbs are commonly involved. Mostly unilateral may be bizarre. There is overgrowth of the skull with thick and areas of rarefaction and sclerosis.

Albright syndrome is a triad of fibrous dysplasia, café-au-lait spots and precocious puberty (Fig. 13.12a–e). The structural strength of the bones with fibrous dysplasia is reduced as a result of the poorly organized trabecular pattern and the thinned cortex. The weakness of the bones leads to the deformities usually present. Angular deformity should be straightened and the bone augmented with cortical bone graft with or without internal fixation. Cortical bone graft is preferred to cancellous bone because it is less likely to be remodelled by the host.

13.5 Osteofibrous Dysplasia

Osteofibrous dysplasia is a rare variant of fibrous dysplasia which usually occurs in tibia and fibula of young children. It is a non-healing non-ossifying fibroma. The

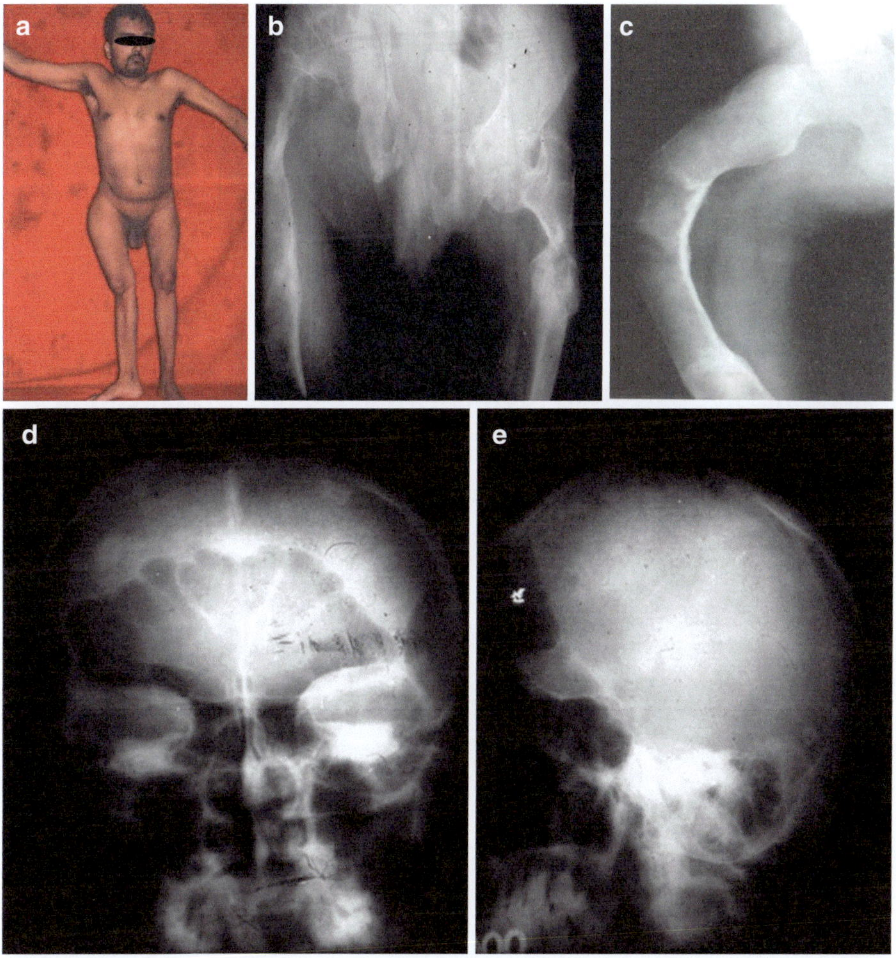

Fig. 13.12 (**a**) A 26-year-old male with McCune-Albright Syndrome. (**b**) Roentgenogram showing grossly deformed pelvis and the femora. (**c**) Roentgenogram revealing shepherd's crook deformity in a different case. (**d**, **e**) Roentgenograms show involvement of the skull along with the orbital-fossae in a case of polyostotic fibrous dysplasia

patients usually are brought to the physician's attention by a parent who has noticed an anterior bowing or mass in the tibia. Osteofibrous dysplasia, on the other hand, arises from the cortex and involves the medullary canal late. It usually is associated with a bowed tibia.

Radiologically the diaphyseal anterolateral cortex contains one or more lucent areas which have a sclerotic margin which often separates the cortical lesion from the underlined medulla.

It usually requires a resection for control. There have been few studies of patients with osteofibrous dysplasia with adequate follow-up, but the evidence is that most are eventually resected.

If the lesion progresses, resection is suggested. If the patient presents after the age of 10 years, especially if the lesion is large or has aggressive features on the plane radiographs, a biopsy is suggested to rule out an adamantinoma. An adamantinoma probably requires a wider resection than does an enlarging osteofibrous dysplasia.

13.6 Idiopathic Osteolysis

This is a heterogeneous group of disorders which is characterized by osteolysis principally affecting the distal extremities, especially the ends of digits. The other sites may be affected on the basis of areas where the bone absorption occurs.

13.6.1 Massive Osteolysis (Gorham-Stout Disease/Disappearing Bone Disease)

Massive osteolysis (Gorham-Stout disease, disappearing bone disease) is a rare resorptive vascular disease which may involve a single bone or a few adjoining bones (Fig. 13.13a–d). The pathogenesis is unknown but it may represent a form of

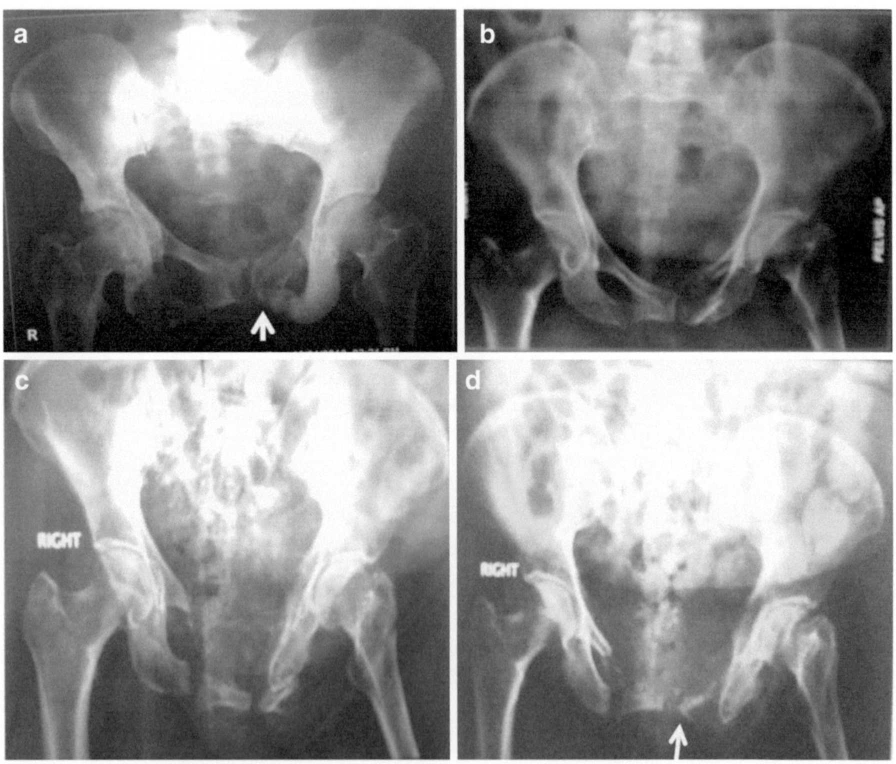

Fig. 13.13 (**a**) Roentgenogram reveals areas of osteolysis in the pubic rami (arrows). (**b**) Roentgenogram reveals progressive absorption of the pubic rami and ischial area. (**c, d**) Roentgenograms reveal progressive absorption of the pubic rami and adjoining areas in the same patient. Note the complete disappearance of the pubic rami on both sides

aggressive angiomatosis. However, the exact etiology is unknown. The disease usually occurs in younger age group. Classically a 'Sucked-Candy' appearance may be observed with tapering of bone at the advancing edge of osteolysis. Rarely the disease may spread across the joints. There may be pain and fracture. The bones of the pelvis and shoulder are most frequently involved. Visceral involvement is rare. The radiograph reveals extensive dissolution of both, cortical and cancellous bone.

13.6.2 Hajdu-Cheney Type of the Osteolysis

Patients have short stature with deformities. There are distinctive facial appearances, premature of loss of teeth (Fig. 13.14a, b). Skull my show evidence of Wormian bones and delayed closure of fontanelles. The oribital roof is demineralised resulting in prominent eyes. There is uniform acro-osteolysis of all the terminal phalanges (Fig. 13.15). These changes are comparable to those resulting in distal phalangeal osteolysis. The tubular bones are thin, osteopenic and show evidence of old fractures (Fig. 13.16a, b). The distal ends of the ulnae and carpal bones may also be involved. The outcome is variable and dependent on the extent of osteolysis. This case had all the features with synostosis basis of fourth and fifth metacarpals on both sides.

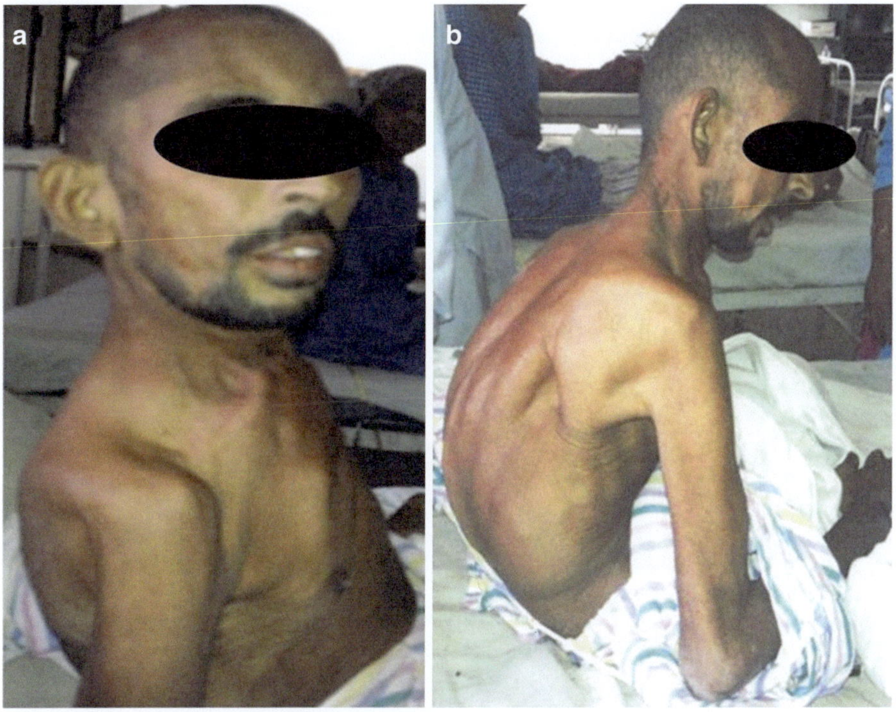

Fig. 13.14 (**a**) A 40-year-old man with Hajdu-Cheney disease showing long narrow face and prominent eyes. (**b**) Note the gross wasting of the muscles around the shoulders

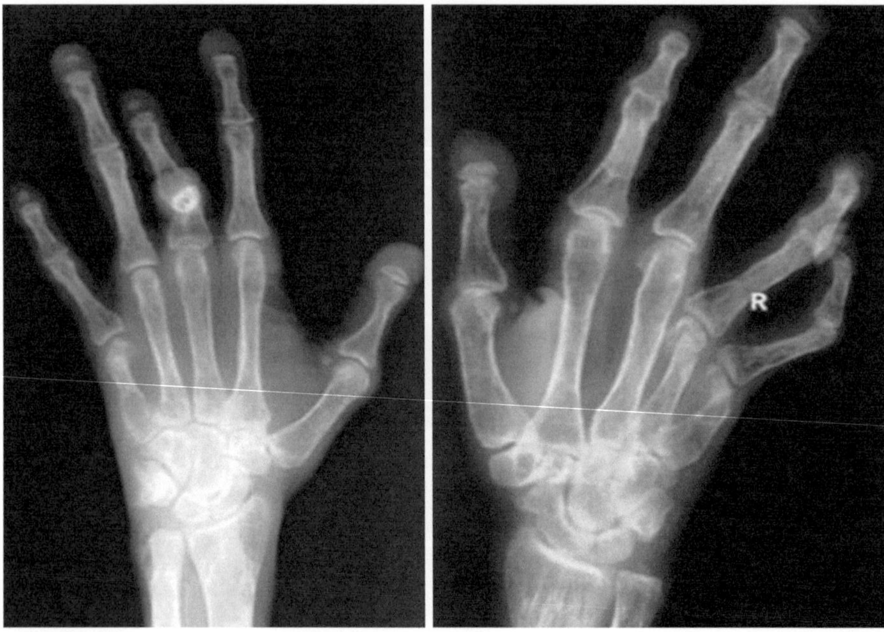

Fig. 13.15 Radiograph of a patient reveals progressive acro-osteolysis has occurred in the distal phalanges. There is evidence of old fracture of the proximal phalanx of the left middle finger and fracture of the neck of fourth and fifth metacarpals on the right side. The basis of fourth and fifth metacarpal are fused on either side. The bones are osteoporotic

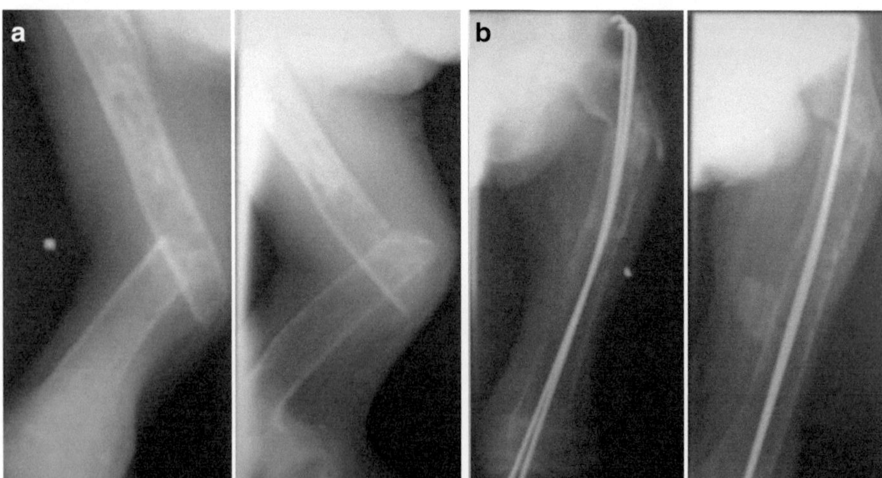

Fig. 13.16 (**a**) Old un-united fracture of the femoral shaft with evidence of gross osteoporosis. (**b**) An attempt has been made to achieve stabilization of the fragments by two long K-wires

13.7 Arthrogryposis Multiplex Congenita

Arthrogryposis multiplex congenita (Amyoplasia or Multiple congenital contracture syndrome) is a congenital non-progressive neuromuscular anomaly characterized by joint contractures, muscle weakness and fibrosis. It is not a specific disorder but a symptom complex of joint contractures associated with both neurogenic and myopathic disorders. The deformities are more neurogenic than myopathic. Disorders associated with dysgenesis of anterior horn cells are known to be common pathologic type. The limbs are creaseless and tubular. The muscle mass is reduced. The major joints include shoulder, wrist, hip, knee and elbow. These children are born with multiple non-progressive soft tissue contractures and restriction of joint movements. Muscle biopsy is usually the most important diagnostic procedure. The intelligences are usually normal.

The shoulder is adducted and internally rotated, elbows are extended and the wrists severely flexed (Fig. 13.17). The hips are flexed, abducted and externally rotated. The knees are usually extended because of quadriceps fibrosis (Fig. 13.18a–g). The joint motion is restricted. In some patients soft tissue webs may be identifiable on the flexor aspect of knees and elbows (pterygia syndrome). The deformities

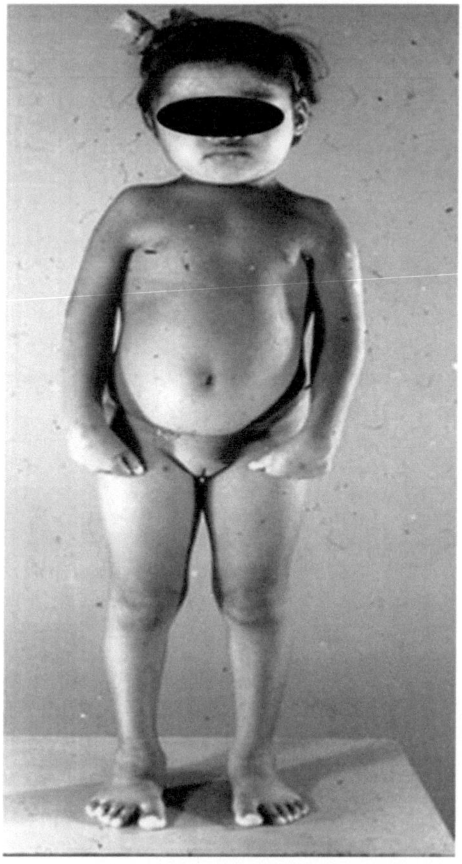

Fig. 13.17 A 6-year-old girl shows adduction of the shoulders, straight elbows lacking flexion creases and severe wrist flexion

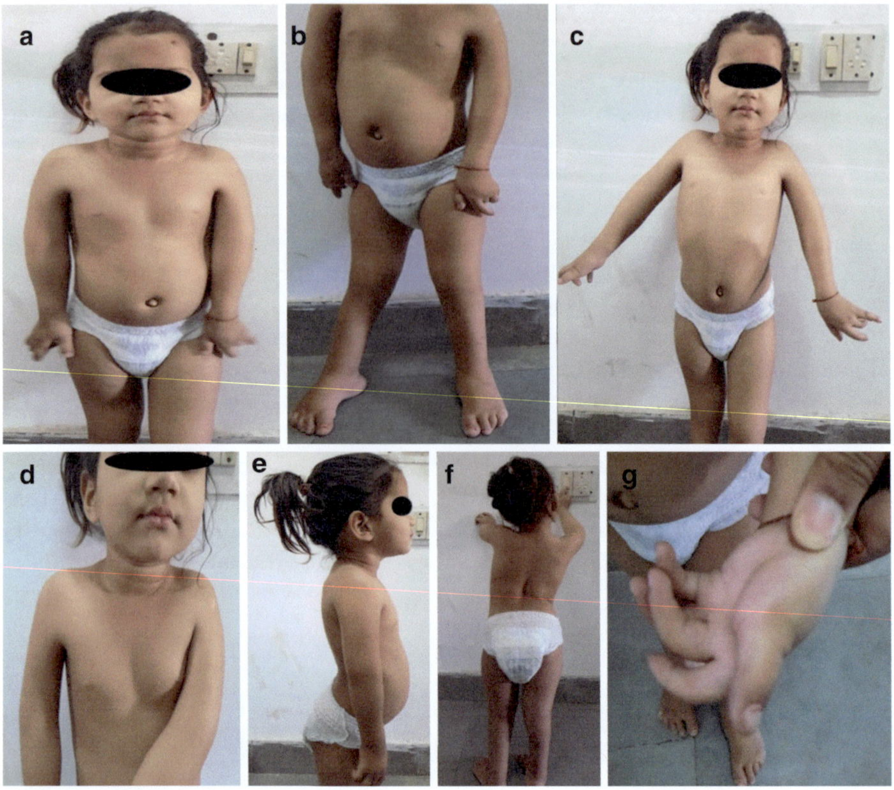

Fig. 13.18 (**a–g**) A 6-year-old girl suffering from arthrogryposis multiplex congenital is showing adducted shoulders, straight elbows and creaseless flexion contracture of the fingers. Note the severe bilateral quadriceps fibrosis and lumbar lordosis

are more severe and rigid distally. The hips may be dislocated. Scoliosis, growth retardation, facial deformities, respiratory problem and abdominal hernias may also be associated. Patients with arthrogryposis are weak and continue to be so later in life.

Radiologically there may be loss of subcutaneous fat and muscles. The joint changes are observed. Adaptive changes occur as a consequence of the fixed position of the joints (Figs. 13.19a–c and 13.20a–c).

The object of treatment is to make the patient functionally oriented. The management of each joint presents a unique opportunity for orthopaedic intervention. The basic needs of the patient have to be addressed. The overall goals are lower limb alignment and stability for ambulation and upper extremity motion for self-care. Surgery should be attempted in early stages, before the adaptive joint changes take place. The correction of contractures should be attempted with care to avoid fractures. The treatment depends on overall involvement of the body joints. The aim of the treatment is to help the patient for independent ambulation by releasing the contractures, perform osteotomies and arthrodesis as and when indicated. The upper extremities are made mobile for self-help like feeding and toileting and if possible to use crutches.

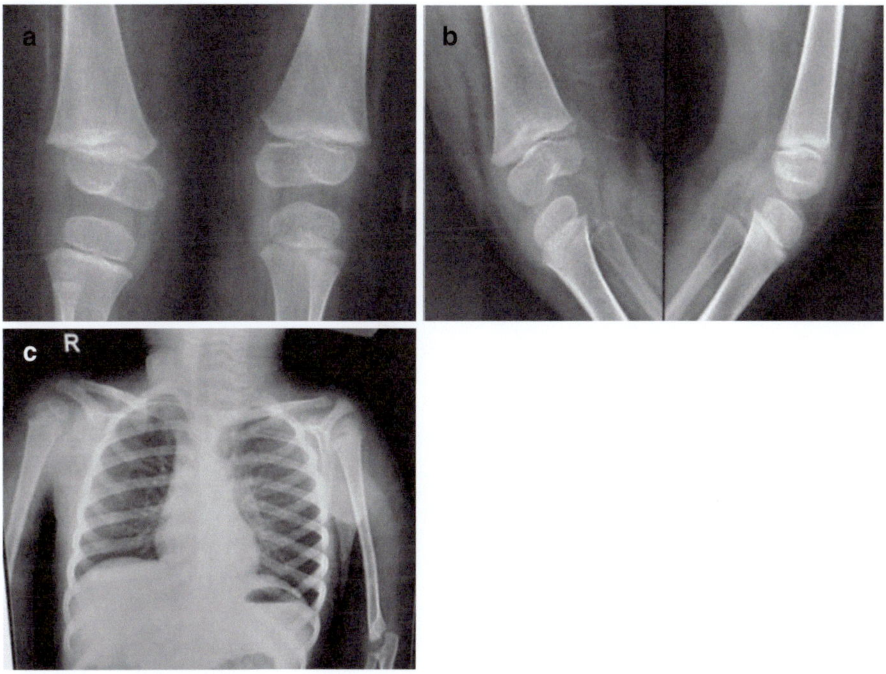

Fig. 13.19 (**a**, **b**) Radiograph of the knees with the absence of the patella on either side. (**c**) Radiograph reveals adduction of the shoulders and straight elbows

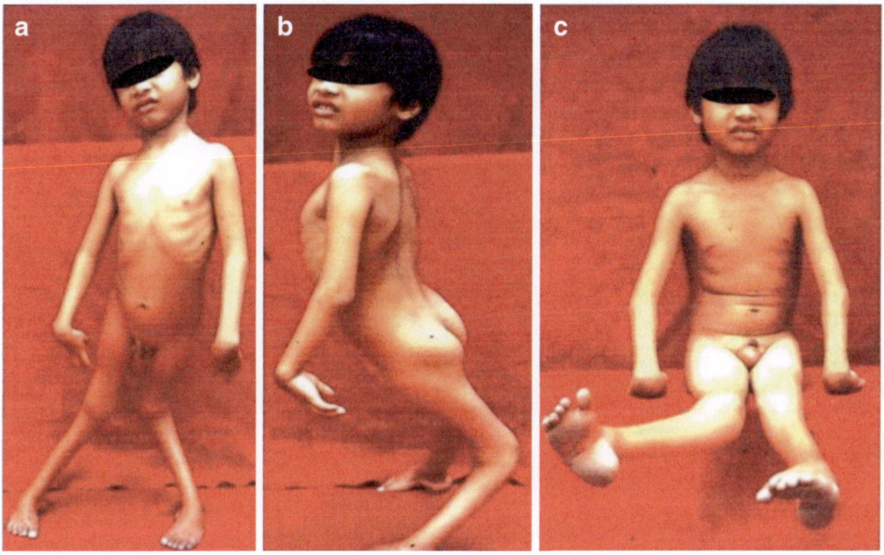

Fig. 13.20 (**a**, **b**, **c**) An 8-year-old boy with multiple deformities had adducted upper limb with severe flexion at the wrists. Note the severe bilateral recurvatum deformity at the knees with compensatory lumbar lordosis. However, the patient has managed to stand. The right knee is in valgus position

The natural history and long-term follow-up of the disease are not well known. There is paucity of description of adults in the exiting literature (Fig. 13.21a–k). The contractures worsen and the joints become stiffer. Nearly 25–30% of the cases is nonambulators and is limited household walkers.

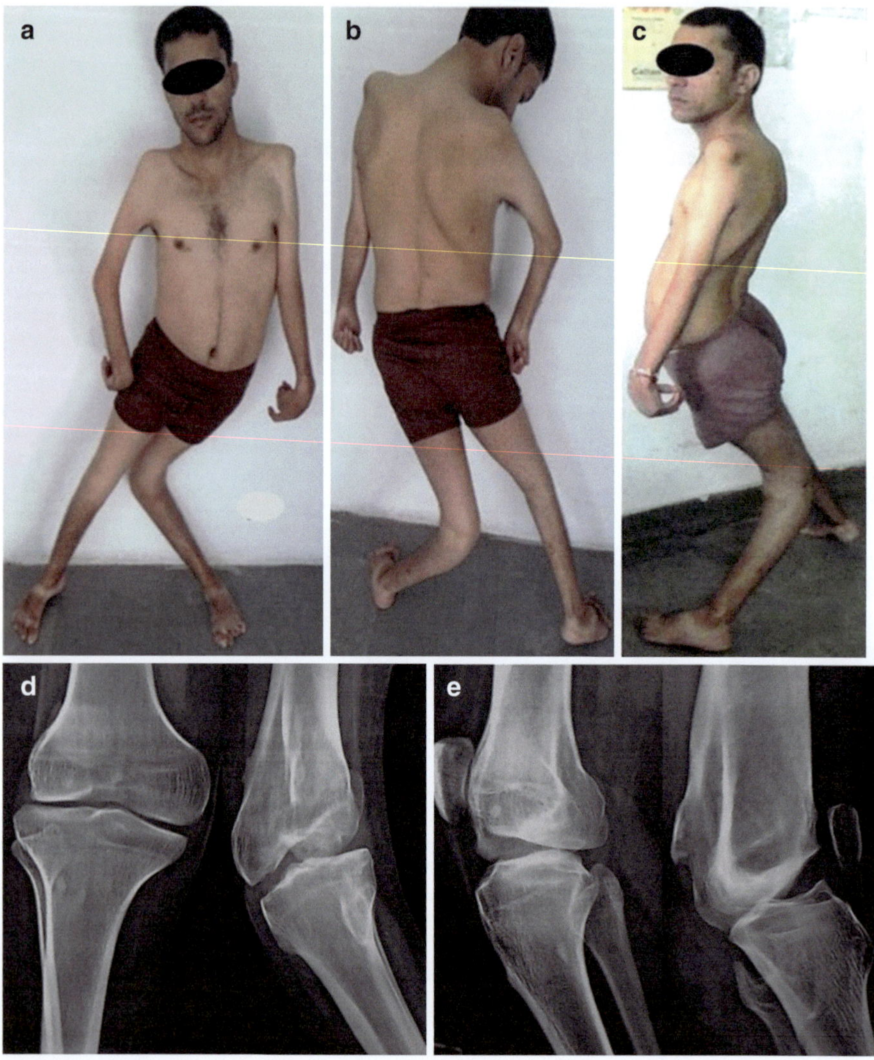

Fig. 13.21 (**a**) A 25-year-old male came with classical clinical signs of Arthrogryposis Multiplex congenita. The upper limbs are straight with severe flexion at the wrists. The left knee is abducted. (**b**) The left knee reveals severe recurvatum. (**c**) Note the severe compensatory lumbar lordosis. (**d**) The roentgenogram of the left knee revealed adaptive changes with abduction and recurvatum. (**e**) The deformed bone ends of the left knee reveal signs of early osteoarthrosis. (**f**) Roentgenogram showing lateral view of both the wrists with flexion deformity of the fingers. (**g**) Roentgenogram of the pelvis with both the hips shows bilateral coxa-valga. (**h**) Post-operative roentgenogram of the left knee after an arthrodesis. The Charnley Clamp is in position stabilizing the arthrodesis. (**i**) Roentgenogram revealing a sound knee arthrodesis. (**j**, **k**) The patient showing complete correction of the left knee deformity after an arthrodesis. The patient can now put on shoes and can walk in an acceptable manner

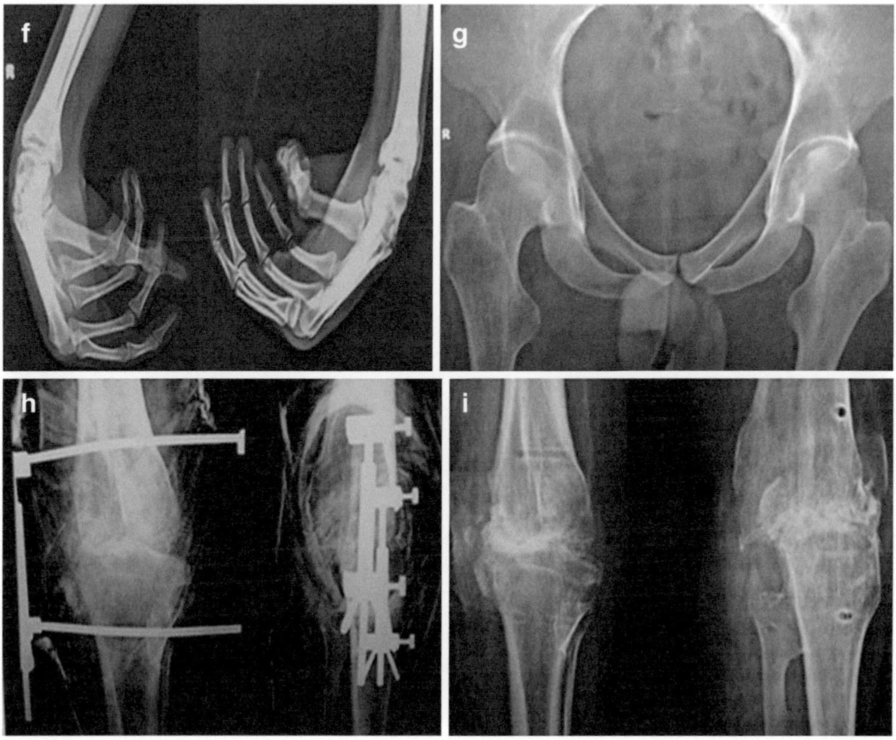

Fig. 13.21 (continued)

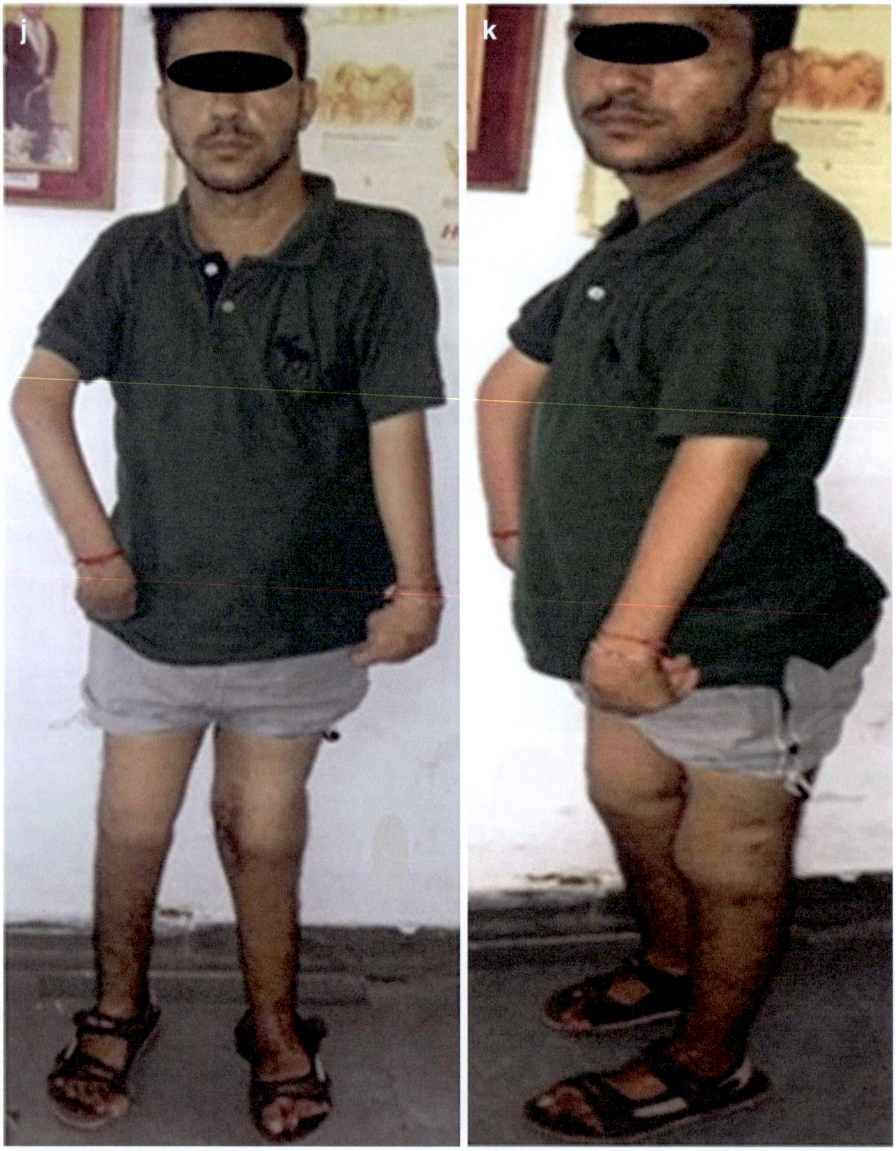

Fig. 13.21 (continued)

13.8 Muscular Dystrophies

The muscular dystrophies are a group of noninflammatory inherited disorders, with progressive degeneration and weakness of skeletal muscle without apparent cause in the peripheral or central nervous system. These have been divided by clinical distribution, severity of muscle weakness and pattern of genetic inheritance.

13.8.1 Duchenne Muscular Dystrophy

The Duchenne muscular dystrophy (Pseudo-hypertrophic muscular dystrophy) is most common. This is clinically evident around the age of 5 years. At the onset the child is not able to achieve independent ambulation and is hesitant to walk or run. There is progressive weakness is proximal muscles group which is symmetrical. The common muscles involved are the gluteus maximus, the quadriceps and tibialis anterior. Pseudohypertrophy of the calf muscles is common which may be due to degeneration of muscles and accumulation of fat. The calf muscles thus show apparent hypertrophy (increase in girth) (Fig. 13.22a–d). The child has difficulty in getting up from squatting position and by climbing upon his lower limbs and by supporting the body weight with his hands and his legs, knees and thighs (Gower's sign) (Fig. 13.23a–c). The abdominal muscles are weak. Subsequently the shoulder girdle is also affected. Muscle biopsy reveals replacement of muscle fibres by fibro-fatty-tissue. The laboratory conformation is increased by serum creatinine phosphokinase (CPK) level.

The gait is slow and the patient develops compensatory changes as the disease progresses. The patient tries to maintain the centre of gravity to avoid a fall. Thus the head is kept posterior and there is compensatory lumber lordosis. Tendo-Achilles contractures develop and the patient walks on the fore foot. Hence, the patient gradually develops wide-based and waddling gait. Such a presentation in the absence of sensory deficit makes the diagnosis which should be confirmed by biopsy. The treatment modalities include physical therapy, orthoses and care of complications.

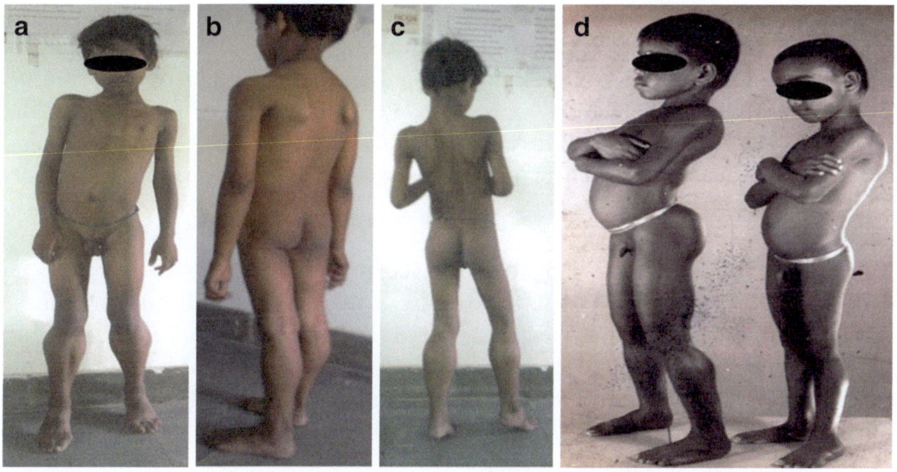

Fig. 13.22 (**a**, **b**, **c**) Clinical picture of patients with Duchenne muscular dystrophy. There is pseudohypertrophy of the calves. (**d**) Clinical picture of two brothers revealing Duchenne muscular dystrophy. Note the lumbar lordosis and backward position of the head to avoid a fall

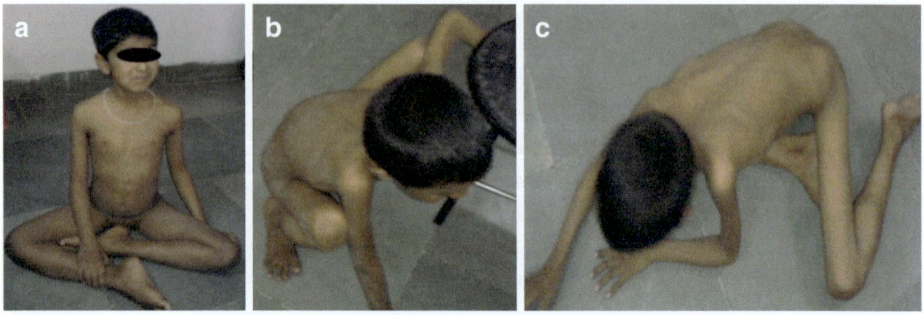

Fig. 13.23 (**a**, **b**, **c**) Note the child has difficulty in getting up from squatting position and by climbing upon his lower limbs and by supporting the body weight with his hands on his knees (Gower's sign)

13.8.2 Becker Muscular Dystrophy

Becker muscular dystrophy presents in a similar fashion as the Duchenne muscular dystrophy but is of less severe nature and less common. The patients often remain ambulatory until early adulthood. The Gower manoeuvre may occur as the weakness progressive. Pseudohypertrophy of the calf is common and gradually equinus and cavus foot deformities occur. There may be a family history of atypical muscular dystrophy. Life expectancy may be a little more as the pulmonary problems are less severe. The management depends on the individual patients and is mostly surgical.

13.8.2.1 Limb-Girdle Muscular Dystrophy

Limb-girdle muscular dystrophy is more common and benign than the other forms. The age of onset is variable, may be in second or third decade of life (Fig. 13.24a, b). The initial muscle weakness involves either the pelvic or the shoulder girdle and the progress is usually slow. There are soft tissue contractures. However, the patients may remain ambulatory for many years.

13.8.2.2 Facio-Scapulo-Humeral Muscular Dystrophy

Facio-scapulo-humeral muscular dystrophy is manifest by muscular weakness of the shoulder girdle and upper arm. These results in scapulothoracic wasting with scapular winging and a marked decrease in the shoulder movements (Fig. 13.25a–d). The onset may occur in late childhood. It is common in females. Life expectancy is relatively good. The treatment depends on the stage of the disease and is usually surgical.

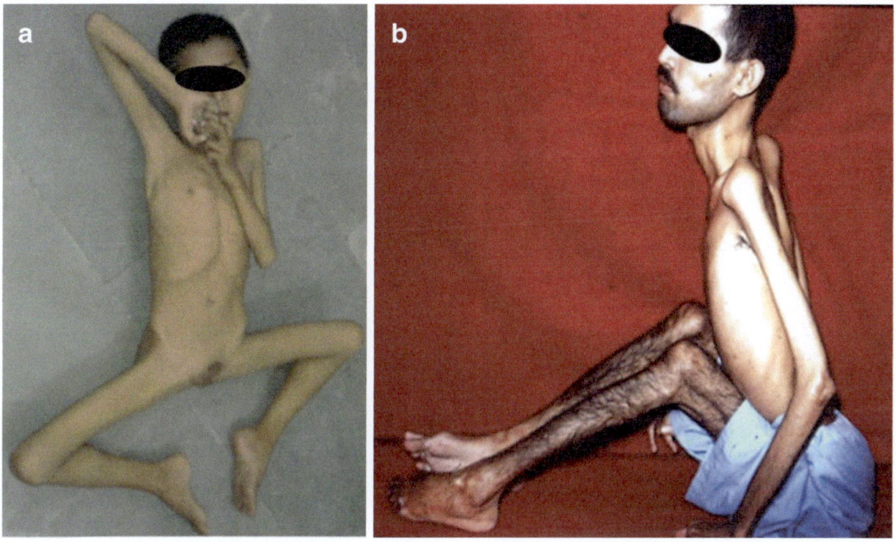

Fig. 13.24 (**a**, **b**) Clinical picture of advanced muscular dystrophy in both these patients where they have become totally dependent. The patients usually die of respiratory paralysis

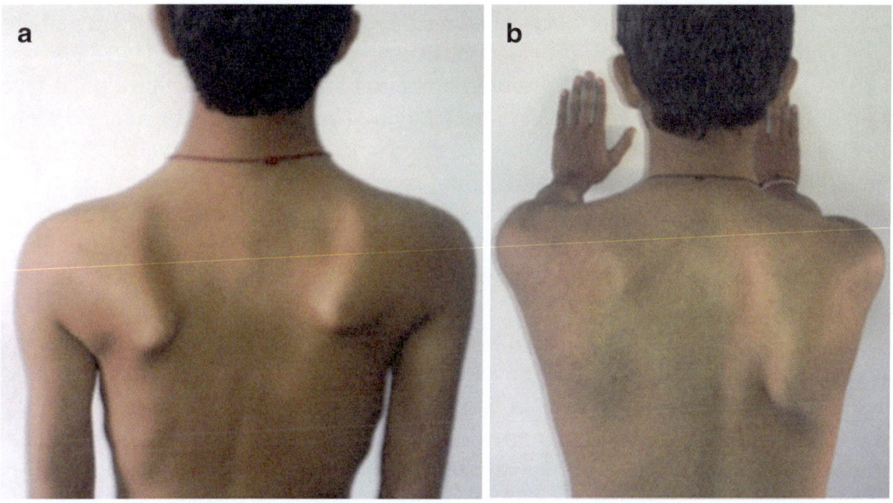

Fig. 13.25 (**a**) Clinical photograph of a patient with facio-scapulo-humeral muscular dystrophy with winging of the scapulae. Note the high placement of the right scapula. (**b**) The winging of the scapula is more prominent while pushing the wall. (**c**, **d**) The front and back view of a 20-year-old patient with abduction contracture of the shoulders. Note the winging of the scapulae

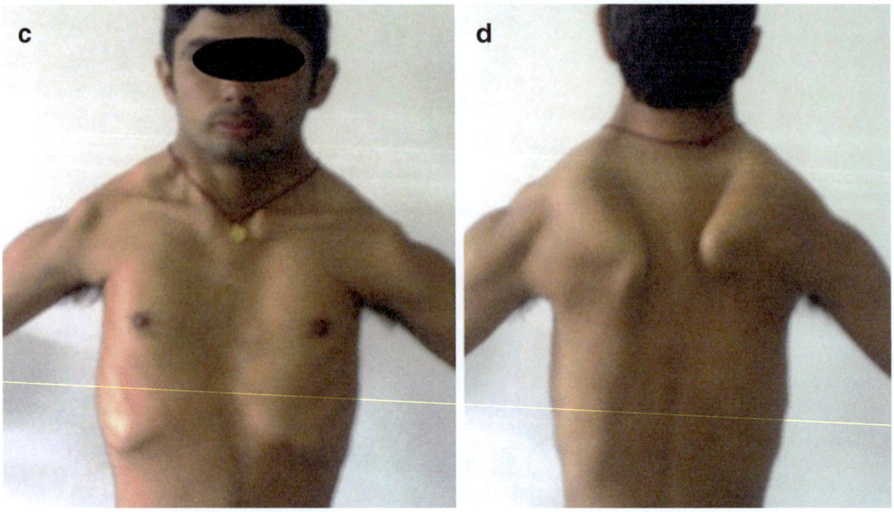

Fig. 13.25 (continued)

13.9 Larsen's Syndrome

Larsen's syndrome is a rare heterogeneous syndrome is characterized by marked joint laxity and multiple joint dislocations may be noted at birth (Fig. 13.26a–d). The hip, knee and elbow joints are predominantly affected. Infants have a flat face with a low nasal bridge and widely spaced eyes. There is delayed dental development and there may be a cleft palate or uvula. Hand abnormalities include cylindrical fingers, short metacarpals and spatulate thumbs. There may be associated abnormalities of epiphyseal development. The spine may also be abnormally curved. Secondary osteoarthritis occurs as a consequence of the marked join instability. Survival is normal unless there is severe spinal involvement and neurological complications.

Joint subluxation and dislocation is seen in multiple joints. There may be associated abnormalities of epiphyseal developmental. The calcaneum contains an extra ossification centre which appears in late infancy or early childhood. The spine may show congenital segmentation anomalies, particularly in the cervical and thoracic regions. The spine may also be abnormally curved. Hands show typical sausages shape fingers with irregular lengths of metacarpals and phalanges. There may be many accessory carpal bones. Supernumerary or abnormally—shaped carpal and phalangeal bones may be noted.

No characteristic pathological changes are seen in joint tissues. The treatment consists of close reduction of various joints. In unsuccessful attempt an open reduction may be necessary (Fig. 13.27a–r). Secondary osteoarthritis occurs as a consequence of the marked joint instability. Survival is normal unless there is severe spinal involvement and neurological complications. Differential diagnosis includes other causes of joint laxity.

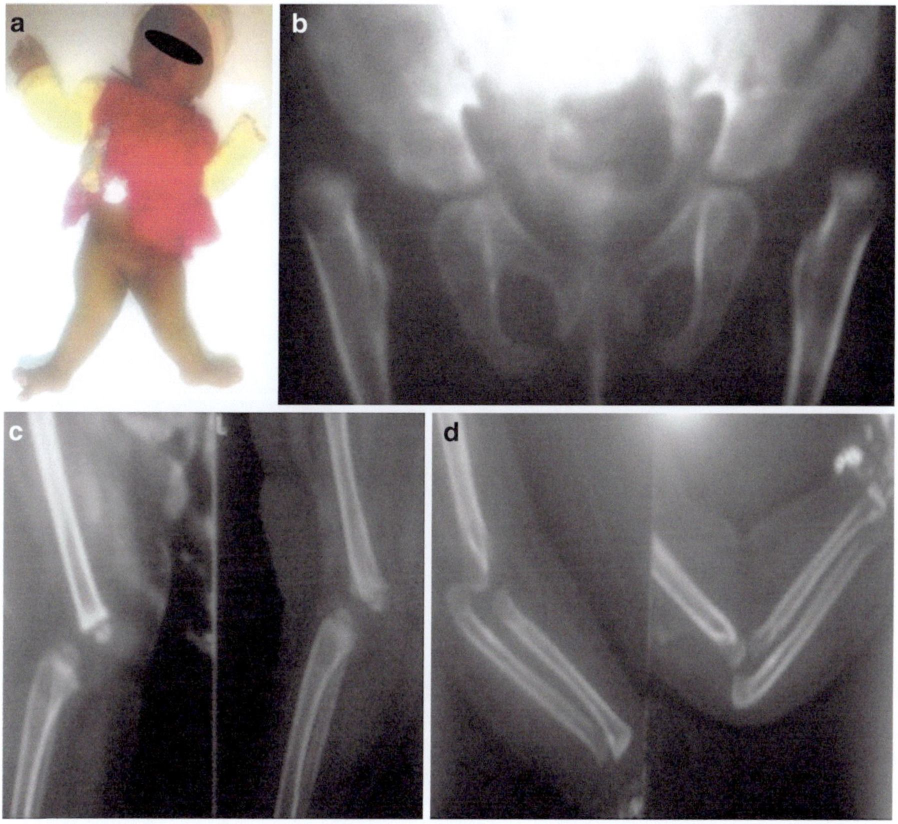

Fig. 13.26 (**a**) A child with Larsen's syndrome showing bilateral genu-recurvatum. (**b**) Roentgenogram reveals bilateral high-dislocation of hips. (**c**) Roentgenogram showing bilateral anterior dislocation of the knees. (**d**) Roentgenogram revealing posterior dislocation of the elbows

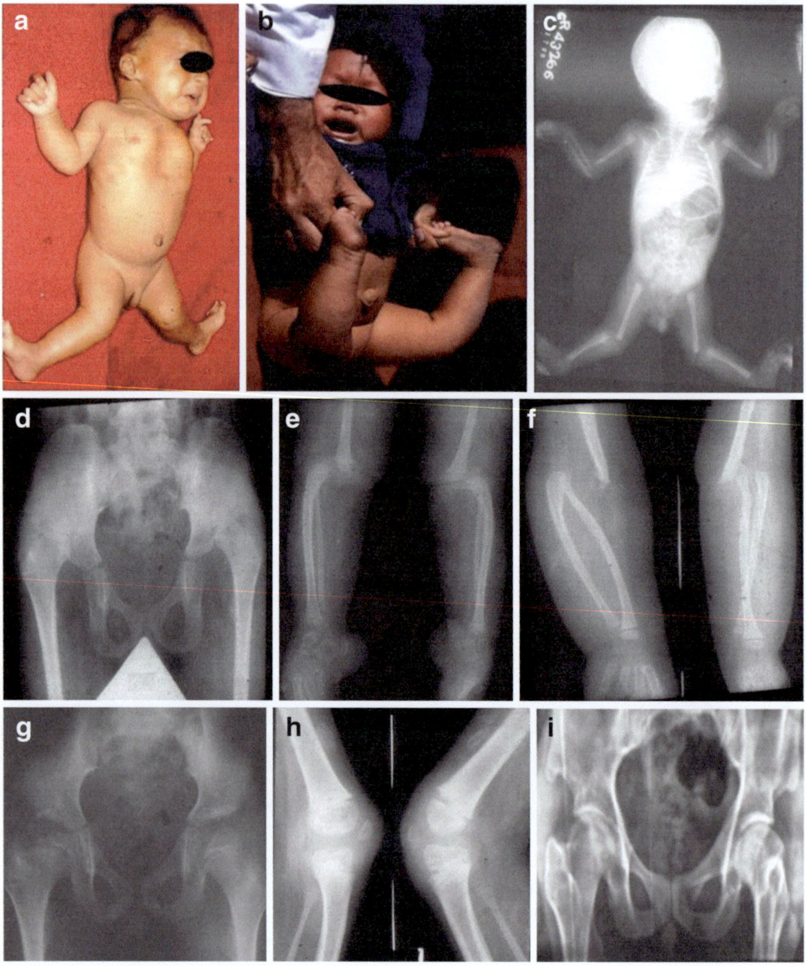

Fig. 13.27 (**a**) A newly born male child with bilateral genu-recurvatum and pectus excavatum. (**b**) Note the dislocation at both the knees. (**c**) Roentgenogram revealing dislocations of various joints of the upper and lower extremities. (**d**) Roentgenogram showing high riding bilateral dislocation of the hips. Note the pelvic abnormity, particularly the anomaly of both the iliac crests and the pubic rami. (**e**) Roentgenogram revealing bilateral dislocation of the knees. (**f**) Roentgenogram reveals disorganized and dislocated elbows. (**g**) Roentgenogram at the age of 5 years revealing both hip joints after open reduction. (**h**) Lateral radiographs revealing location of both the knees after open reduction. (**i, j**) Roentgenogram of the hips and the knees at the age of 24 years. The various joints have started showing signs of early degeneration. (**k**) Roentgenogram of the spine also reveals signs of enhanced lumbar lordosis. (**l**) Roentgenogram of the hands showing abnormalities of the carpal bones. The interphalangial joints reveal subluxations and shortening of the distal phalanges. (**m**) The trasal bones, metatarsals and the phalanges appear abnormal and reveal deformities due to subluxation at the metacarpo and inter phalangeal joints. (**n**) The axis deformity of the toes and flat feet can be observed which may be due to ligamentous laxity. (**o**) Both the elbows remain dislocated. (**p**) Roentgenogram of both the elbows reveal complete dislocation but are functional. The bone ends have lost their normal anatomical configuration—a rare presentation. (**q**) Clinical picture of the patient after 18 years, showing dislocation of both the elbows. However, he had good range of movement. (**r**) The same patient at the age of 29 years. He is a Software engineer

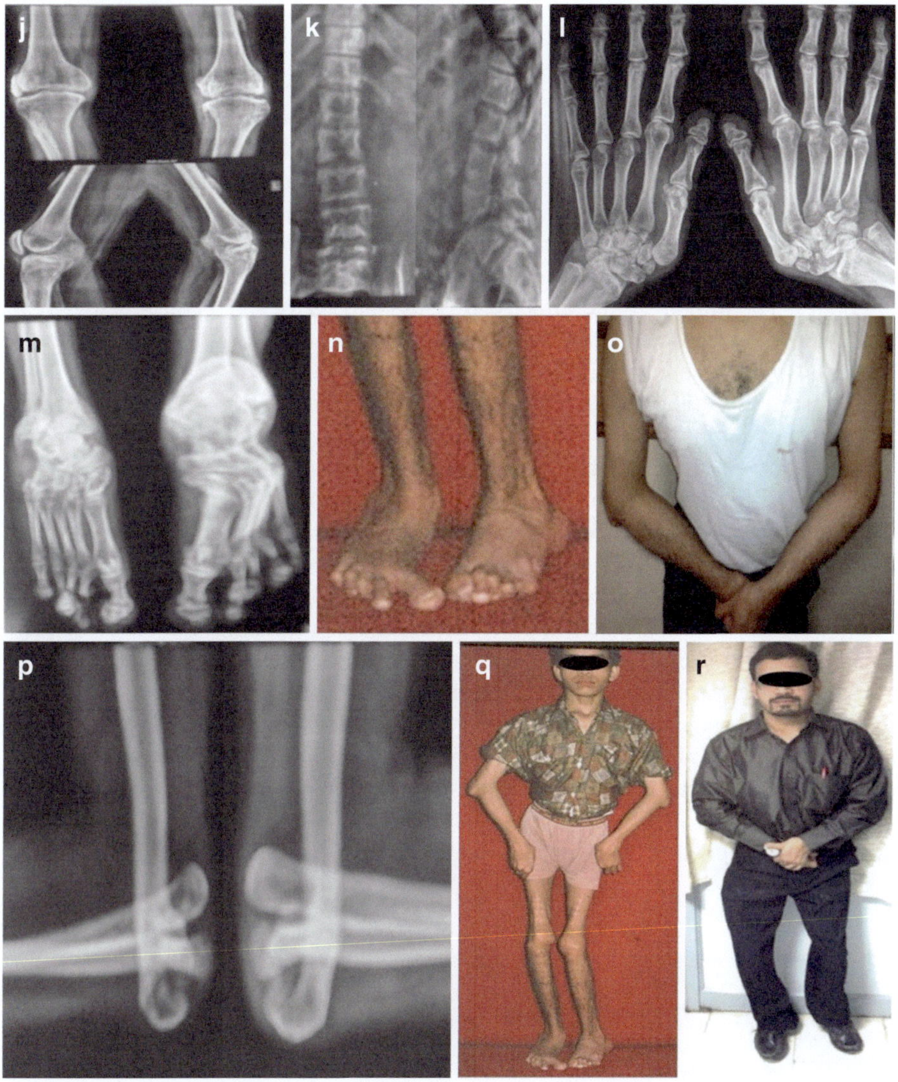

Fig. 13.27 (continued)

13.10 Infantile Digital Fibromatosis (Congenital Digital Fibromas)

Infantile digital fibromatosis of infancy is a rare condition. However, when present, they are common in females. They arise from the fingers and toes and are often multiple in distributions. They start as small nodules which may increase to an enormous size, affecting mainly the dorsolateral aspect of the distal parts of the adjoining fingers (Fig. 13.28a, b). The covering skin may become reddish in colour and fixed to the tumour. The swellings are painless and often adhere to the deeper tissues sparing the bones. The tumours quite often regress spontaneously and a conservative treatment has been recommended. Recurrence has been reported even after excisions.

An 11-month-old female infant was brought with a history of progressive swellings of the distal parts of the fingers, one on the left and four on the right fingers, associated with bilateral deformed toes (Fig. 13.29a, b). On X-rays the four radial metacarpals on the left side were short and deformed. There was bilateral absence of the second and hypoplasia of the third metatarsals (Fig. 13.30a, b). One week after birth a small swelling was noticed on the left ring finger. Subsequently similar swellings appeared one after another on other fingers. The tumours involved the terminal phalanges sparing the nails. The tumours were firm and painless. The overlying skin was red and shining. There were flexion contractures of the right ring and four left fingers. Bilateral syndactyly and flexion contracture of the second toe were also present.

Arteriography showed a vascular nature of the tumours. This was confirmed during excision of one of the tumours, no bleeding occurring during surgery. The specimen removed was a nonencapsulated greyish white fibrous mass. Histopathological examination showed that the dermis was replaced by interlacing bands of fibrous connective tissue with an abundance of collagen, similar to dermato-fibromas with no sign of malignancy. A section of deformed bone showed a normal histopathological appearance.

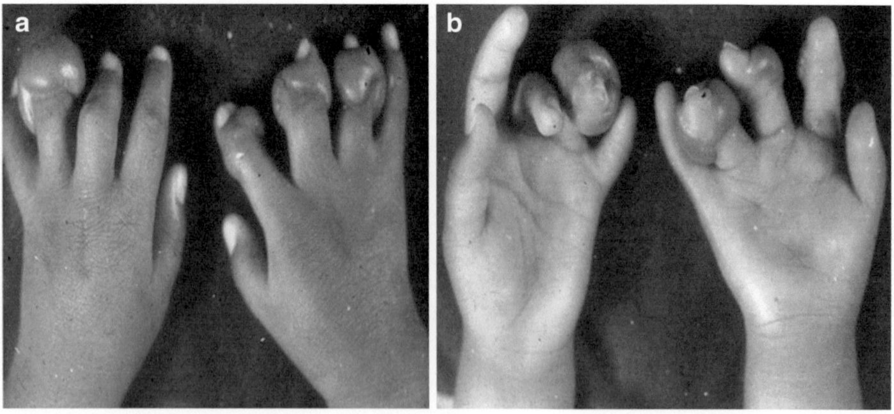

Fig. 13.28 (**a, b**) Dorsal and palmar views of an 11-months-old girl with infantile digital fibromatosis reveals the lesions on the distal ends of the fingers

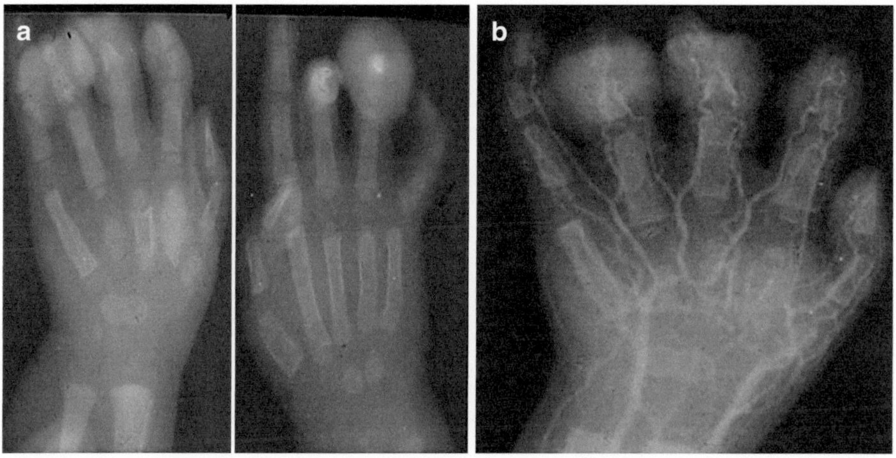

Fig. 13.29 (**a**) Roentgenogram reveals soft tissue lesions with pressure effect on the distal phalanges. (**b**) An angiogram shows digital arteries with normal vasculature

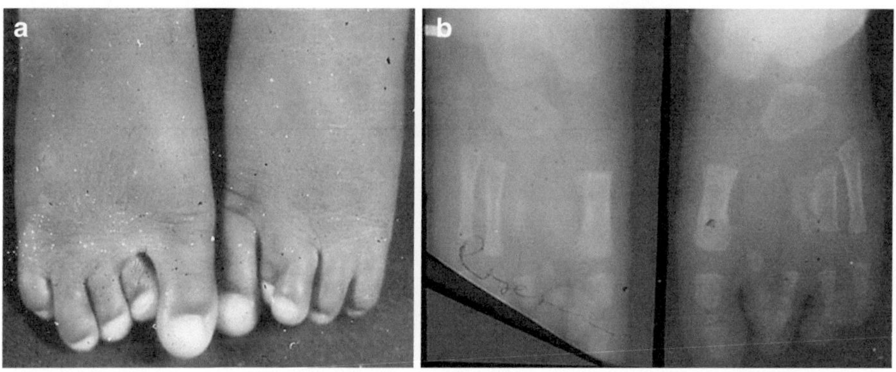

Fig. 13.30 (**a**) The second and third toes appear smaller on the right side and short second toe on the left side. (**b**) Roentgenogram reveals disorganized second and third metatarsals on the right side and second metatarsal on the left side which has resulted in the shortening of the toes

13.10.1 Poland Syndrome

There is Symbrachydactyly-short digits with Syndactyly. There may be ipsilateral absence of Pectoralis minor along with the sternal portion of the Pectoralis major (Fig. 13.31a–c). There may also be absence/hypoplastic nipple and areola. 10% cases may have cutaneous syndactyly. On the affected side the hand may be hypoplastic.

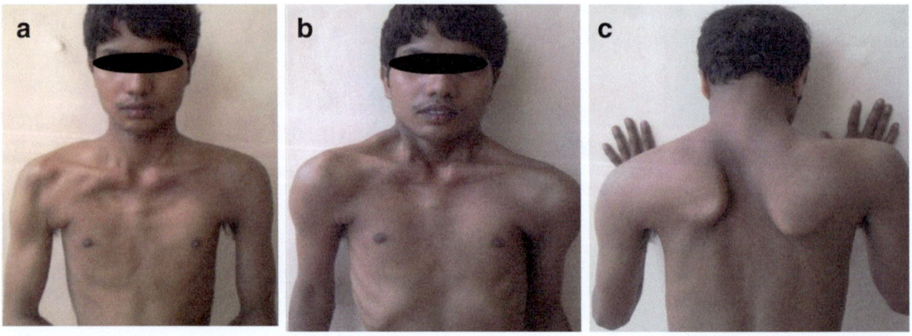

Fig. 13.31 (**a, b**) A 25-year-old patient showing bilateral absence of pectoralis minor and the sternal portion of the pectoralis major. Note the prominent medial half of the clavicles. There is pectus excavatum. The axillary folds are transverse. (**c**) In the same patient there is bilateral winging of the scapulae. Note the hypoplasia of the left scapula

13.10.2 Nail-Patella Syndrome

Nail-patella syndrome includes dysplasia of nails, hypoplasia of patella, elbow dysplasia and may be iliac horns. The nails of the fingers than toes are dystrophic, hypoplastic and even may be absent (Fig. 13.32a–d). The nails of the fingers may be more involved than toe nails. The patellae are also abnormally placed and are hypoplastic or may be absent which may result in valgus or varus alignment (Fig. 13.33). The head of the radius may be dislocated and the elbow joint may be dysplastic with abnormalities in the capitellum and trochlea. Iliac horns and bony exostoses on the posterior surface of the ilium may be present.

The patients may have short stature, shoulder girdle dysplasia, foot deformity and contractures of various joints may be noted.

13.10.3 Proteus Syndrome

There may be overgrowth of some parts with disfigurement to the extent that a description of 'Elephant-Man' has been made. There may be bizarre array of abnormalities. The deformity increases with age. A differential diagnosis of the following can be considered:

1. Hamartomatous overgrowth.
2. Idiopathic hemihypertrophy.
3. Maffucci syndrome.
4. Neurofibromatosis.
5. Macrodactyly-partial gigantism.

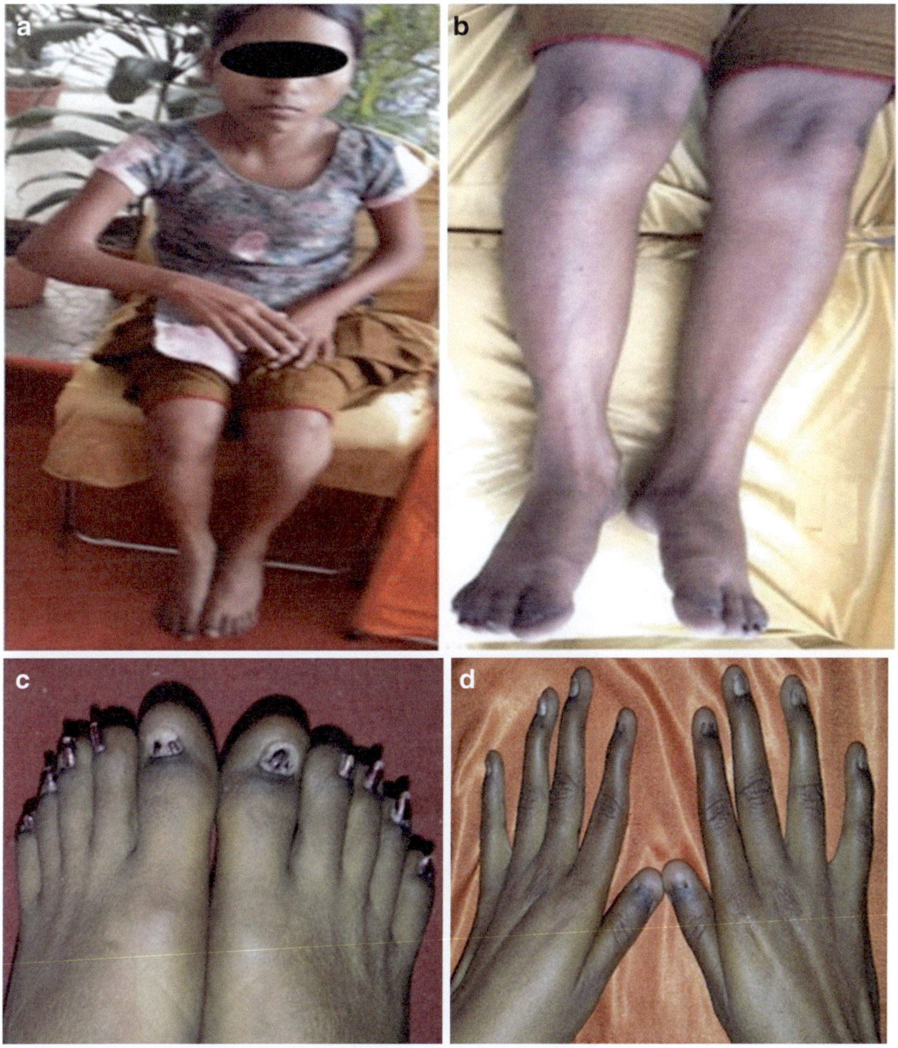

Fig. 13.32 (**a**) A 10-year-old girl revealing hypoplastic patellae and nails. (**b**) The patient revealing hypoplastic patellae and dystrophic nails. (**c**, **d**) The nails of the fingers and the toes are hypoplastic and dystrophic

13.10.4 Holt-Oram Syndrome

There are upper limb anomalies with heart disease. The thumb is almost always affected. It ranges between total absences to tri-phalangeal thumb (Fig. 13.34). The most extreme cases may be associated with Phocomelia which may be bilateral or unilateral (Fig. 13.35).

Fig. 13.33 Roentgenogram shows hypoplastic and abnormally placed patella

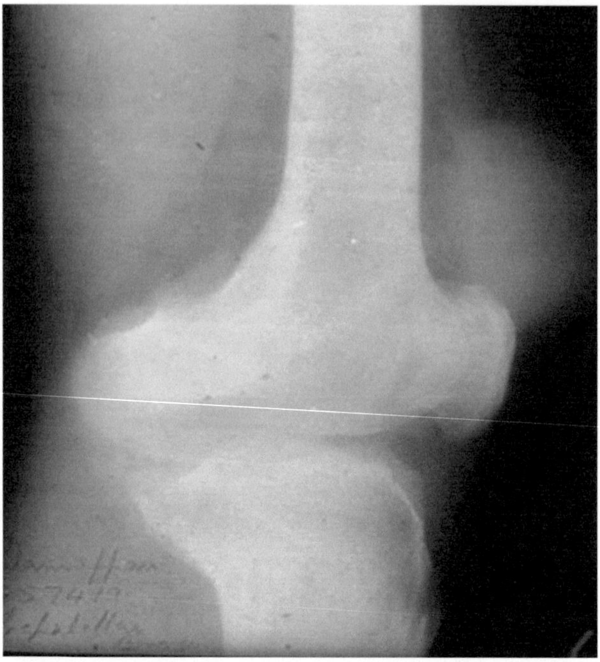

Fig. 13.34 Roentgenogram revealing bilateral tri-phalangeal thumb. Note the absence of middle phalanges in the index and the ring fingers. The middle phalanx in both the middle fingers is small

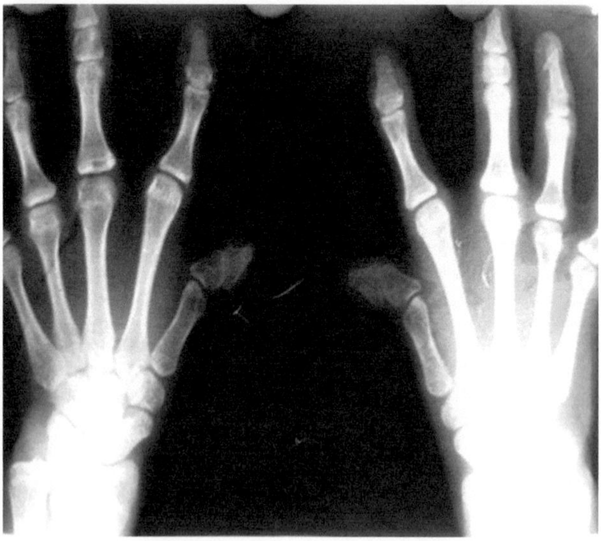

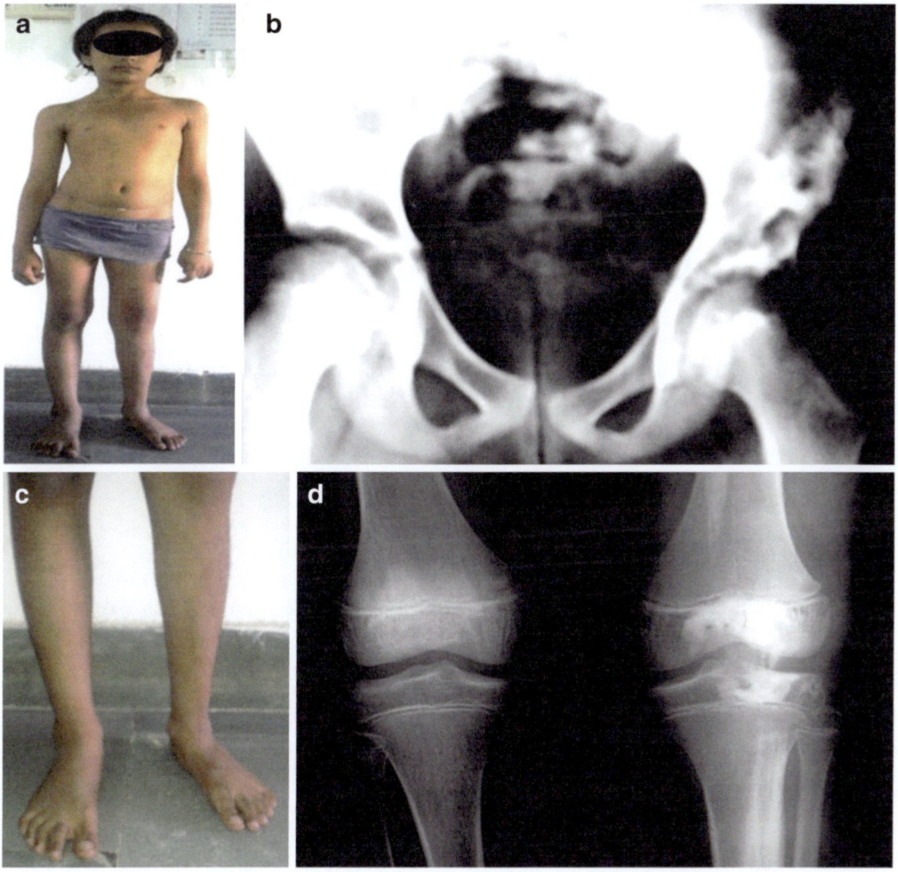

Fig. 13.35 (**a**) Clinical picture of an 8-year-old girl came with a gradually increasing limp on the left side. (**b**) The roentgenogram revealed lytic and sclerotic changes in the supra-acetabular area. The patient has been under observation. (**c**) She had occasional pain in the left knee. (**d**) Roentgenogram of the left knee revealed sclerotic changes in the lower end of the femur and upper end of the tibia in the physeal region

13.10.5 Pterygia Syndrome

This rare congenital anomaly includes soft tissue contractures with webbing of the skin in the flexor areas of the joints particularly at the knee. As a result the patient cannot extend the knee joints and walk with half flexion gate. The joints are otherwise not involved.

13.10.6 Mild Dysplasias

It is a common observation that mild variety of skeletal dysplasias goes un-noticed till they cause the secondary changes in the form of osteoarthrosis. There may also be other Osteoarticular changes in late life.

13.10.7 Lethal Skeletal Dysplasias

The earlier authors have proved that there are many lethal types of dysplasias. Their identity has made after birth and on post-mortem. These mainly include Achondrogenesis, Fibrochondrogenesis and Thanatophoric dysplasia. There are many others which may be listed in this category.

13.10.8 The Unknown Skeletal Dysplasias

There are many ill-understood and unrecognizable patterns of such human malformations which are yet to be classified. We have to understand the exact pathophysiology of both, Dwarfism and Gigantism and the others.

Suggested Readings

Levine MA. The McCune-Albright syndrome. N Engl. J Med. 1991;325:1738–9.

Listernick R, Charrow J. Neurofibromatosis type I in childhood. J Pediatr. 1990;116:845–53.

Martuza RC, Eldridge R. Neurofibromatosis 2 (bilateral acoustic neurofibromatosis). N Engl J Med. 1988;113:39–52.

Dawe C, Wynne-Davies R, Fulford GE. Clinical variation in dyschondrosteosis: a report on 13 individuals in 8 families. J Bone Joint Surg (Br). 1982;64:377.

Crawford AH, Bagamery N. Osseous manifestations of neurofibromatosis in childhood. J Pediatr Orthop. 1986;6:72–88.

Banker BQ. Neuropathologic aspects of arthrogryposis multiplex congenital. Clin Orthop. 1985;194:30.

Drummond DS, Siller TN, Cruess RL. Management of arthrogryposis multiplex congenital. In: AAOS instructional course lecture, vol. 23. St Louis: CV Mosby; 1974. p. 79.

Athanasou NA. Pathological basis orthopaedic and rheumatic disease. London: Edward Arnold; 2001.

Hansen OM. Surgical anatomy in treatment of patients with arthrogryposis. J Bone Joint Surg Am. 1961;43:855.

Lloyd-Roberts GC, Lettn AWF. Arthrogryposis multiplex congenital. J Bone Joint Surg Am. 1961;43:855.

Lloyd-Roberts GC, Lettn AWF. Arthrogryposis multiplex congenital. J Bone Joint Surg (Br). 1970;52:494.

Thompson GH, Bilenker RM. Comprehensive management of arthrogryposis multiplex congenital. Clin Orthop. 1985;194:6–14.

Hall JG. Arthrogryposis multiplex congenital: etiology, genetics, classification, diagnostic approach and general aspects. J Pediatr Orthop B. 1997;6(3):159–66.

Larsen B, Reimann I, Becker-Andersen H. Congenital calcaneal valgus. Acta Orthop Scand. 1974;45:145.

Bowen JR, Ortega K, Ray S, et al. Spinal deformities in Larsen syndrome. Clin Orthop. 1985;197:159.

Oki T, Terashima Y, Murachi S, et al. Clinical features and treatment of joint dislocations in Larsen's syndrome: report of three cases in one family. Clin Orthop. 1976;119:206.

Steel HH, Kohl EJ. Multiple congenital dislocations associated with other skeletal anomalies (Larsen's syndrome) in three siblings. J Bone Joint Surg Am. 1972;54:75.

Laville JM, Lakermance P, Limouzy F. Larsen's syndrome. Review of the literature and analysis of 38 cases. J Ped Orthoped. 1994;14:63–73.

Lowan DCR, Takayama N, Flatt AE. Poland's syndrome: a review of 43 cases. J Bone Joint Surg Am. 1976;58:52.

Wilson MR, Louis DS, Stevenson TR. Poland's syndrome: variable expression in associated anomalies. J Hand Surg [Am]. 1988;13:880.

McKibbin B. Anatomical factors in the stability of the hip in the newborn. J Bone Joint Surg. 1970;52B:148.

Wynne-Davis RW. Acetabular dysplasia and familial joint laxity. Two aetiological factors in congenital dislocation of the hip. A review of 589 patients and their families. J Bone Joint Surg. 1970;52B:704–416.

Udell J, Schumacher HR, Kaplan F, Fallon MD. Idiopathic familial acroosteolysis. Histomorphometric study and literature review of the Hajdu-Cheney syndrome. Arthritis Rheum. 1986;29:1032–8.

Guidera KJ, Satterwhite Y, Ogden JA, Pugh L, Ganey T. Nail-patella syndrome. A review of 44 orthopaedic patients. J Pediatr Orthop. 1991;11:737–42.

Henry A. A monostotic fibrous dysplasia. J Bone Joint Surg (Br). 1969;51:300.

Parekh SG, Donthineni-Rao R, Ricchetti E, Lackman RD. Fibrous dysplasia. J Am Acad Orthop Surg. 2004;12:305–13.

Crawford AH, Schorry EK. Neurofibromatosis update. J Pediatr Orthop. 2006;26:413–23.

Part II

General Considerations

General Considerations

14

In general, limb malformations are classified as reduction (deficiency), excess and fusion or malsegmentation deformities. Aplasia and hypoplasia are the common abnormalities; in the major long bones. They occur in the following bones in the descending order of frequency: fibula, radius, femur, ulna and humerus. When individual bones are affected, alterations in adjacent structures are frequently present. The classification of primary constitutional disorders of skeletal development is complex. Current classification is in terms of longitudinal or transverse orientation and complete or partial extent.

Skeletal dysplasias usually present clinically with disproportionate short stature, skeletal deformity and functional abnormality. In contrast, secondary disorders of bone development result in proportionate short stature. Some malformations that develop postnatally are present but unexpressed at birth as in the case of certain skeletal dysplasias of late onset. Primary disorders of skeletal development should be distinguished from retarded or abnormal skeletal development which occurs secondary to a known cause, such as a nutritional deficiency, endocrine or chronic debilitating disease (e.g., cardiac or liver failure) of childhood.

The age of onset of the condition should be noted and the obstetric, perinatal and family history carefully evaluated. A detailed examination should be carried out to determine whether other organ systems are involved. A full set of skeletal radiographs should be obtained in order to determine the parts of the skeleton affected, the pattern of bone involvement and any other specific features.

Cytogenetic and molecular genetic studies can be employed to identify a specific gene abnormality. Where indicated, biochemical abnormalities associated with a specific primary (or secondary) disorder of skeletal development should also be sought.

S. S. Yadav, *Skeletal Dysplasias & Congenital Malformations*,
https://doi.org/10.1007/978-981-99-5679-1_14

This is not only because a very large number of different forms of congenital malformations have been described but also because the extent of heterogeneity among these conditions makes them difficult to identify and describe adequately. Many developmental abnormalities are rare and some are unique.

Suggested Reading

Swanson AB. A classification for congenital limb malformations. J Hand Surg. 1976;1:8.

15 Congenital Disorders of Neck and Chest

15.1 Klippel-Feil Syndrome

Klippel-Feil Syndrome is a rare developmental disorder characterized by a short neck, limitation in the motion of the cervical spine with low posterior hairline (Fig. 15.1). The incidence is two per 100,000 births. The etiology is not known. This disorder is classified in into three types:

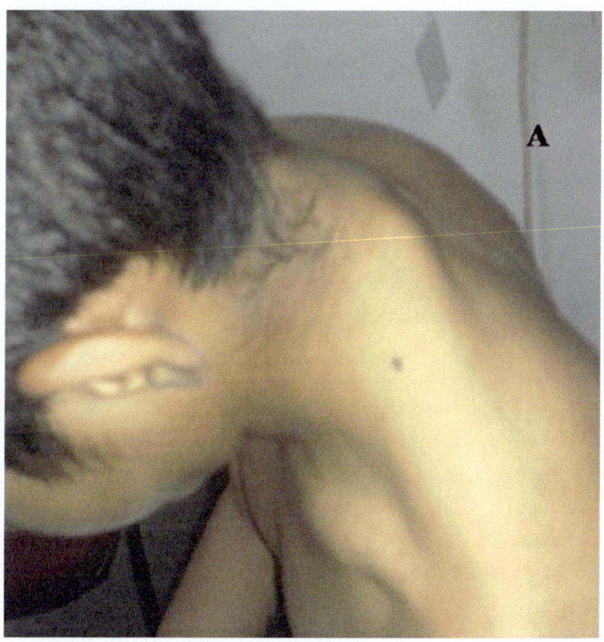

Fig. 15.1 A patient with Klippel-Feil syndrome. Note the low hairline

S. S. Yadav, *Skeletal Dysplasias & Congenital Malformations*,
https://doi.org/10.1007/978-981-99-5679-1_15

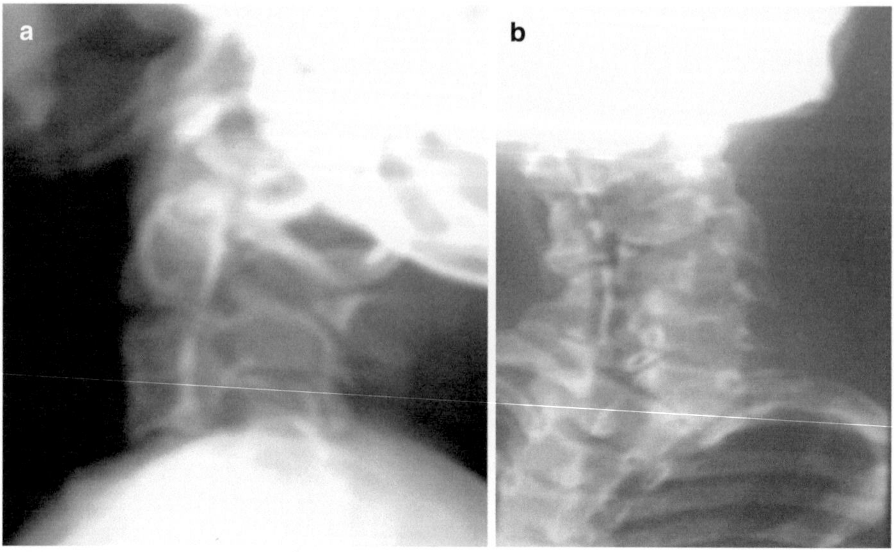

Fig. 15.2 (**a**) Radiograph reveals segmental fusion of the upper cervical vertebrae. (**b**) The cervical spine is deformed showing lateral curve

1. Massive fusion of the cervical vertebrae.
2. Fusion of one or two cervical vertebrae.
3. When the thoracic or lumbar vertebrae are also involved along with the cervical spine.

Nearly one third of the children have an associated Sprengel's deformity. Scoliosis can also occur with some non-skeletal anomalies. The radiograph show varying degrees of vertebral fusion. The common pattern of fusion is C1–C2 and C3–C4 resulting in high degree of instability at the vertebrae above (Fig. 15.2a, b). Flexion and extension radiographs are helpful. The associated anomalies could be Scoliosis, torticollis or congenital heart diseases.

15.2 Sprengel's Deformity

Sprengel deformity (Congenital high scapula) is a rare congenital deformity resulting from the absence of the scapula to descend properly (Fig. 15.3a–d). Affected children have two major problems—cosmetic and functional. The failure of descent of the scapula may be caused by great intrauterine pressure due to variations in the amniotic fluid or abnormal articulations of the scapula with the vertebral column, named as omovertebral bone. The other possible cause may be changes in the shape and size of the scapula (Fig. 15.4a–f). The clinical appearance as a result of elevation and malrotation of the scapula may be worsened by frequent anomalies such as congenital scoliosis, Klippel-Feil syndrome and asymmetry of the chest wall. Dysostosis with prominent axial involvement includes Sprengel shoulder and Klippel-Feil syndrome (Figs. 15.5a–h, 15.6a–h, 15.7a–d, and 15.8a, b).

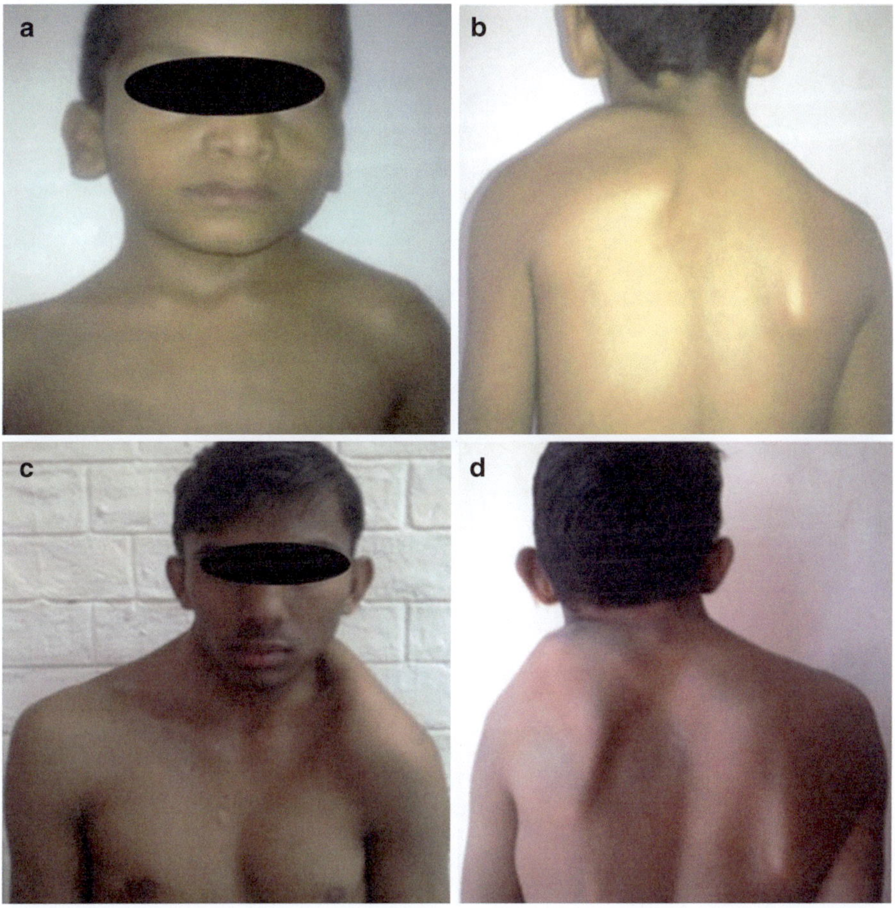

Fig. 15.3 (**a**) A 10-year-old boy showing elevated left shoulder complex. (**b**) The left scapula is considerably elevated. (**c**) A 23-year-old male with severe Sprengel shoulder on the left side. Note the high-riding scapula resulting in severe deformity. (**d**) The left scapula is cephalically placed but not touching the hairline

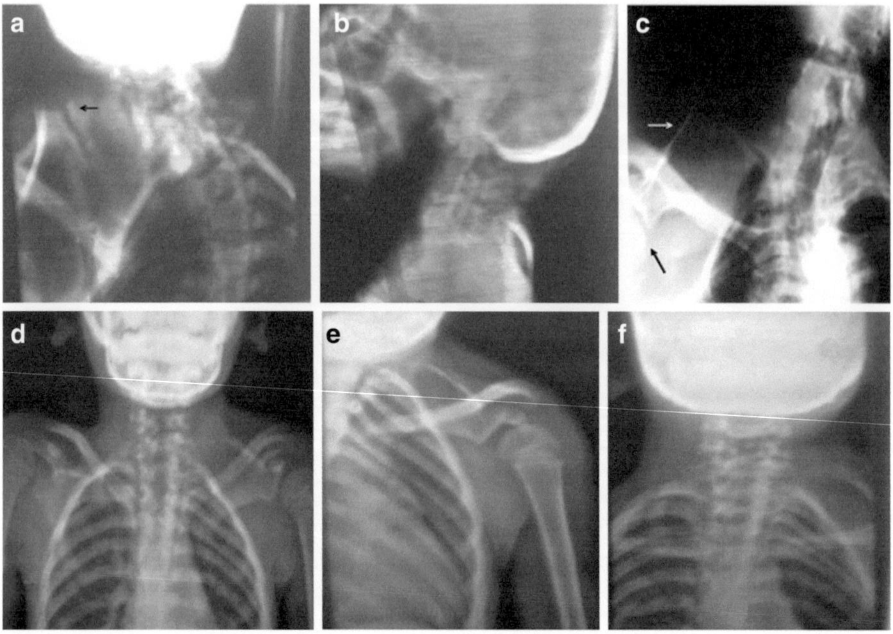

Fig. 15.4 (**a**) Radiograph revealing severe abnormality of the right scapula with segmentation (Arrow). (**b**) The cervical spines are fused which cannot be defined. (**c**) The radiograph shows abnormal shape of the right scapula in a different case (White arrow). The cervical spine is segmentally fused. The gleno-humeral joint can also be seen (Black arrow). (**d**) Roentgenogram reveals elevated scapula on the left side. (**e**) The direction of the left scapula is changed due to elevation. (**f**) Roentgenogram shows elevated scapula on the left side in another patient

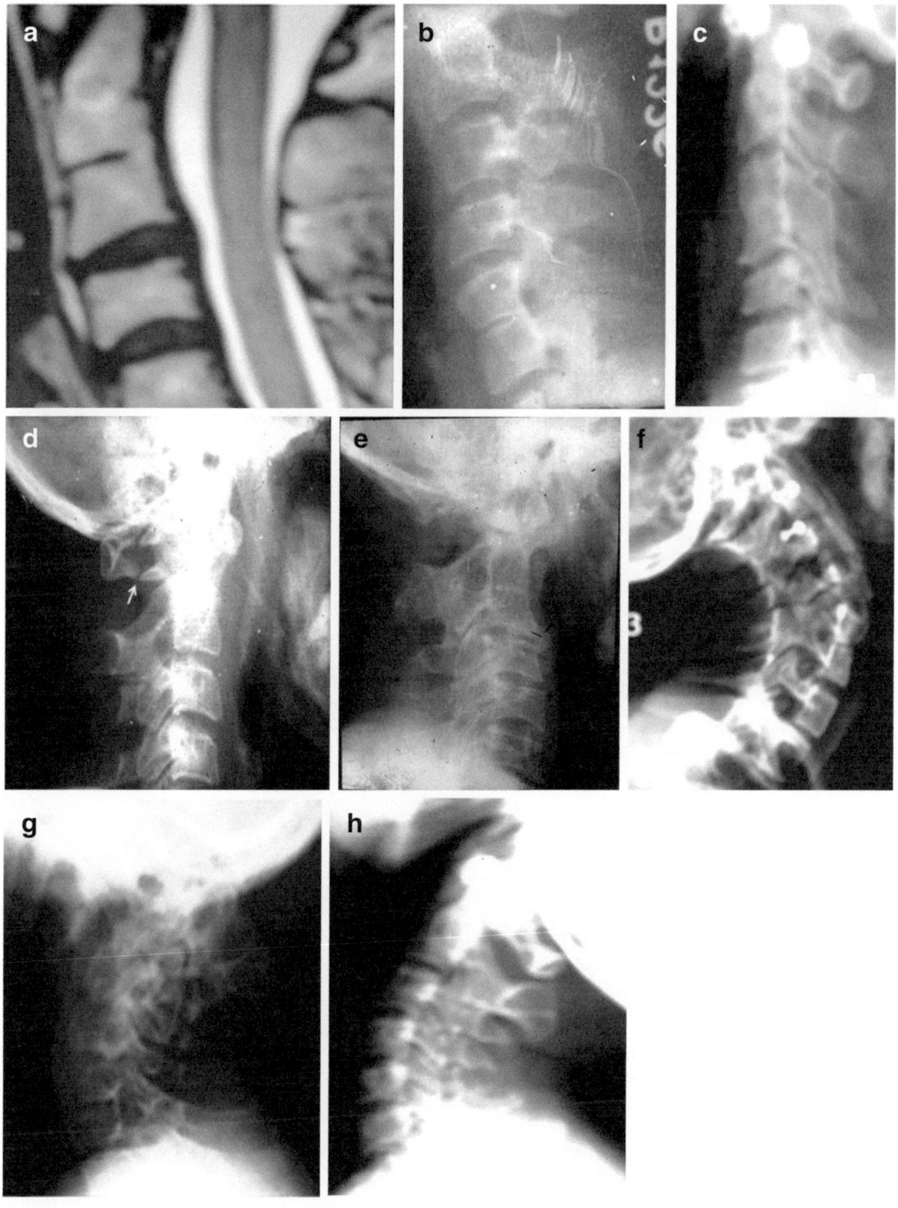

Fig. 15.5 (**a**) MRI reveals congenital synostosis between third and fourth vertebrae. (**b**, **c**) Radiographs revealing congenital fusion of C 3,4 and 5,6 in two different cases. (**d**) Congenital non-union of both the arches of the atlus—a rare anomaly (Arrow). (**e**) Radiograph revealing congenital absence of the posterior arch of the atlus. The axis and the third cervical vertebra reveal block fusion. (**f**) Roentgenogram revealing congenital fusion of C5–C6. (**g**) Radiograph revealing fusion of upper four cervical vertebrae. (**h**) Fusion of the spinus processes of the second and third cervical vertebrae

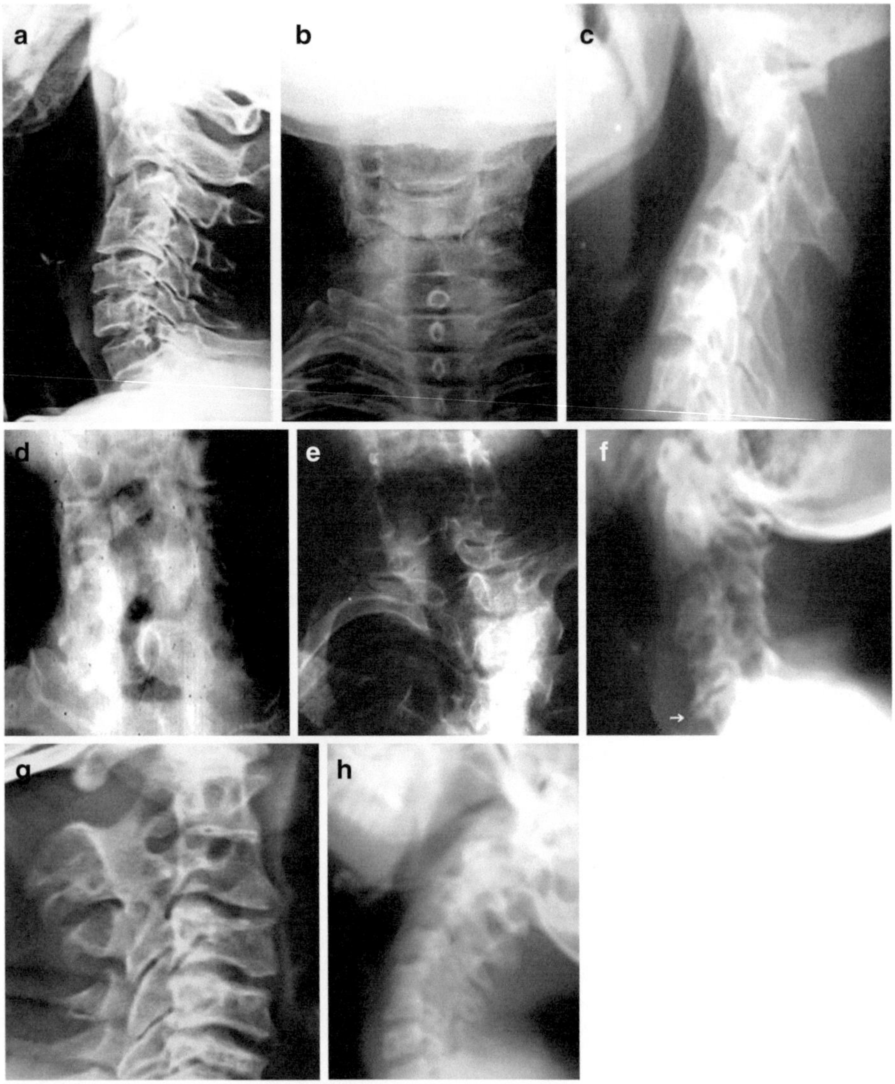

Fig. 15.6 (**a**, **b**) Roentgenogram reveals vertebral fusion of C3 and C4 resulting in change of the curve and pain in the neck. (**c**) Radiograph shows anterior subluxation of the head at the C1–C2 level. This may result in neurological presentations. Note the fusion between second and third vertebrae posteriorly. (**d**) Radiograph reveals spina bifida occulta of C7. (**e**) Note the lack of fusion of the laminae of the lower cervical vertebrae. (**f**) Radiograph revealing multiple congenital anomalies of upper cervical spine resulting in cervical spondylosis (arrow). (**g**) Roentgenogram reveals congenital fusion of C2 and 3 with prominent second cervical spine. (**h**) Complete congenital fusion of the fourth and fifth cervical spine

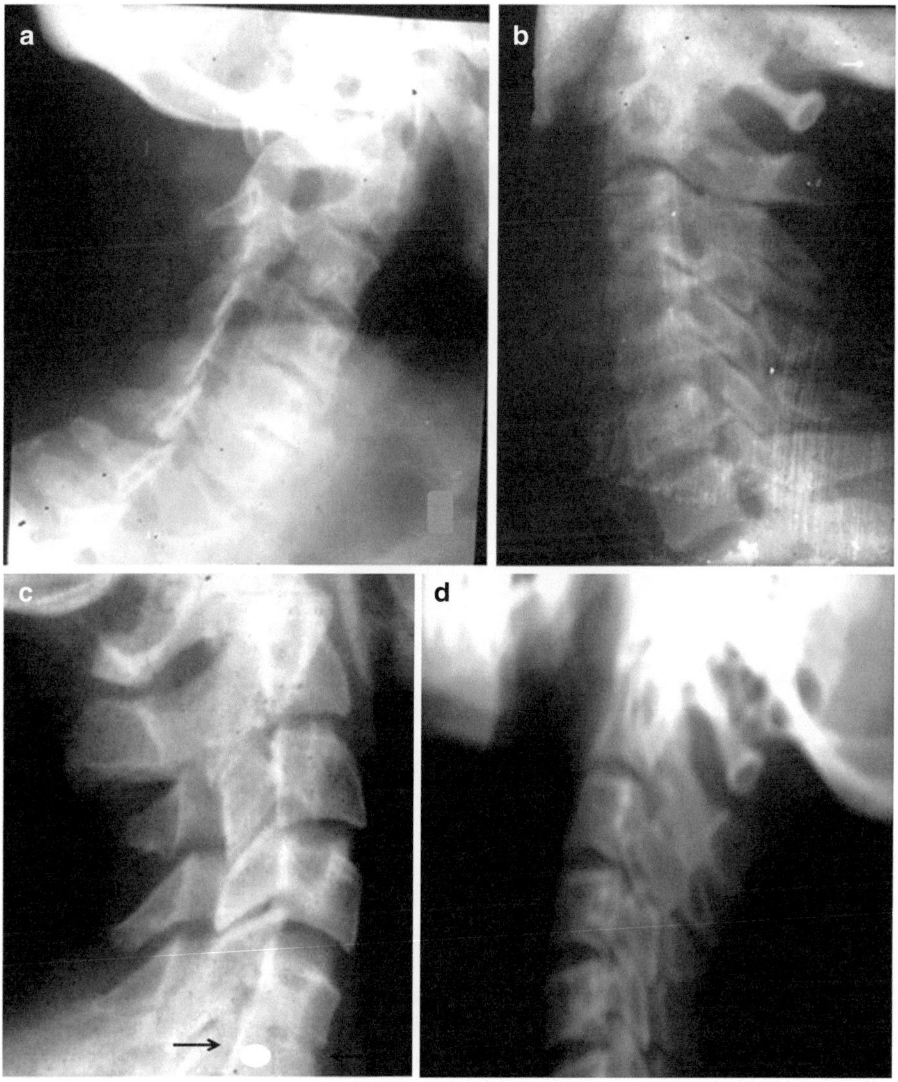

Fig. 15.7 (**a**) Roentgenogram revealing fusion of C2–C3. (**b**) Roentgenogram revealing fusion of C3–C4. (**c**) Roentgenogram shows congenital fusion of C5–C6 (arrow). (**d**) Roentgenogram revealing congenital fusion of C2-3 posteriorly

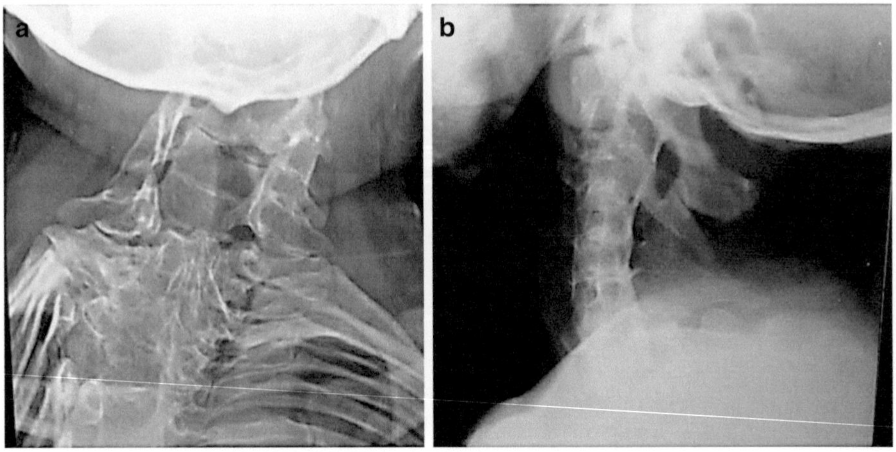

Fig. 15.8 (**a**) Roentgenogram reveals congenital absence of the laminae of the cervical spine with thoracic scoliosis. (**b**) Note the fusion of the whole cervical spine and malformed upper spinous processes

15.2.1 Other Congenital Anomalies of the Cervical Spine

There are other congenital anomalies which are rarely associated with specific names.

15.3 Cervical Rib

Cervical rib is a supernumerary rib which can be detected arising from the seventh cervical vertebra and could be a cause of compression of the neurovascular structure at the root of the neck (Fig. 15.9a–c). Clinically it may present as thoracic outlet syndrome (TOS). In most of the cases the brachial plexus is affects. However, the subclavian vein and subclavian artery may also be involved. It is classified as osseous and fibrous. When symptomatic, the treatment is conservative. In a few cases it may also require excision of the rib.

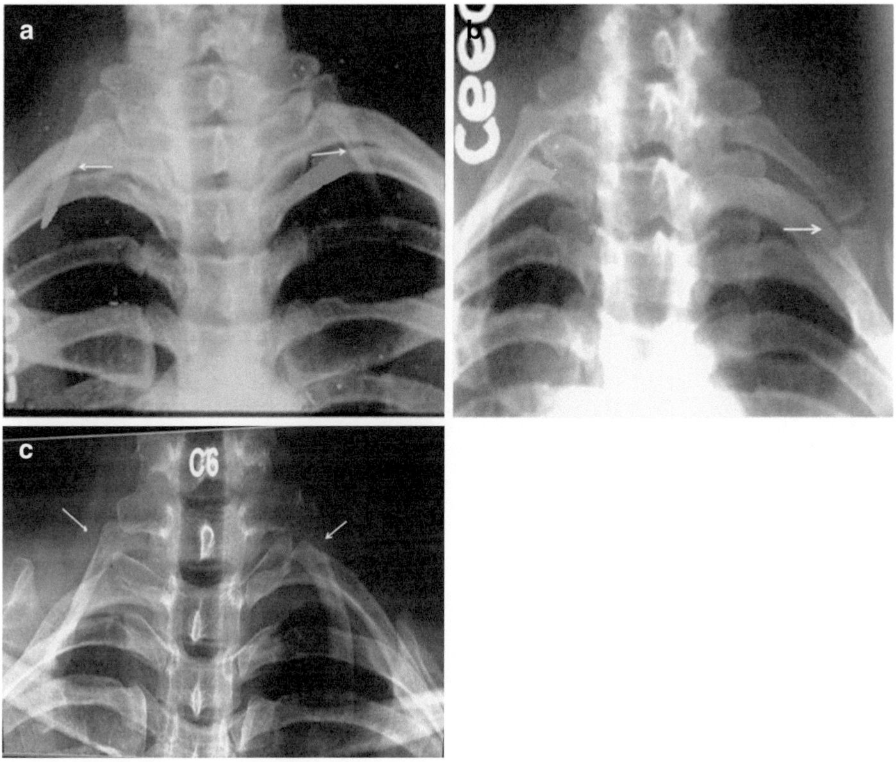

Fig. 15.9 (**a**) Bilateral osseous cervical ribs arising from the C7 vertebra. (**b**) Left sided cervical rib with pseudoarthrosis (arrow). (**c**) Roentgenogram reveals bilateral cervical rib

15.4 Congenital Muscular Torticollis

Torticollis (wry neck) is a congenital anomaly characterized by limited neck motion in which the head is tilted towards the involved side and the chin on the opposite side (Fig. 15.10a, b). In most of the cases it is unilateral. However, bilateral cases have been observed (Fig. 15.11). There is shorting of the sternocleidomastoid muscle. In most cases this may be a consequence of the injury during birth. As a result a firm and non-tender mass may be felt over the muscle which gradually disappears resulting in fibrosis.

Radiographs of the cervical spine always should be obtained to ensure that the deformity is truly a muscular torticollis and is not associated with congenital vertebral lesions of the cervical spine.

Most of the cases can be managed conservatively by active and passive starching excercises. Surgery is recommended when the deformity persists even after starching measures. Surgical treatments include a unipolar or a bipolar release. If left untreated, will result in permanent changes, particularly in the bones and asymmetry of the face.

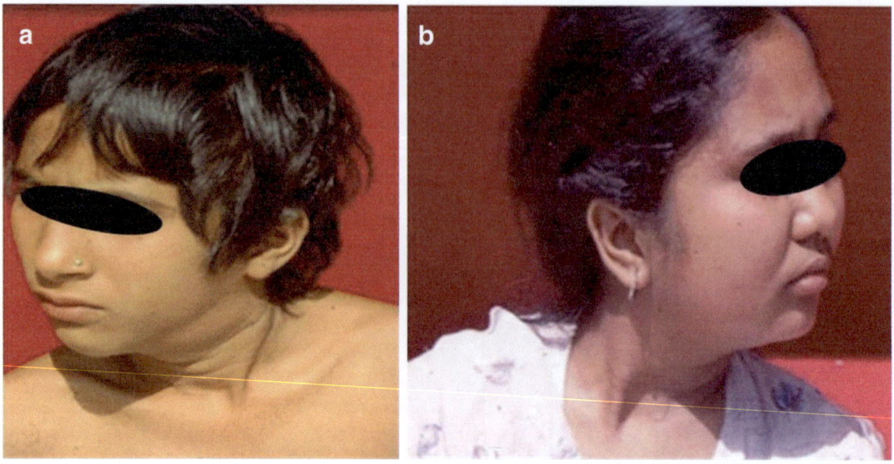

Fig. 15.10 (**a**) Patient revealing torticollis on the left side. Note the fibrosed sternocleidomastoid muscle. (**b**) Torticollis on the right side

Fig. 15.11 A rare presentation of bilateral muscular torticollis

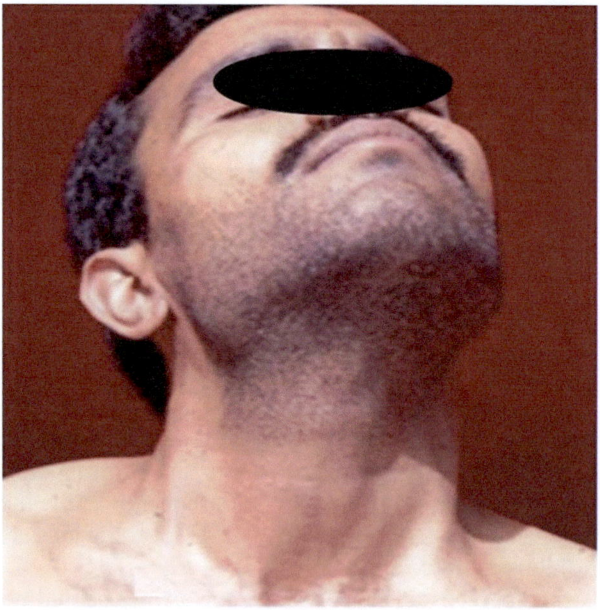

15.5 Erbs Palsy/Birth Palsy

The Erbs Palsy is sustained by the newborn mainly during a difficult delivery. The child has internal rotation of the limb at the shoulder and the palm is facing backward-Policeman's tip (Fig. 15.12). The lesion is at C5–C6 level of the brachial plexus at the Erbs point. In majority of the early cases recovery takes place in about 12 months. The incident has become considerably more infrequent in recent years because of the modern obstetric management. Early diagnosis of Erbs Palsy in a newly born child is usually not difficult. The appearance is characterized by adduction and internal rotation and contracture of the shoulder with loss of complete extension of the elbow. Conservative treatment has been recommended. Many soft tissue operations have been described. However, in neglected cases a derotation osteotomy of the humerus may be indicated.

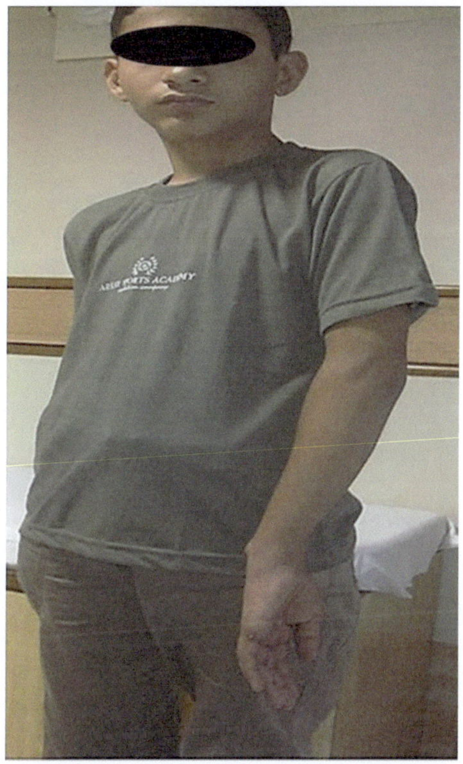

Fig. 15.12 A neglected case of Erbs palsy on the left side. Note the severe internal rotation deformity of the extremity (Policeman's tip)

15.6 Pectus Excavatum Congenitum

Pectus excavatum (depressed chest) is a common chest deformity affecting males more than females (Fig. 15.13a–d). The appearance of the said abnormality varies from mild to severe and worsens in the teenagers. It can be seen as a single disorder

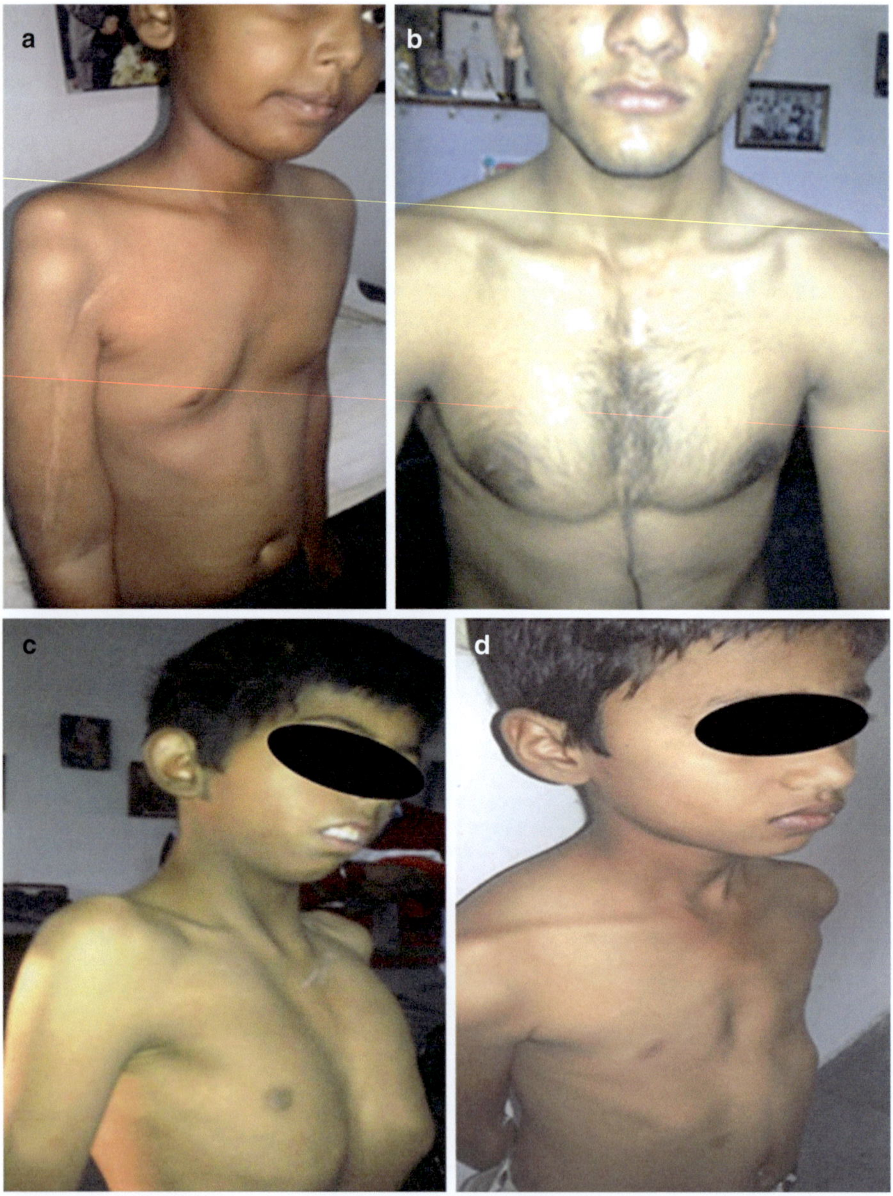

Fig. 15.13 (**a**, **b**) Patients with pectus excavatum Congenitum. (**c**, **d**) Both these patients reveal evidence of pectus excavatum which has been symptomless

or as a part of Marfan's and Poland syndromes. Family occurrences can be seen in one third cases. However, the etiology of disease is not known. The condition is usually asymptomatic except raising cosmetic questions. Rarely there may be restrictive pulmonary impairment.

15.7 Pectus Carinatum Congenitum

Pectus carinatum is commonly known as bird-chest. There is protrusion of the anterior chest wall (Fig. 15.14a–c). The males are affected four times more frequently than females. The disease can occur in isolation our in association with other abnormalities. This is a symptomless syndrome. It is rarely symptomatic resulting in progressive emphysema and respiratory tract infections. It may be associated with other conditions such as kyphosis, scoliosis or occasionally associated with Morquio's syndrome.

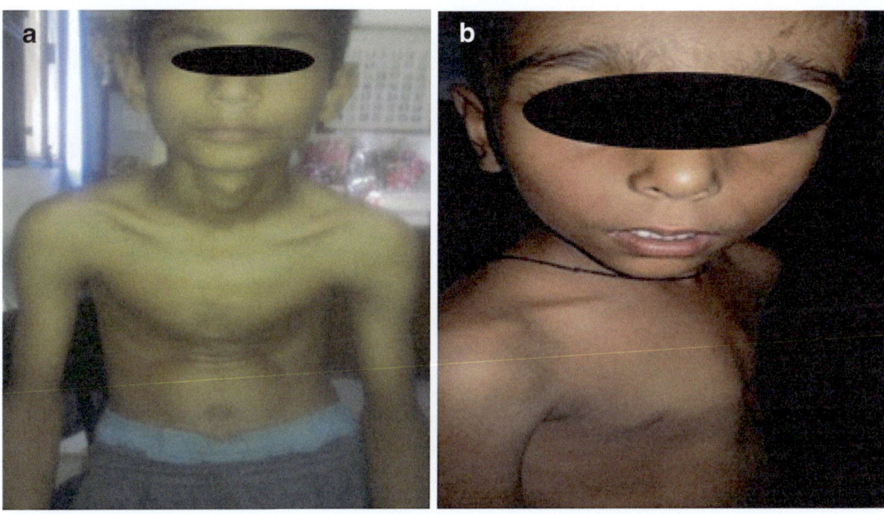

Fig. 15.14 (**a**) Patient revealing Pectus carinatum congenitum (Pigeon chest). (**b**) Patient showing elevated chest-pectus carinatum. (**c**) Patient showing elevated chest-pectus carinatum

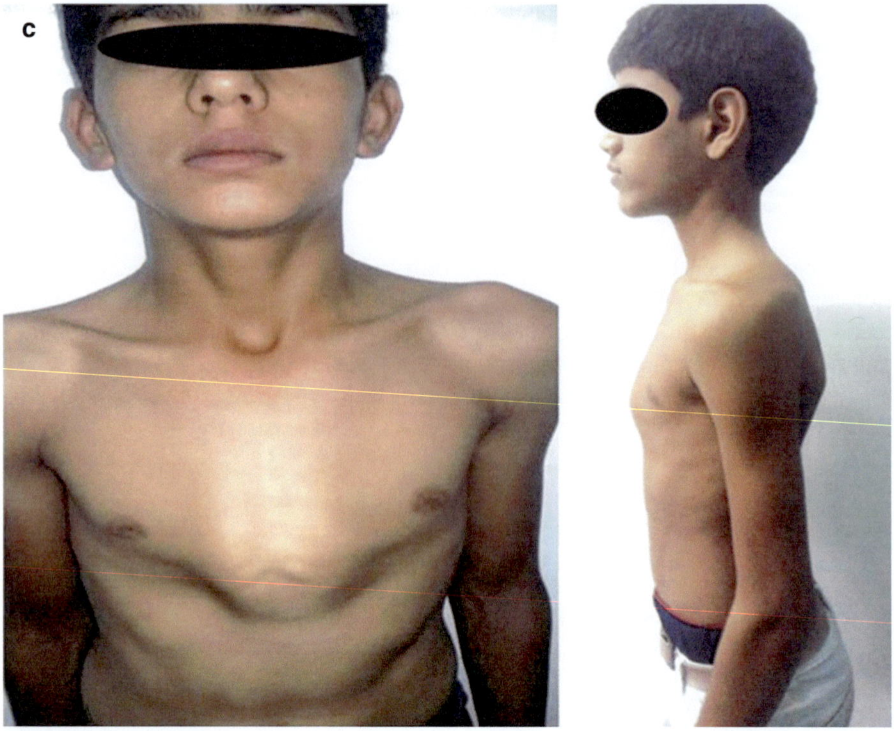

Fig. 15.14 (continued)

Suggested Readings

Bonola A. A surgical treatment of the Klippel-Feil syndrome. J Bone Joint Surg (Br). 1956;38:440.

Shoul MI, Ritvo M. Clinical and roentgenological manifestations of the Klippel-Feil syndrome (congenital fusion of the cervical vertebrae, brevicollis). Report of eight additional cases and review of the literature. Am J Roentgenol. 1952;68:369.

Adler JB, Patterson RL. Erb's palsy: long time results of treatment in eighty-eight cases. J Bone Joint Surg Am. 1967;49:1052.

Armstrong D, Pickrell K, Fetter B, Pitts W. Torticollis: an analysis of 271 cases. Plast Reconstr Surg. 1965;35:14.

Canale ST, Griffin DW, Hubbard CN. Congenital muscular torticollis: a long-term follow-up. J Bone Joint Surg Am. 1982;64:810.

Coventry MB, Harris LE. Congenital muscular torticollis in infancy. Some observations regarding treatment. J Bone Joint Surg [Am]. 1959;41:815.

Tse P, Cheng J, Chow Y, Leung PC. Surgery for neglected congenital torticollis. Acta Orthop Scand. 1987;58:270.

Congenital Disorders of the Spine 16

16.1 Congenital Scoliosis

Congenital scoliosis occurs due to abnormal vertebral development which is present at birth. The identification of the curvature is observed much earlier in life than idiopathic scoliosis (Fig. 16.1a–f). The congenital curves tend to be rigid and resistant to correction. To check the increase in such curves, even early fusion may be necessary.

The classification of such a vertebral anomaly has been described as segmentation, formation and mixed problems, i.e., anterior, posterior, lateral, anterolateral and postrolateral (Fig. 16.2a, b). A vertebra that has not been formed fully is called a hemi-vertebra. Hemi-vertebrae may be single or multiple and may balance the deformity. The presence of a single hemi-vertebra is common and may not cause much progression of the deformity. When balanced, the contra lateral hemivertebrae do not progress and thus do not require any treatment (Fig. 16.3a, b). When separated by several segments, a double curve is produced and both the curves may progress to require fusion. This is because of anomalous growth in the vertebral ring. Depending on such anomalies, the resultant deformity can be scoliotic, kyphotic or a combination of all these. The magnitude of deformity is proportional to the growth potential of the vertebrae, the maximum changes can be observed up to the age of 16–18 years (Fig. 16.4a, b). Diastematomyelia is a form of dysraphism which may be associated with congenital scoliosis.

Orthotic Treatment Because the primary deformity in congenital scoliosis is in the bones rather than the soft tissues, the curves tend to be rigid. They are not as amenable to orthotic treatment as idiopathic or paralytic curves. Never the less, there are definite indications for orthotic treatment of congenital spine deformities. The orthosis of choice is the Milwaukee brace which can be tried.

S. S. Yadav, *Skeletal Dysplasias & Congenital Malformations*,
https://doi.org/10.1007/978-981-99-5679-1_16

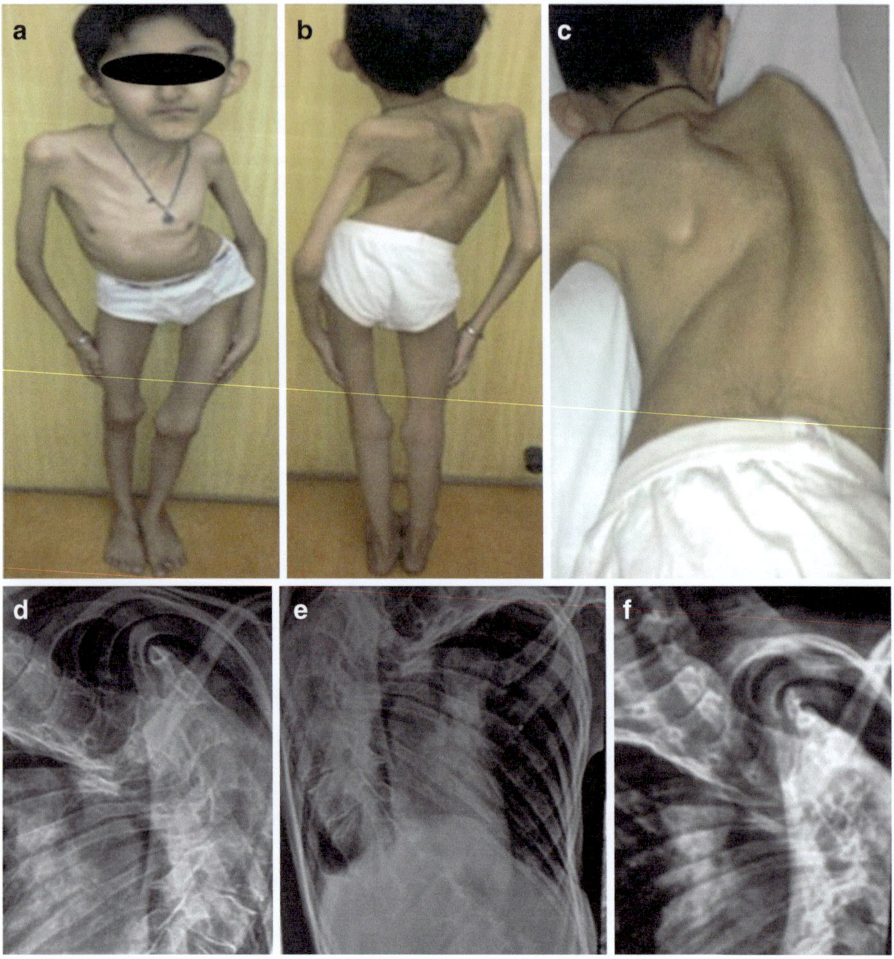

Fig. 16.1 (**a**) A 12-year-old boy with severe congenital scoliosis. (**b**) Note the trunk asymmetry, decompensated scoliosis and pelvic tilt is visible. (**c**) A prominent rib hump occurs due to forward bending on the convex side of the deformity. (**d**, **e**, **f**) Radiographs of the same patient with massive hump reveal severe scoliosis with extensive rib anomalies

Surgical Treatment Surgery is the most customary treatment of severe or progressive congenital scoliosis. Several types of operative procedures can be tried and two fundamental questions arise: What is the best procedure? What is the best age for surgery? The answer has not been simple. There are four basic procedures for the surgical treatment of congenital scoliosis: posterior fusion, anterior and posterior fusion, convex growth arrest (anterior epiphysiodesis and posterior hemiarthrodesis) and hemi-vertebra excision. The operative treatment chosen must be tailored to that specific patient. It depends on the age of the patient and the type of deformity.

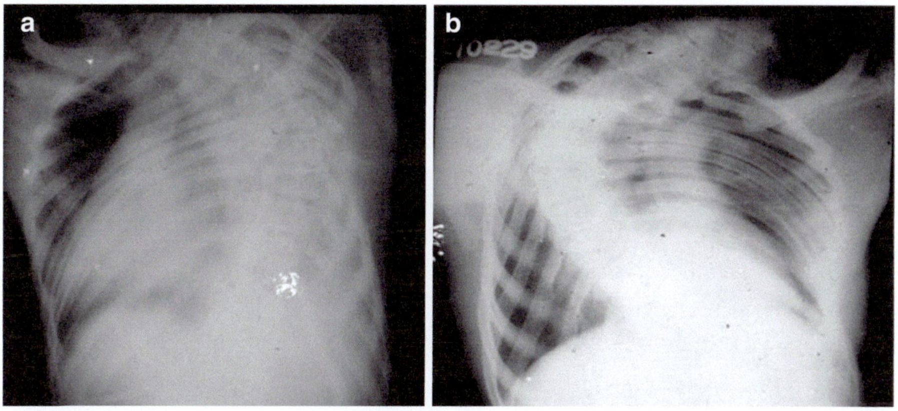

Fig. 16.2 (**a**, **b**) Radiographs revealing severe kyphoscoliotic deformity

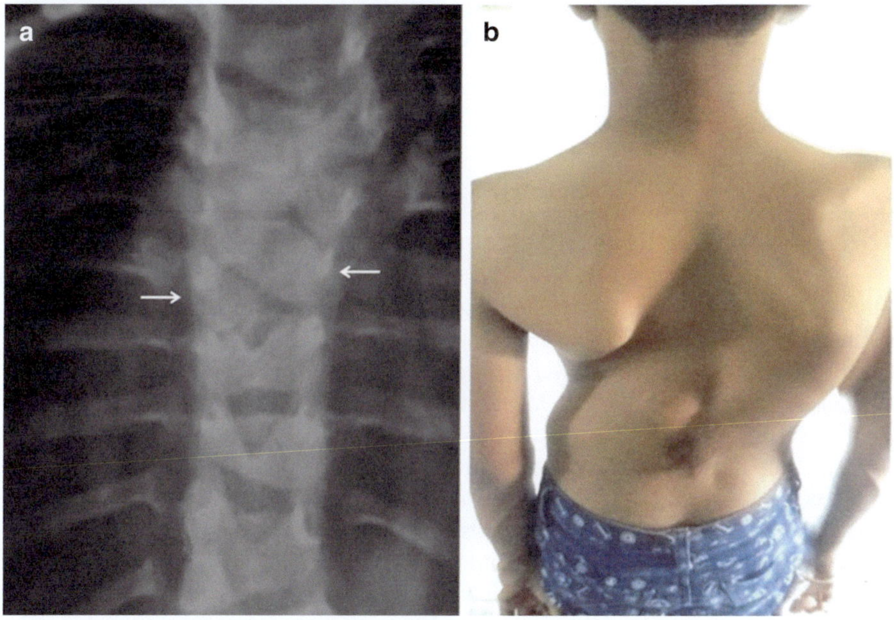

Fig. 16.3 (**a**) Roentgenogram reveals double hemi-vertebrae on left and right (arrows). The result is straight spine. No treatment was given. (**b**) A 20-year-old boy had thoracic scoliosis with a prominent left scapula and a hair patch in the lumbar area

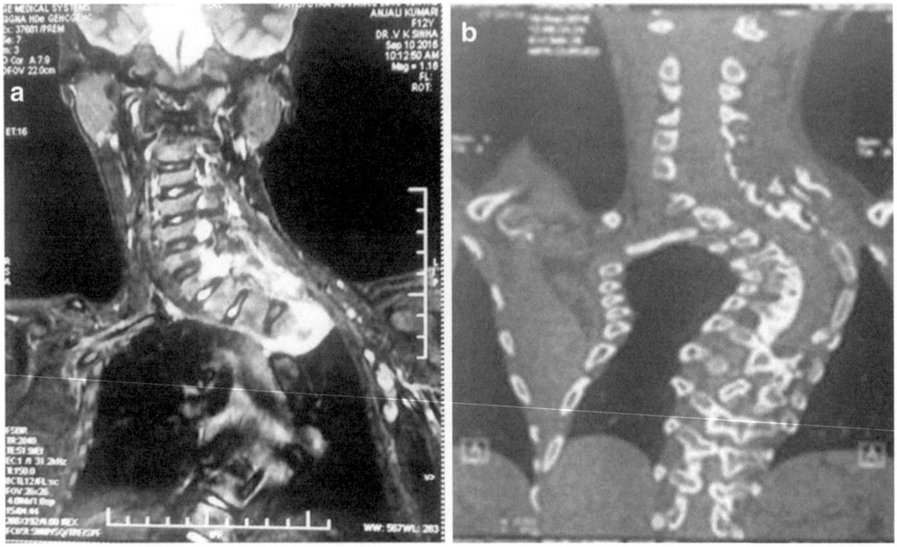

Fig. 16.4 (**a**) MRI of a 12-year-old boy reveals severe scoliosis of the upper dorsal spine. The patient had occasional neurological symptoms in the lower limbs. (**b**) CT scan presentation of the same patient

16.2 Idiopathic Scoliosis

Idiopathic scoliosis is the lateral bending of the spine in combination with vertebral rotation and torsion. These are classified as infantile (0–3 years), juvenile (4–10 years) and adolescent (over 10 years) forms. The curves are usually progressive and located at the thoracic level with convexity and combined with kyphosis (Fig. 16.5a–c). The most frequent curves are at the thoracic level. The deformity is S shaped and occurs less frequently at the thoracolumbar level (Fig. 16.6a, b).

Growth has a definite role in idiopathic scoliosis and knowledge of normal spinal growth is essential in treating all patients with spinal deformities. There is a relation between curve increase and growth and there is a role for growth in the etiology of idiopathic scoliosis. No single causative factor has been found for idiopathic scoliosis and it appears that the etiology is probably multifactorial.

When the child is still growing, the treatment depends on the magnitude of the curve. The choice available for treatment is non-operative and operative. Observation is considered a non-operative treatment. All curves in the adolescent over 45–50° are treated operatively.

Non-operative Treatment The aim of non-operative treatment is to control the curve and prevent it from progressing and to obviate the needs for surgical stabilization. In addition, the cosmesis of the child should be improved whenever possible. The choices for non-operative therapy include orthotic treatment and exercises.

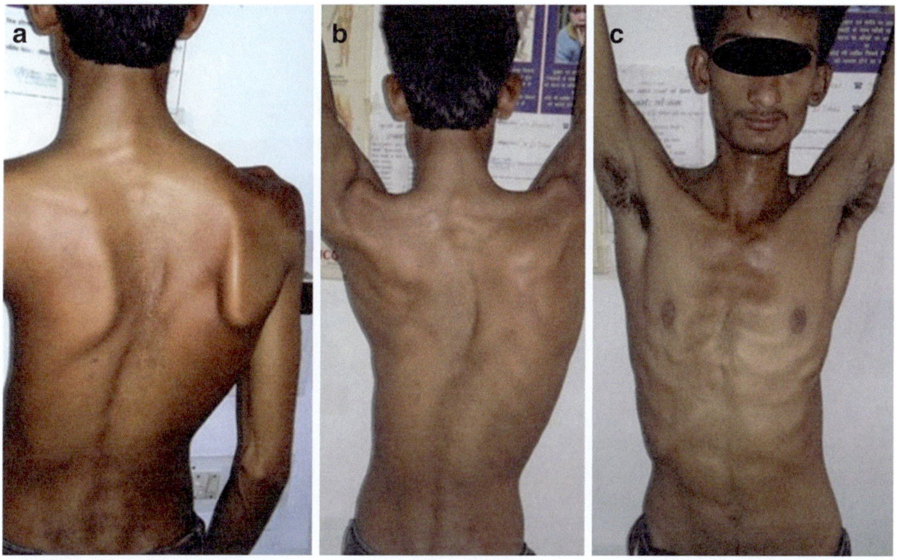

Fig. 16.5 (**a**) The clinical manifestation of a lumbar (left side) and thoracic (right side) scoliosis. (**b**, **c**) Asymmetry of the trunk and the curve of the spine are shown when the shoulders are lifted (back and front view)

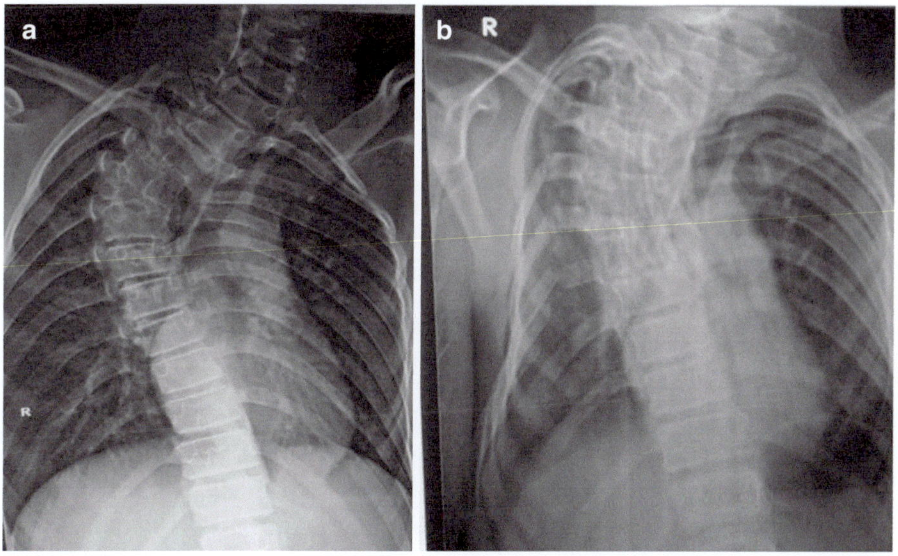

Fig. 16.6 (**a**, **b**) Roentgenograms show moderate to severe kypho-scoliotic presentation of the dorsal spine

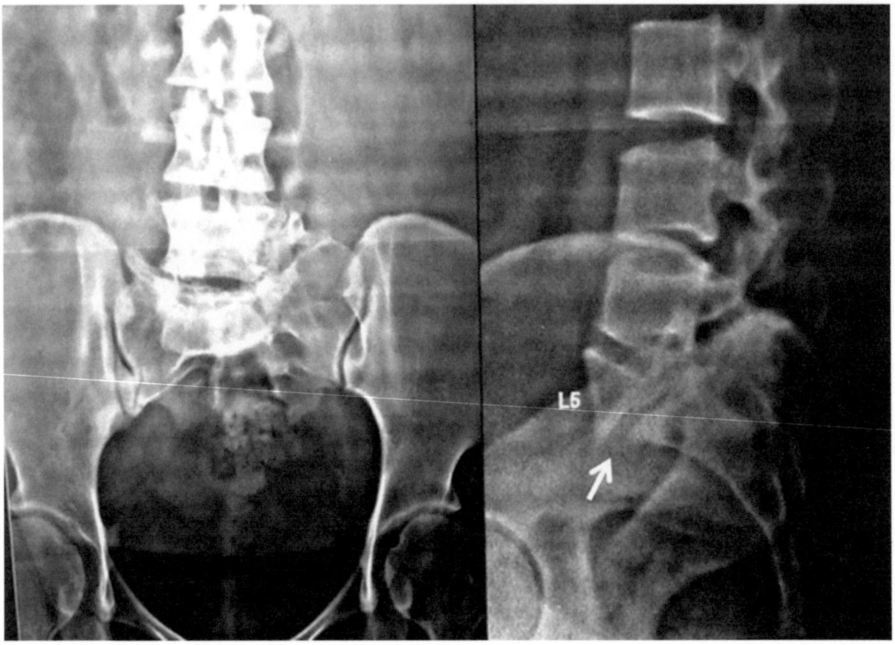

Fig. 16.7 Roentgenograms reveal posterior wedging of L5 vertebra

Operative Treatment Despite the best non-operative treatment some curves progress. In addition, curves present that are too large or too late for non-operative treatment. These cases are considered for surgery. The decision to perform surgery is made with knowledge of natural history. Indications for operative treatment of idiopathic scoliosis in the adolescent are: (1) A child in the rapid growing phase who presents with curve of over 40–45°. (2) Progression to a curve of 40–45° while under non-operative therapy, either in adolescent idiopathic scoliosis or in the juvenile at the onset of the adolescent growth spurt. (3) Curves over 50–60° in a mature adolescent. The decision for surgery must not be based on the cobb measurement alone. The age and maturity of the child and the growth potential and possibility of progression are important (Fig. 16.7).

16.3 Kyphosis

It is a curvature of the spine in the segital plane in which the convexity of the curve is directed posteriorly. Normally the thoracic spine and the sacrum are kyphotic and the cervical spine and the lumbar spine are lordotic. The ranges of such normal curves usually keep on increasing with age. Normally these curves balance each other so that the head is centred over the pelvis. In the upright position, the spine is subjected to the forces of gravity and several other structures are known to maintain its stability which includes all the constituents of the disc complex. Any alteration of function in children usually causes progressive kyphotic deformity.

Kyphotic deformity is also seen in Achondroplasia, Pseudochondroplasia, Spondyloepiphyseal Dysplasia, Diastrophic Dwarfism, Mucopolysaccharidosis, Marfan syndrome, Tuberculosis of Spine, Neurofibromatosis, Post-traumatic and Juvenile and Senile osteoporosis.

16.4 Congenital Kyphosis

Congenital kyphosis occurs due to abnormal development of the vertebrae which occurs due to failure of developing segments of the spine (Fig. 16.8a, b). Patients with congenital kyphosis often have other congenital anomalies like Klippel-Feil syndrome and other non-skeletal anomalies may be present.

In the early childhood mild cases of congenital kyphosis may be over-looked. A kyphotic deformity in the thoracolumbar junction may be observed. A thorough physical and radiological assessment should be made. A CT scan can identify the nature of bony deformity. An MRI is essential to access intra spinal abnormalities.

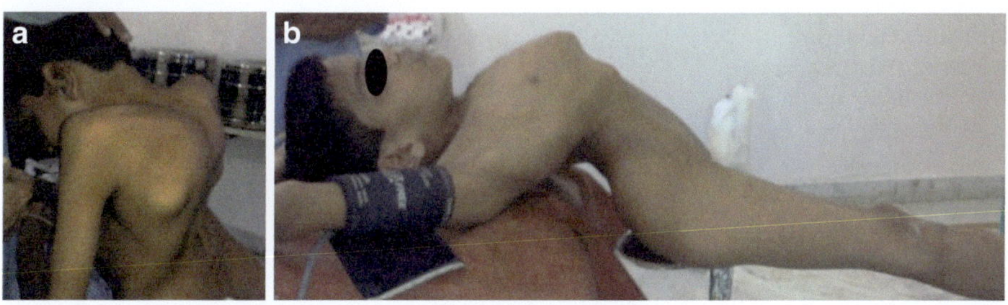

Fig. 16.8 (**a**) Patient with severe kyphotic deformity of the thoracic spine with lordosis in the dorsolumbar area. He had bilateral flexion deformity of the hips. (**b**) This is how the patient was taken up on the O.T. table for hip release surgery

16.5 Scheuermann Disease

Scheuermann disease (Juvenile/adolescent kyphosis) is a common cause of structural kyphosis from cervical to lumbar spine. A typical Scheuermann disease is identified with a rigid thoracic kyphosis in children with its apex located between T7 to T9. Scheuermann disease is usually present in children about the time of puberty or it may develop later in adolescence. The onset of the disease usually occurs around 10 years of age. Males are affected more frequently than females.

The etiology of this condition is still unknown, but it is thought to be associated with a defect in the cartilage end plate, reducing the ability of this structure to cushion the pressure of normal weight-bearing and bending. The vertebral bone and plate is irregular and there is anterior wedging of thoracic vertebrae due to decreased longitudinal growth anteriorly (where the pressure of weight-bearing is greatest). The deformity is known to occur in patients with sudden growth spurt (Fig. 16.9). This is due to the fact that the ring apophysis ossifies around that age and the changes in the vertebral body in the form of wedging irregularity of the end plate can be judged only then. The end plates are narrow and segmentally indented towards the vertebral body. These findings are manifestations of pathogenitic process. Prolapses

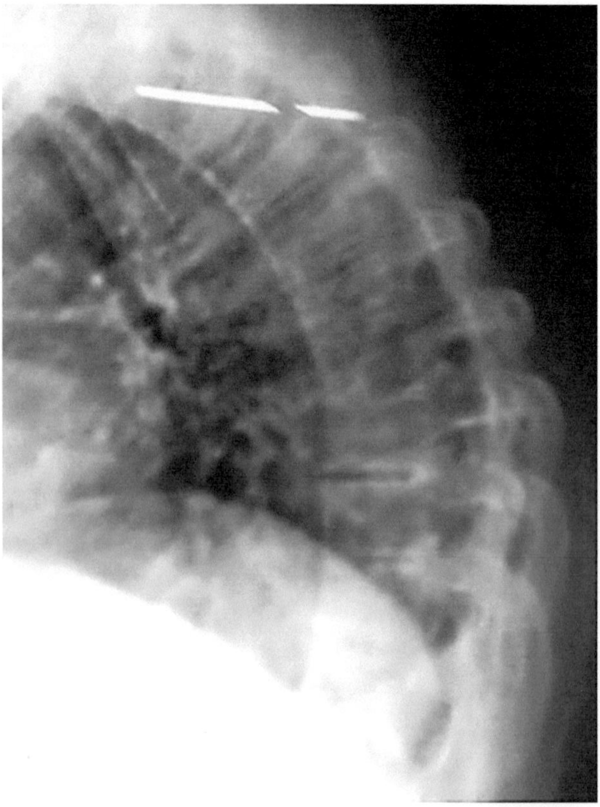

Fig. 16.9 Lateral radiograph of a patient of Scheuermann disease with kyphotic deformity. Note the narrowing of the intervertebral disc spaces and irregularity of the vertebral end plates

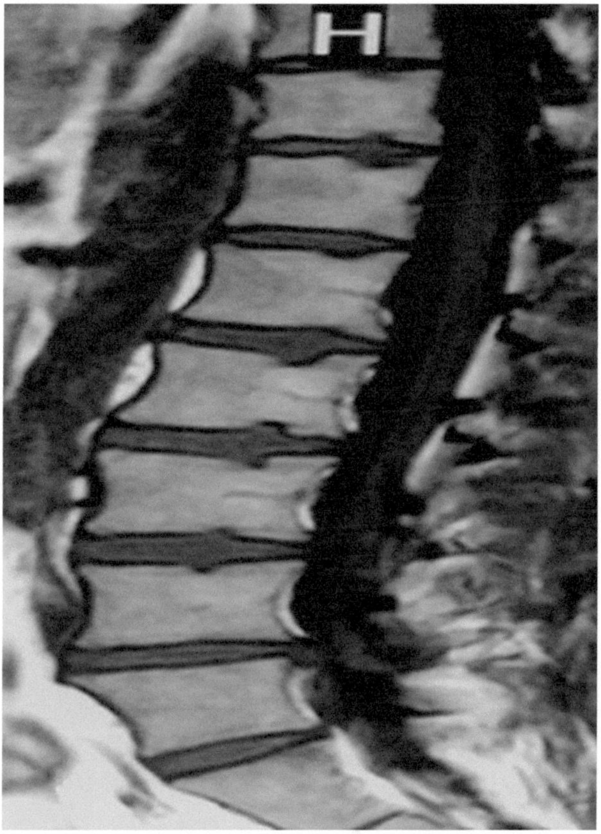

Fig. 16.10 Schmoral Nodes are also seen outside the area of kyphosis and in patients who did not have kyphotic deformity

of the disc tissue through gaps in the end plates may be secondary to loss of mechanical strength in these defective areas. Schmorl nodes may be observed in patients with Scheuermann disease (Fig. 16.10). Endocrine abnormalities are known to be associated with Scheuermann disease. Osteoporosis and scoliosis is also known to occur in such cases.

Along with the kyphotic deformity the patient also complains of pain over the deformity which is aggravated by standing and during physical activity. In adults the patients may complain of low backache caused by exaggerated lumbar lordosis.

16.6 Postural Kyphosis

Postural kyphosis is a flexible deformity of the spine and commonly occurs in juvenile and adolescent patients. Postural kyphosis results in a round back while bending forward. Postural kyphosis should be differentiated from pathologic types of kyphosis such as Scheuermann disease and Pott's spine. Radiographs are necessary for confirming the diagnosis.

16.7 Diastematomyelia

Diastematomyelia is bifurcation of the spinal cord separated by a fibrous band or bony septa/spur. It is a form of dysraphism associated with congenital scoliosis. It may be associated with hair patches, dimples and even skin pigmentation. It may be associated with clubfeet, cavus feet, flat feet and vertical talus. It is possible that these intraspinal anomalies may not have any symptoms.

In diastematomyelia the MRI findings are suggestive of split cord malformation with syringomyelia, low-lying cord and dural ectasia (Fig. 16.11a–d). There is tethering and thickening of filum terminale with incomplete fusion of posterior elements of L3/L4 vertebral body with lipoma or transitional L5–S1 segment with sacralization of L5 vertebral body.

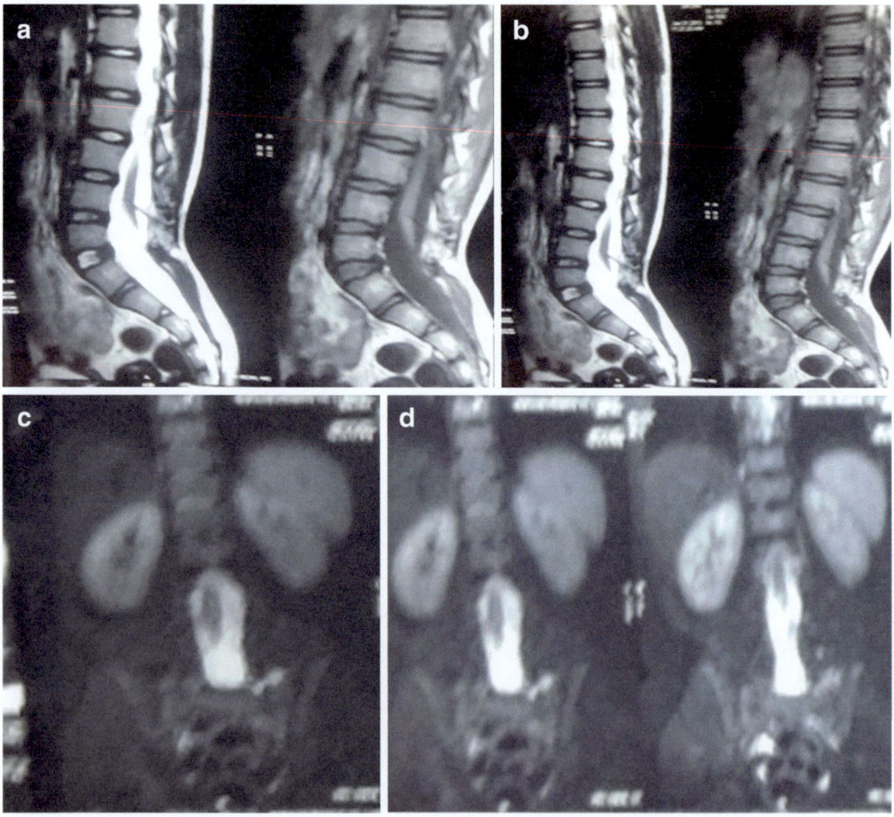

Fig. 16.11 (**a**, **b**) MRI of lumbar spine in a case of diastematomyelia reveals bifurcation of the spinal cord. (**c**, **d**) MRI shows bifurcation of the cord in the lumbar spine in a case of diastematomyelia

16.8 Spondylolysis and Spondylolisthesis

A defect in the neural arch of the spine resulting in narrowing due to bony bridge results in Spondylolysis. Majority of such cases develop at L5–S1 level followed by L4–L5. The defects may be classified as congenital, isthmic, traumatic, degenerative and pathological Spondylolysis. Since the defect is bilateral it may result in forwarded shift of the vertebra above and results in Spondylolisthesis. According to Meyerding classification such a shift has been graded (1–4) as per the distance covered by the shift. **Grade-1 (25% shift), Grade-2 (50% shift), Grade-3 (75% shift) and Grade-4 (Complete slipping of the vertebra)** (Figs. 16.12a–c, 16.13, 16.14a–c, 16.15a, b, and 16.16a, b). Although Spondylolysis may be symptomless, Spondylolisthesis commonly is symptomatic in the form of low backache, paraspinal muscle spasm and radiation of pain in the lower limbs. In severe cases there may be a 'step-off' sign present at the level of Spondylolisthesis (Fig. 16.17). Depending on the observations, reduction of the displaced vertebra and a post spinal fusion may be necessary.

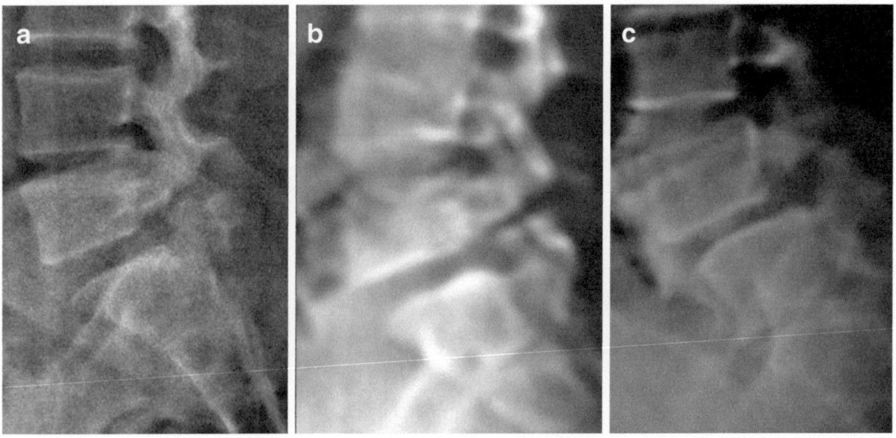

Fig. 16.12 (**a**, **b**) Radiographs showing spondylolysis with defect in the pars interarticularis. (**c**) There is onset of spondylolisthesis Gr–I

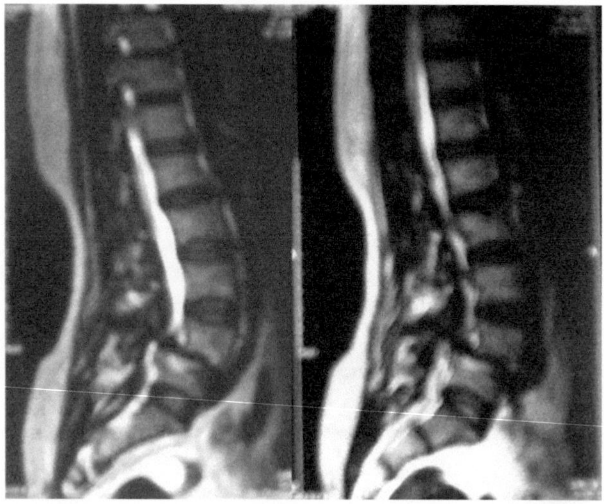

Fig. 16.13 M.R.I. reveals Gr–I spondylolisthesis with moderate stenosis of the canal

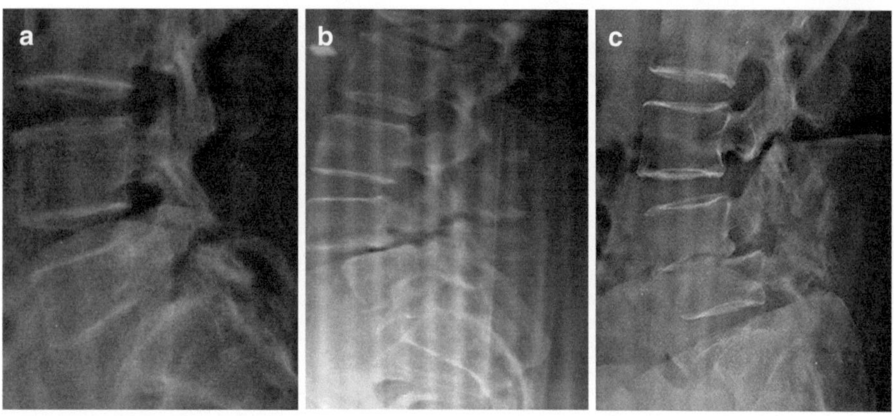

Fig. 16.14 (**a**, **b**) Roentgenogram showing Gr-I spondylolisthesis at L5, S–1 and L4–5 level. (**c**) Radiograph shows Gr-I spondylolisthesis at L5, S-1 level

16.9 Complications

The complications may be observed in the form of pseudoarthrosis at the site of the fusion. There may also be neurological complications after surgery particularly when a reduction of the slip of the vertebra is attempted. Even with immediate decompression in these cases, permanent neurologic deficits can persist. A postoperative increase in slippage has also been noted, especially with decompression, which tends to remove the midline stabilizing structures, thereby increasing the instability. It may also occur in higher degrees of spondylolisthesis.

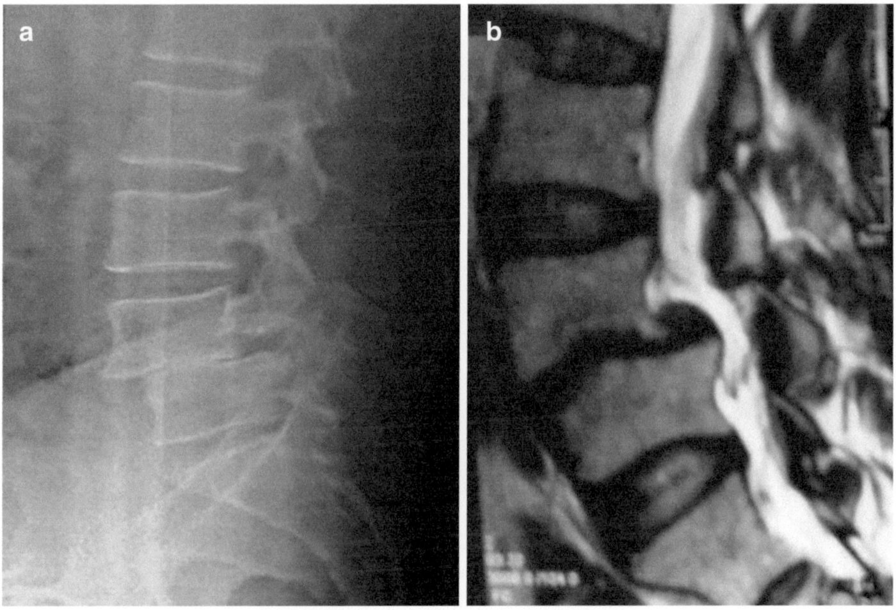

Fig. 16.15 (**a**) Roentgenogram shows Gr–II spondylolisthesis at L4–5 level. (**b**) MRI reveals Gr–II spondylolisthesis with PIVD at L4–5 level

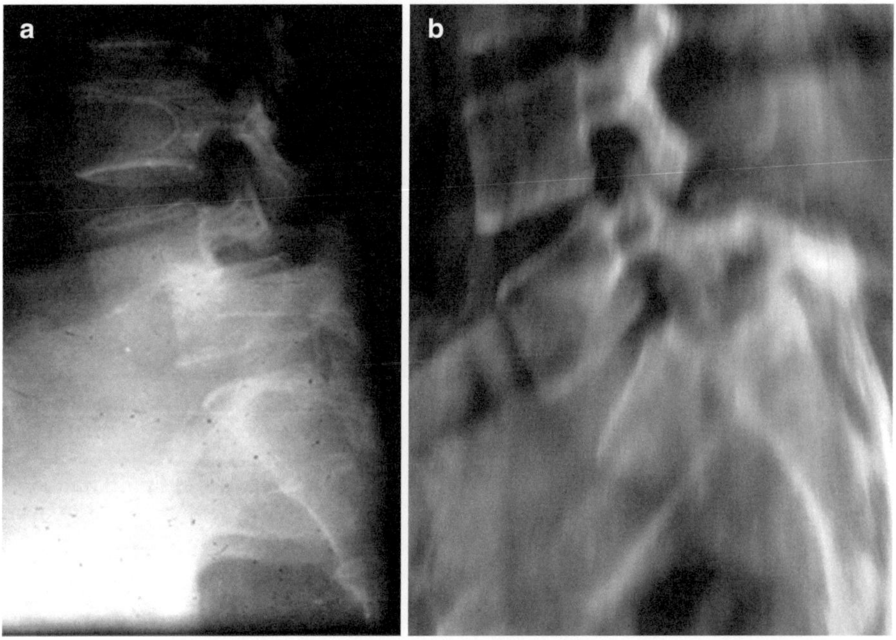

Fig. 16.16 (**a**) Radiograph shows spondylolisthesis Gr–III at L4–5 level. (**b**) Radiograph shows Gr–IV (total) spondylolisthesis at L5, S1 level

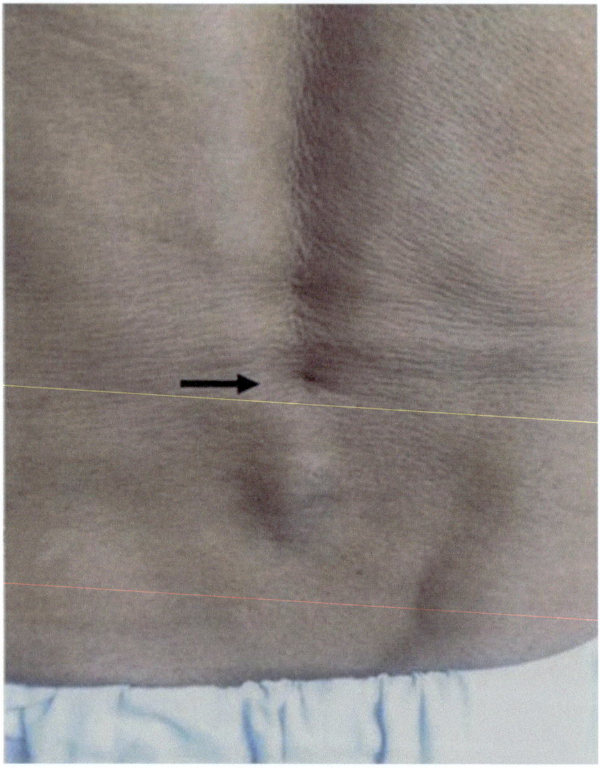

Fig. 16.17 Clinically a 'step-off' can be felt at the site of the defect (arrow)

16.10 Bertolotti Syndrome (Mega Transverse Process L5)

A chronic, persistent low backache along with buttock pain is a common presentation. There may also be ridiculer pain. It is characterized by sacralization of the fifth lumbar vertebra and lumbarization of the first sacral vertebra. It involves unilateral or bilateral fusion of the transverse process of L5 to S1 leading to formation of a transitional fifth vertebra (Fig. 16.18a, b). It may result in pain generating fourth lumbar disc with sciatic type of pain corresponding to the fifth lumbar nerve root. Scoliosis may be frequently associated with this syndrome.

Physical therapy may be helpful in early cases. If not relived it can be treated surgically with excision and posterolateral fusion of the transitional articulation.

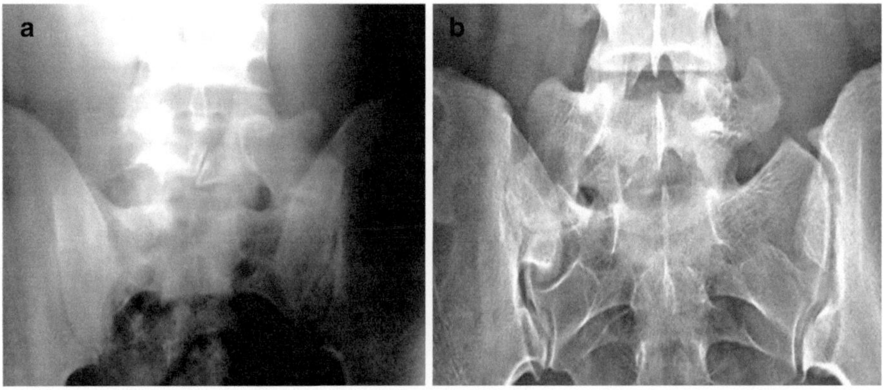

Fig. 16.18 (**a**) Roentgenogram reveals enlarged transverse process of L5 vertebra on the left side, one of the causes of low backache. (**b**) Roentgenogram reveals mega transverse process of L5 on the right side. Patient had severe low backache with radiation

16.11 Spina Bifida Occulta

The bony defect is present at the arch of vertebral body but the lesion is covered by skin. There is no meningeal herniation through the bony defect. The lumbar and lumbo-sacral regions are the most common sites for this lesion. In majority of the cases there are no symptoms. Foot deformities in the form of CTEV may be present. Occasionally there may be neurological symptoms in the form of loss of sensation and radiculopathy. A tuft of hair is usually present at the site (Fig. 16.19a–d). Spina bi-fida occulta in most cases is symptomless and is an accidental finding on roentgenogram (Fig. 16.20a–d).

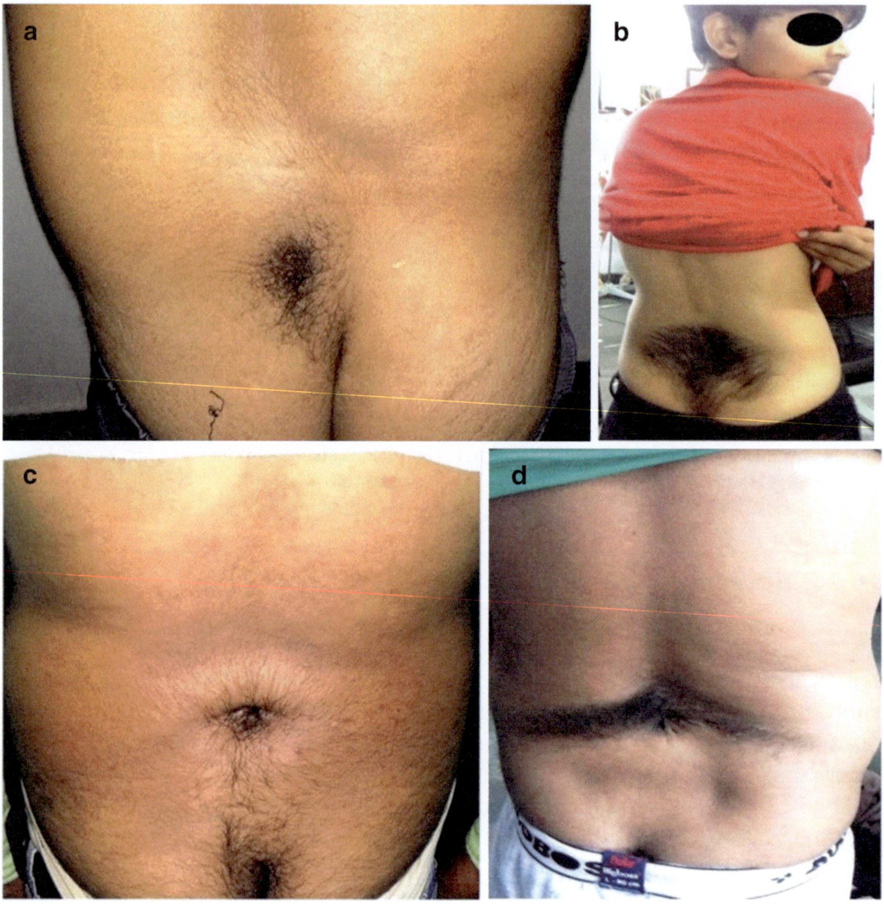

Fig. 16.19 (**a**, **b**) Patients with spina bifida occulta showing tuft of hair at the site of the defect. (**c, d**) Clinical presentation of tuft hair in cases with spina bifida occulta

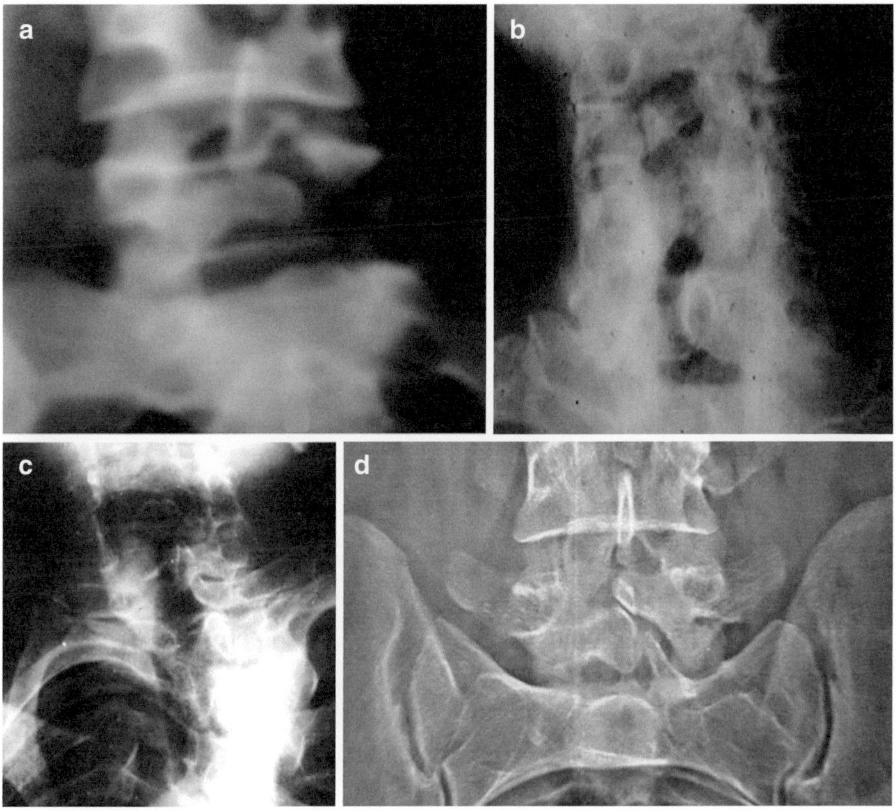

Fig. 16.20 (**a**) Roentgenogram revealing Spina Bifida occulta at L5 level. (**b**) Radiograph reveals spina bifida occulta of C7. (**c**) Multiple lack of fusion of the laminae of the lower cervical vertebrae. (**d**) Spina bifida occulta L5. Note the osseous spike from S1

16.12 Meningocele and Myelomeningocele (Spina Bifida Cystica)

Neural tube defects are classified under the generic terms myelodysplasia or spinal dysraphia and spina bifida aperta. Neural tube defects can be classified into four sub types:

1. Meningocoele (A cyst involving only the meninges)
2. Myelomeningocele (Cyst with abnormal neural element)
3. Lipomeningocele (Sac contains lipoma)
4. Rachischisis (Complete absence of skin and sac with dysplastic S. cord).

Meningocoele/myelomeningocele are the most frequent subgroups of neural defect with sensory and motor defects in the lower extremities along with disturbed

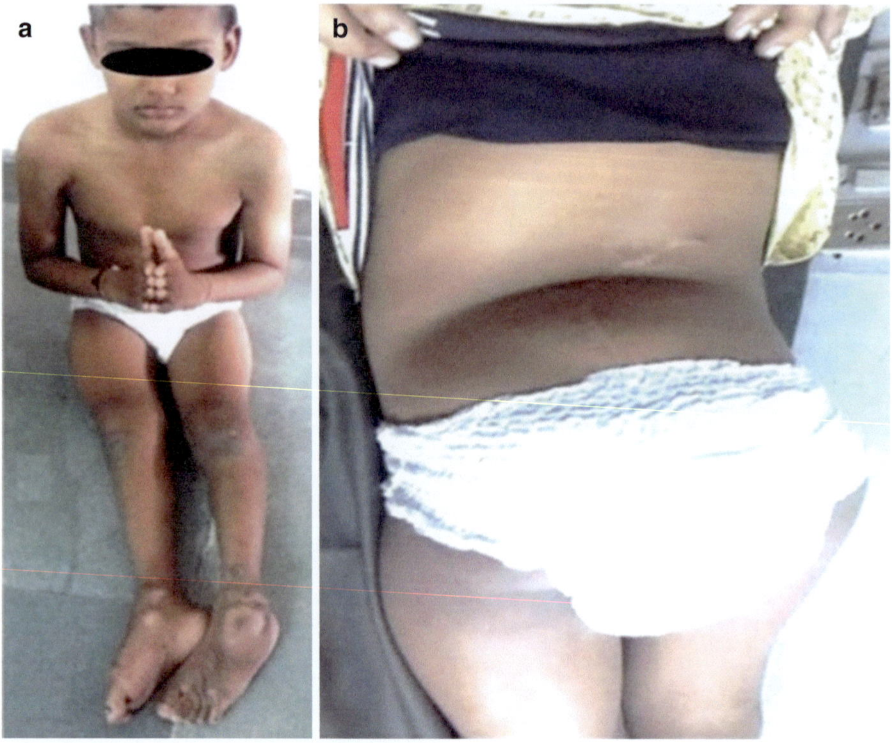

Fig. 16.21 (**a**) A 12-year-old boy with myelomeningocele reveals distal paralysis of the lower limbs. Note the equinovarus deformity on the right side. (**b**) The scar of the initial operation is visible. The boy needs to wear nappy because of the accompanying incontinent

sphincteric control (Fig. 16.21a, b). In meningocele the CSF only come out of the defect and the swelling is purely cystic whereas in myelomeningocele the neural fibres are also herniated into the sac along with the CSF and result in plenty of orthopaedic complications including paralysis of the lower limbs. The radiograph reveals loss of vertebral arches at the site (Fig. 16.22a–c).

The management of such a complex situation has been challenging. The protruded sac is excised after neural fibres are reduced back in their normal anatomical position. However, it may not be possible to carry out this procedure completely and there are leftover neurological symptoms. The tethered cord syndrome (TCM) is another major cause of decreased neural function. It is due to adhesions that the open neural elements do not find an anatomical placement in the canal. Moreover, majority of such children have some degree of tethered cord at the time of birth. After the sac is closed there is a tendency for the herniated neural fibres to get adhered to the line of repair. This process keeps on increasing with age hence these patients always have variable neurological symptoms (Fig. 16.23a, b).

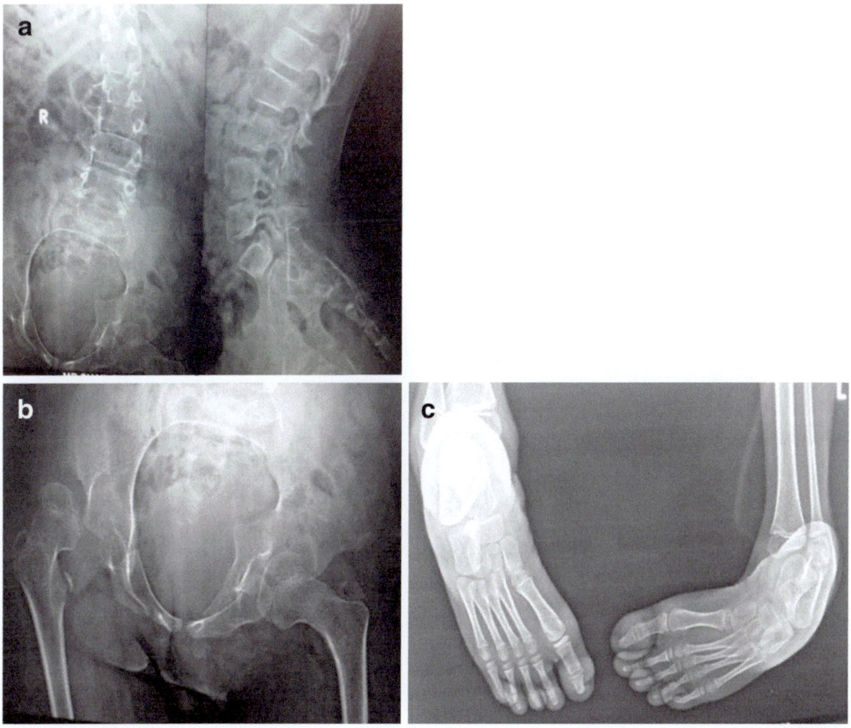

Fig. 16.22 (**a**) AP roentgenogram of the same boy showing unclosed vertebral arches. The lateral view reveals the loss of posterior vertebral structures. (**b**) Roentgenogram of the same patient showing paralytic dislocation of the right hip with secondary valgus position of the femoral neck, dysplasia of the acetabulum. (**c**) The roentgenograms of the feet of the same patient showing equinovarus deformity of the right foot which was treated

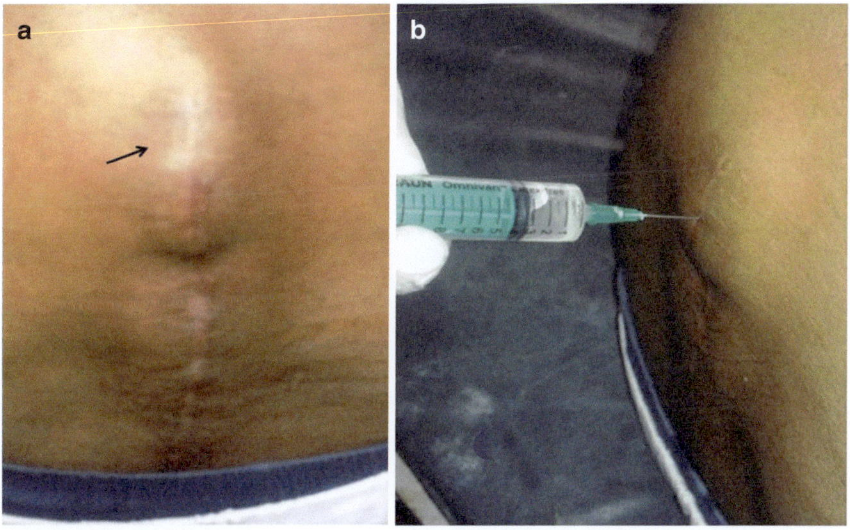

Fig. 16.23 (**a**) A case of meningocele showing subcutaneous extension after surgery (arrow). (**b**) The CSF is being aspirated to confirm the diagnosis before further repair

16.13 Sacral Agenesis

Sacral agenesis is a rare congenital anomaly which could be total or partial (Fig. 16.24a, b). Depending upon the extent of vertebral involvement there may be an evidence of neurological involvement. Such children may have symptoms like myelomeningocele. The sensory symptoms may also be revealed below the affected vertebrae. In severe cases a part of lumbar vertebrae may also be affected (Fig. 16.25a, b).

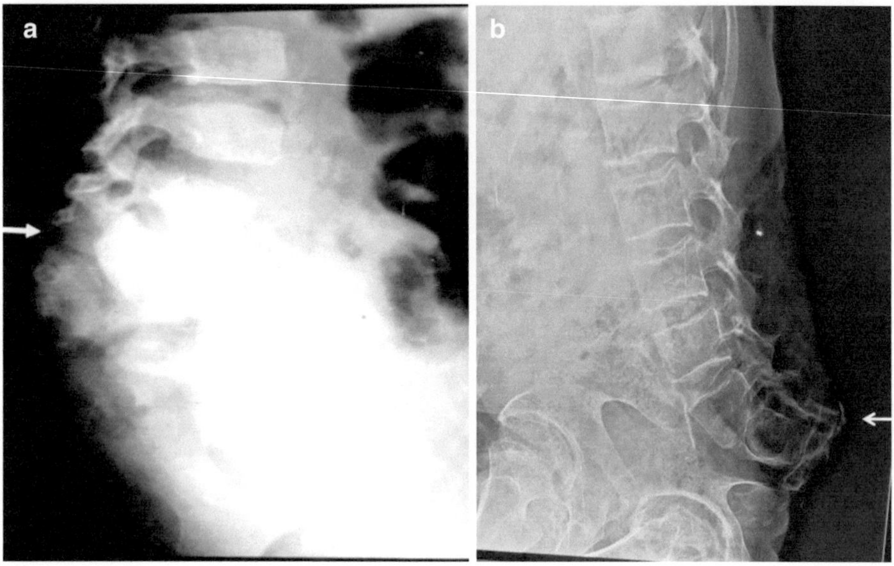

Fig. 16.24 (**a**) Roentgenogram revealing complete absence of the sacrum and the coccyx. (**b**) Roentgenogram shows partial absence of the sacrum

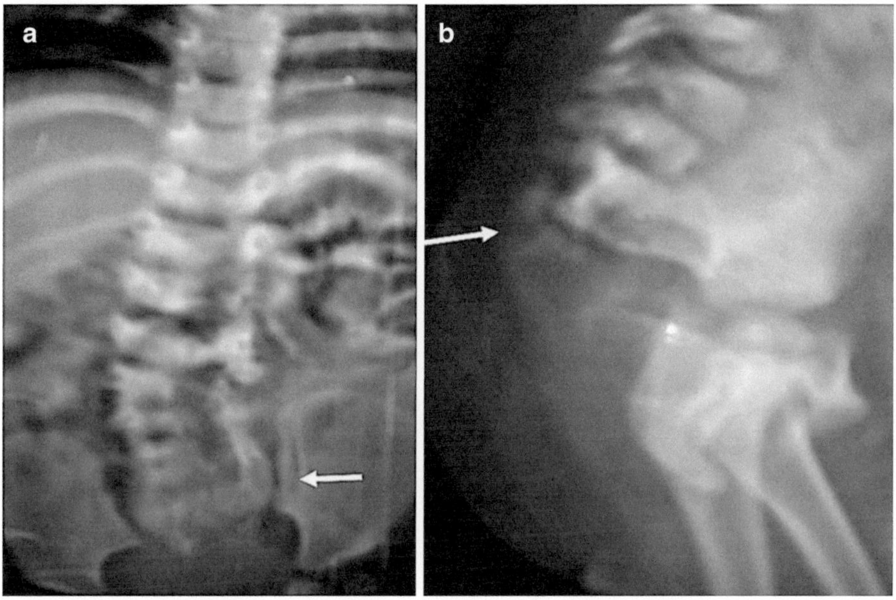

Fig. 16.25 (**a**) Roentgenogram revealing spina bifida and absence of the distal sacral vertebrae (arrow). Note the bony bar on the left side of the lumbar vertebrae. (**b**) Lateral view of the same patient showing complete absence of the sacrum (arrow) and maldevelopment of L5 vertebra

Suggested Readings

Bunnell W. A study of the natural history of idiopathic scoliosis. Spine. 1986;11:773.

Ceballos T, Ferrer-Torrelles M, Castillo F, Fernandez-Paredes E. Prognosis in infantile idiopathic scoliosis. J Bone Joint Surg Am. 1980;62:863.

Figueiredo UM, James JI. Juvenile idiopathic scoliosis. J Bone Joint Surg (Br). 1981;61

Goldstein L. Surgical management of scoliosis. J Bone Joint Surg Am. 1966;48:167.

Lovell and Winter's, Raymond TM, Stuart LW. Pediatric Orthopaedics. 4th ed. Lippincott-Raven, I; 1996.

Aufdermaur M. Juvenile kyphosis (Scheuermann's disease): radiology, histology and pathogenesis. Clin Orthop. 1981;154:166.

Ascani E, LaRossa G. Scheuermann's kyphosis. New York: Reven Press; 1994. p. 557.

Birch JG, Herring JA. Spinal deformity in Marfan syndrome. J Pediatr Orthop. 1987;7:546.

Bradford DS. Juvenile kyphosis. Clin Orthop. 1977;128:45.

Bradford DS, Moe JH, Winter RB. Kyphosis and postural round back deformity in children and adolescents. Minn Med. 1973;56:114.

Bradford DS. Management of spondylolysis and spondylolisthesis. Instr Course Lect. 1983;32:151.

Hensinger RN. Spondylolysis and spondylolisthesis in children and adolescents. (review). J Bone Joint Surg [Am]. 1989;71:1098.

Dolan K. Developmental abnormalities of the cervical spine below the axis. Radiol Clin N Am. 1977;15:167.

Congenital Skeletal Limb Malformations

17

Congenital anomalies are a challenge to the Orthopaedic Surgeon and the Prosthetist. It is difficult to discuss congenital anomalies without classifying them in a way that is universally acceptable to all specialists who treat such problems.

The classification to be presented has been evolved in an endeavour to standardize nomenclature. Malformations are categorized into various groups.

The classification for congenital limb malformations adopted by the American Society for surgery of the hand, the International federation of societies for surgery of the hand and the International societies of Prosthetics and Orthotics is well accepted. This method groups similar pattern of deficiencies according to the parts that have been primarily affected by certain embryological failure, whether the insult involves a total part (skeletal and soft tissue) or only the dermomyofascial structures. The main categories of this classification are (1) failure of formation of parts, (2) failure of differentiation (separation) of parts, (3) duplication, (4) overgrowth, (5) undergrowth, (6) congenital constriction band syndrome and (7) generalized skeletal abnormalities.

Seven descriptive terms are also employed in the classification of these anomalies: amelia, hemimelia, phocomelia, acheiria, apodia, adactylia and aphalangia. The first three terms are derived from the Greek melos, a limb: Amelia, means the absence of a limb and hemimelia means absence of a large part (hemi, a half) of a limb. Phocomelia refers to a flipper-like limb, that is, a hand or foot which is attached more or less directly to the trunk. Acheiria and apodia mean the absence of a hand and a foot, respectively. Adactylia means absence of a digit and is reserved for those cases in which there is absence of the associated metacarpal or metatarsal as well as the digit. Aphalangia naturally means the absence of one or more phalanges.

S. S. Yadav, *Skeletal Dysplasias & Congenital Malformations*,
https://doi.org/10.1007/978-981-99-5679-1_17

Upper Extremity Malformations 18

It is difficult to discuss congenital anomalies of the upper extremity without classifying them in the way that is universally acceptable to all medical specialists who treat such problems in the upper limb. Malformations are categorized into various groups.

A group of deformities categorized as failure of formation of parts is distinguished by failure or arrest of the formation of the limb, which may be complete or partial. These deformities are seldom genetically patterned or inherited. These are subcategorized as transverse and longitudinal.

Transverse arrest is complete absence of the structures distal to a certain point in upper extremity. They are commonly referred to as congenital amputations. Transverse terminal deficiencies may occur at any level from the shoulder down, but they tend to involve the hand and wrist more frequently. Complete absence of the arm at the level of the proximal humerus (Amelia) is rare and occurs during the time of the use of thalidomide or other embryologic toxins. Bilateral presence of this condition is extremely rare. Upper third forearm transverse anomalies are common. The longitudinal arrest is grouped into radial, central, ulnar and intersegmental deficiencies.

Abnormal presentation of additional upper extremity in the form of duplication has also been described.

18.1 Triplication of the Upper Extremity

18.1.1 Triplication of the Right Upper Extremity

A very rare congenital anomaly in the form of triplication of the right upper limb has been recorded by us (Fig. 18.1). All the three components were hypoplastic with there corresponding ill developed scapulae. There was no other congenital malformation in the body and the child was otherwise looking normal.

S. S. Yadav, *Skeletal Dysplasias & Congenital Malformations*,
https://doi.org/10.1007/978-981-99-5679-1_18

Fig. 18.1 Roentgenogram of a 1-month-old girl child with three right upper extremities. Not well developed, all the three revealed anomalies, more so at the distal ends. There were three corresponding rudimentary Scapulae (Unfortunately a clinical picture was not permitted by the parents)

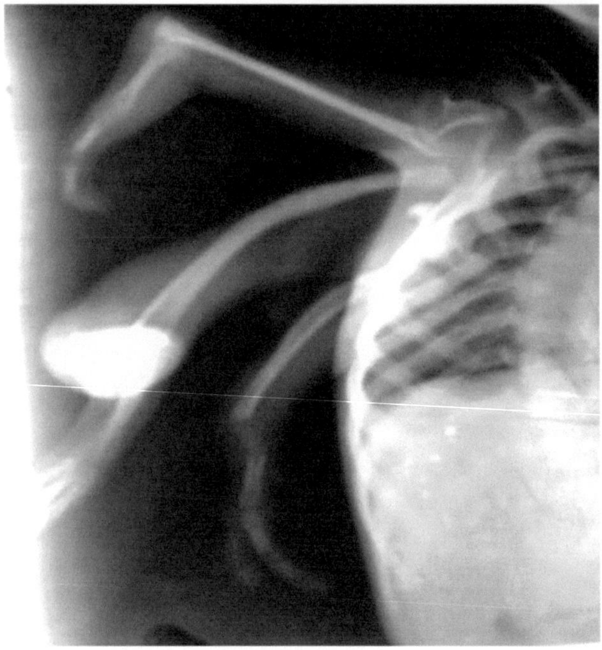

18.1.2 Amelia

Amelia (absent limb) is a rare congenital defect. Bileteral complete absence of the arms at the level of proximal humerus is a very rare anomaly. Generally there is a rudimentary bud on either side (Fig. 18.2a). Children with congenital limb deficiencies make remarkable functional adaptations by using whatever limb components they have to substitute for those that are missing. Patients with bilateral upper extremity absence rarely use prostheses. Certain functional patterns are common to all children with the same defect. They inevitably perform almost all functional activity with the sensory surfaces available to them, even by using their feet (Fig. 18.2b). Radiograph fails to reveal the upper extremities (Fig. 18.3).

18.1.3 Phocomelia (Intersegmental Deficiency)

Phocomelia (short-limb) is a congenital disorder where the terminal part of the limb is attached directly to the trunk. The word phocas comes from the word seal, dramatically describing the shortened limb found in this condition. The forearm bones are short and deficient in either radial or ulnar components (Fig. 18.4a–c). There may also be associated multiple anomalies of the hand (Fig. 18.5a, b).

Phocomelia has been classified into three types. (1) Arm and forearm deficient type, where the hand is attached directly to the shoulder. (2) Arm deficient type, where forearm is attached to the shoulder. (3) Forearm deficient type, where the hand is attached to the arm (Fig. 18.6a–d).

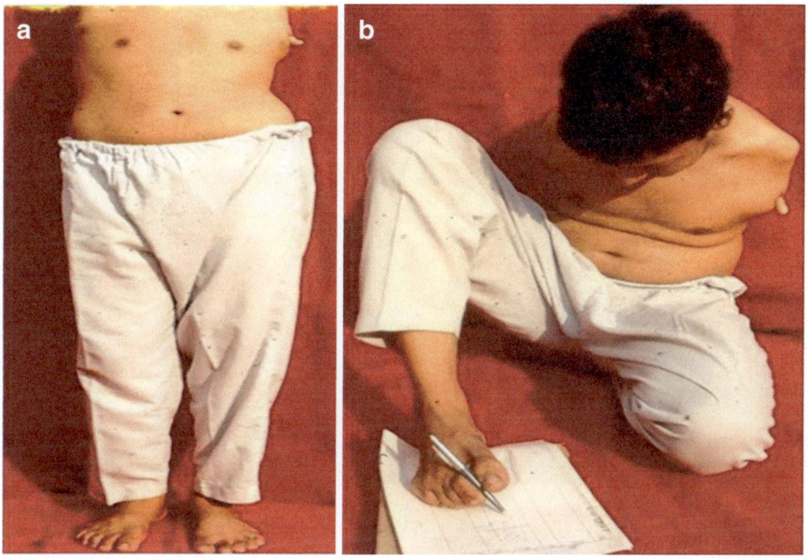

Fig. 18.2 (**a**) A 15-year-old boy with bilateral absence of arms. Note the rudimentary buds on both sides. (**b**) The same boy using his right foot for writing and other functional activities

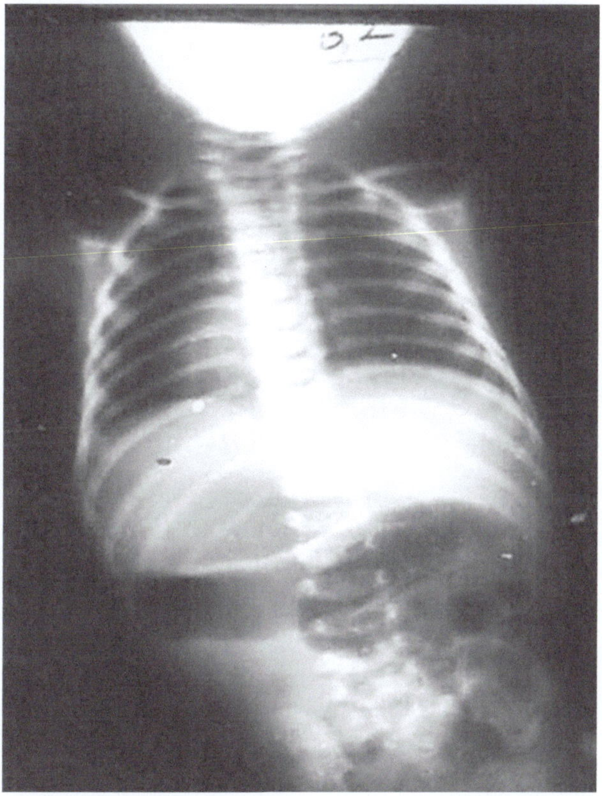

Fig. 18.3 Roentgenogram reveals bilateral complete absence of the upper limbs with normal looking trunk. Note the hypoplastic scapulae

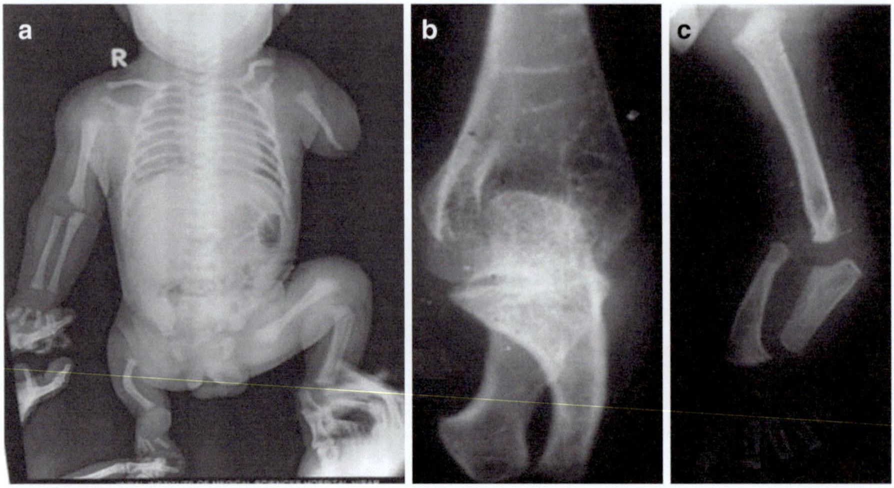

Fig. 18.4 (**a**) Roentgenogram reveals congenital mid arm amputation on the left side and grade-IV proximal femoral focal deficiency (PFFD) on the right side. (**b**) Radiograph of a patient with mid-forearm amputation. The radiograph show upper portion of radius and ulna and a normal elbow. (**c**) Roentgenogram reveals congenital hypoplasia of the radius and ulna with dislocation of the radial head. The elbow joint appears normal

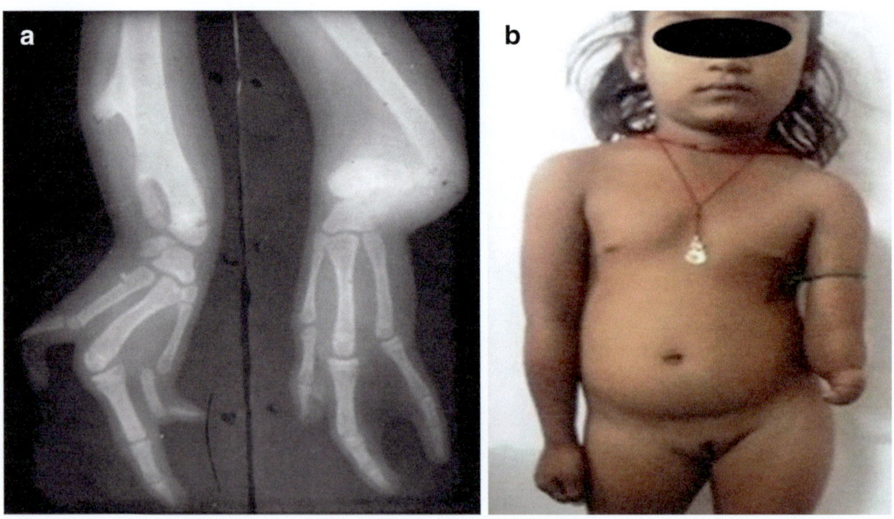

Fig. 18.5 (**a**) Roentgenograms show multiple congenital anomalies of the forearm and hand. Rudimentary ulna is merged with the radius proximally. The carpals are malformed and fused. Note the oligodactylia. (**b**) A 6-year-old girl with a micro hand on the left side attached to the arm (Type C Phocomelia)

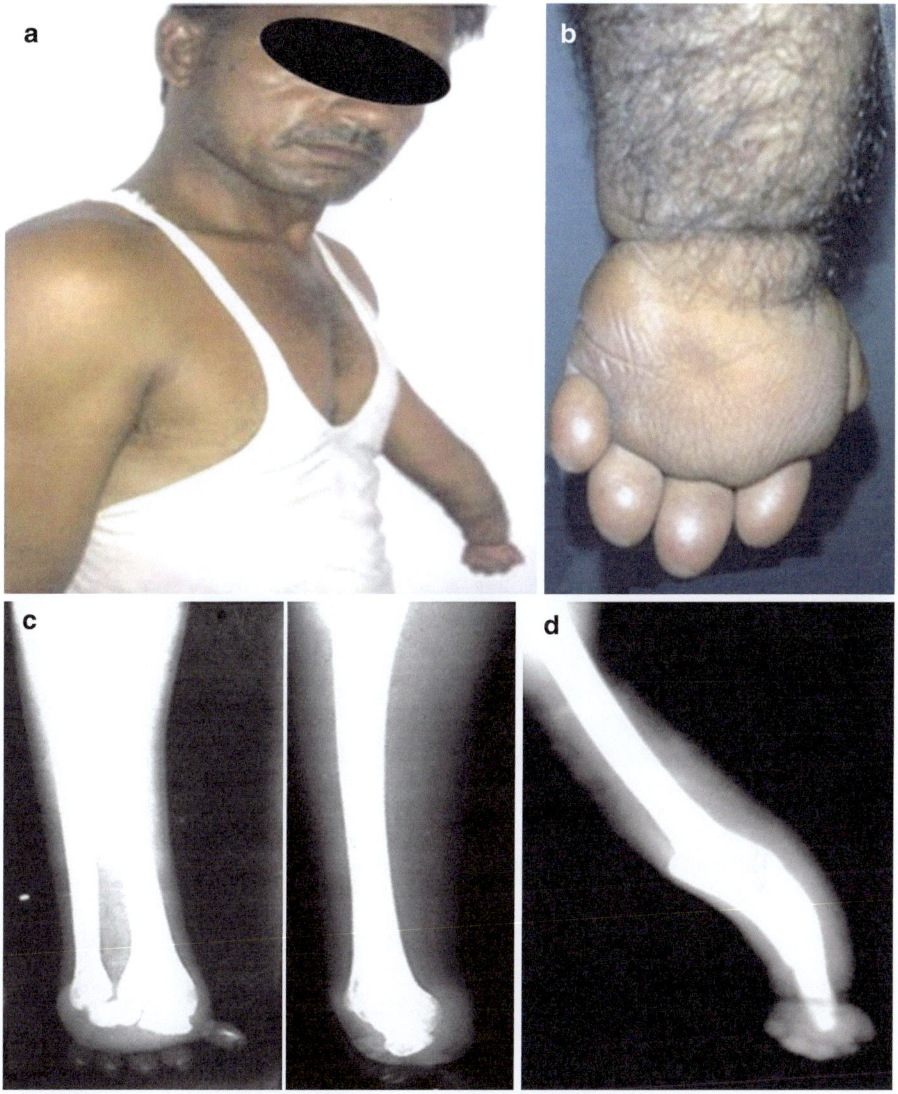

Fig. 18.6 (**a**) A 35-year-old male shows mesomalic and acromalic shortening. Note the congenital non-functional micro hand on the left side which is functionally similar to Acheiria. (**b**) Note the hypoplasia of the hand with small palm and micro digits with no bones (Acheiria). A circular crease on the wrist is evident-constriction band. (**c**) Roentgenogram revealing achromelia. Note two carpals and micro hand with small palm and rudimentary fingers with no metacarpals and phalanges (Adactylia and Acheiria). (**d**) Radiograph revealing hypoplastic radius and ulna. A rare anomaly

Phocomelia may be associated with other congenital defects like- cleft palate and scoliosis (Fig. 18.7a–c). The goal of the treatment is to improve the self-care activities that enable the patient to be as independent as possible.

18.1.4 Congenital Deltoid Contracture

Congenital deltoid contracture is a rare presentation and is often bilateral. The patient is not able to adduct the shoulders fully (Fig. 18.8a, b). The X-rays do not reveal any abnormality. The treatment depends on the severity of the contracture, a release or lengthening of the deltoid is needed (Fig. 18.8c).

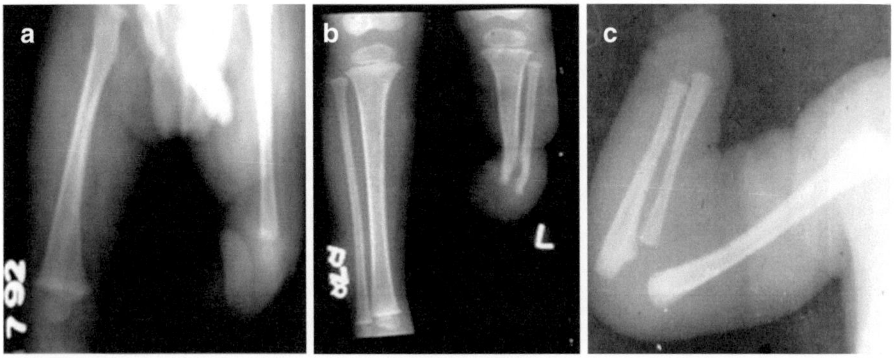

Fig. 18.7 (**a**) Roentgenogram reveals congenital disarticulation at the left knee. (**b**) Note the mid leg congenital amputation on the left side. (**c**) Roentgenogram shows only soft tissues after the distal ends of radius and ulna with complete absence of the hand (Acheiria)

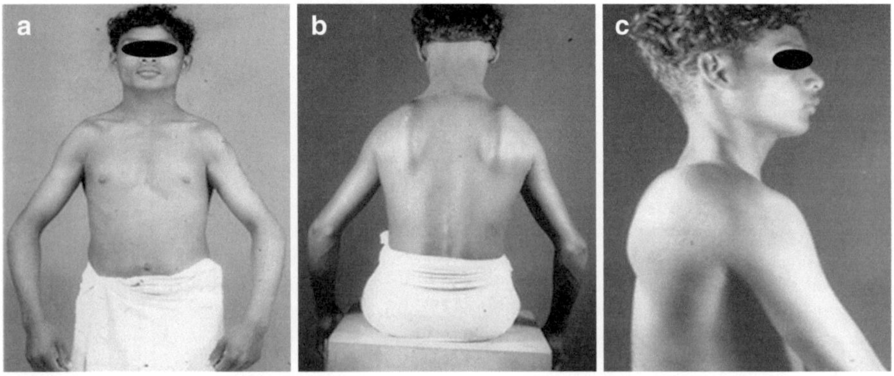

Fig. 18.8 (**a**) A patient with bilateral congenital deltoid contracture anterior view. Note the inability to fully adduct both the arms. (**b**) Posterior appearance with relative winging of the scapulae. (**c**) Post-operative appearance with adducted arms

18.2 Congenital Synostosis

18.2.1 Congenital Humero-Ulnar Synostosis

In this rare defect the lower end of the humerus is fused with the proximal end of the ulna (Fig. 18.9a–c). Mostly unilateral, the distal humeral epiphysis and proximal ulnar epiphysis are absent. Rarely it may occur with radius. Usually the condition is

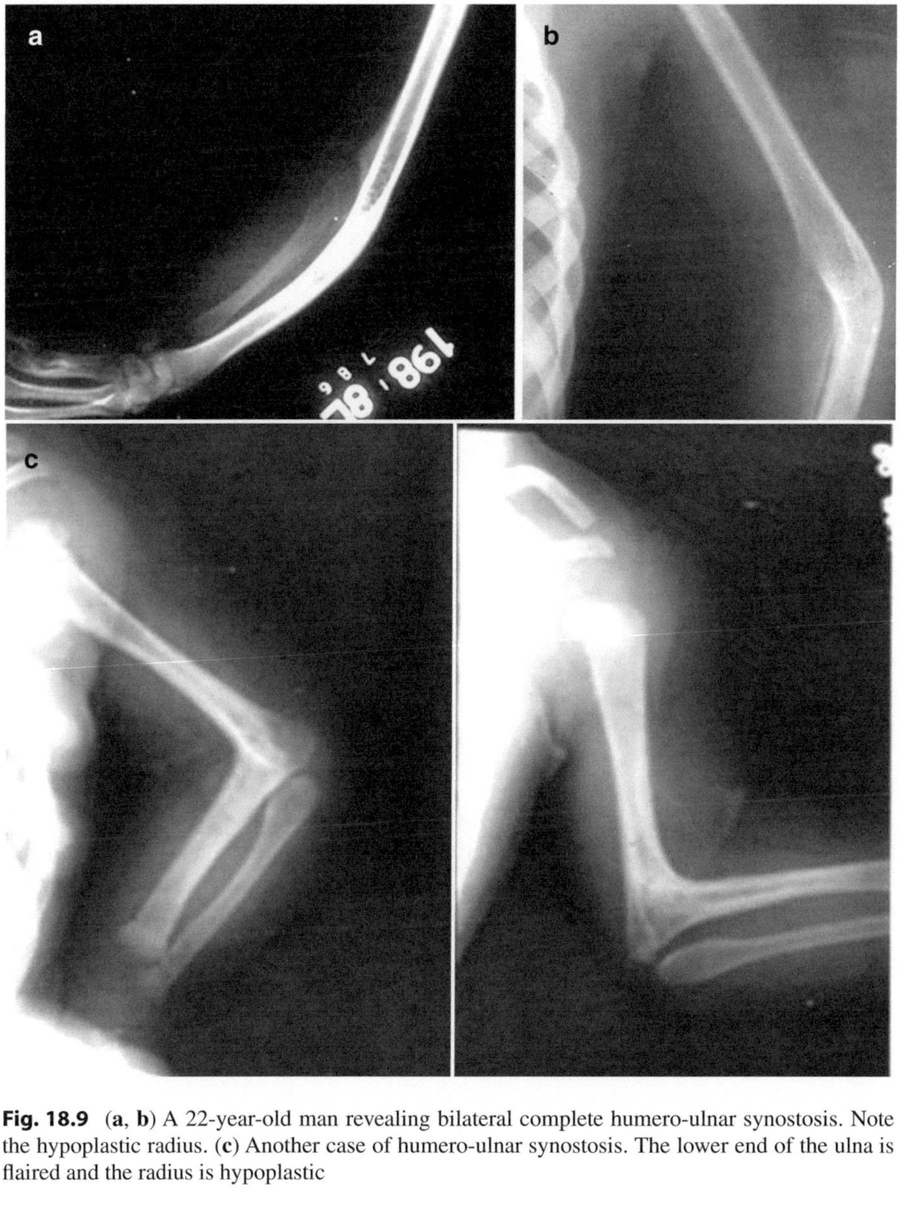

Fig. 18.9 (**a**, **b**) A 22-year-old man revealing bilateral complete humero-ulnar synostosis. Note the hypoplastic radius. (**c**) Another case of humero-ulnar synostosis. The lower end of the ulna is flaired and the radius is hypoplastic

sporadic where the extremity in due course of time becomes short. The associated skeletal abnormalities may occur in the form of syndactyly, polydactyly, clubfoot and Madelung deformity.

18.2.2 Radio-Ulnar Synostosis

This rare congenital synostosis mostly occurs at the proximal ends of the Radius and Ulna (Figs. 18.10 and 18.11a, b). However, distal ends may also reveal synostosis. More than half of the cases are known to be bilateral. Although the abnormality is present at birth, it may be detected at a later stage in the form of loss of pronation and supination. The interosseous membrane also may be abnormal. An occasional fibrous synostosis may allow some moment. There may be atrophy of the muscles in the region. The defect may be associated with other skeletal abnormalities such as syndactyly, polydactyly, clubfoot and carpal coalition.

Two types of radio-ulnar synostosis have been described. Type I is when synostosis occurs at the level of head of the radius and Type II when it occurs below the head of the radius (Fig. 18.12a, b).

In terms of treatment, all attempts to separate the two bones usually fail. Derotation of the non-writing hand by an osteotomy through the proximal fusion mass has been recommended.

18.2.3 Radial Deficiencies (Clubhand)

Various deficiencies of the radius also involve the skeleton on the radial border of the forearm and hand. The lateral rays in the hand may be hypoplastic or absent. The thumb may be absent and when present is defective to some degree. Such a congenital longitudinal deficiency of radius is associated with radial deviation of the hand resulting in severe defect and shortening of the forearm. There may be complete absence, proximal or distal deficiency of the Radius. Although the etiology is not known, it may be multifactorial.

Congenital deficiency is total or partial absence of parts that occur on the radial (preaxial) border of the upper extremity. The deficiency may range from Hypoplasia of the thumb to varying degrees of deficiencies of the radius. Typically the hand is deviated radially, the extent being determined by the degree of absence of the radius.

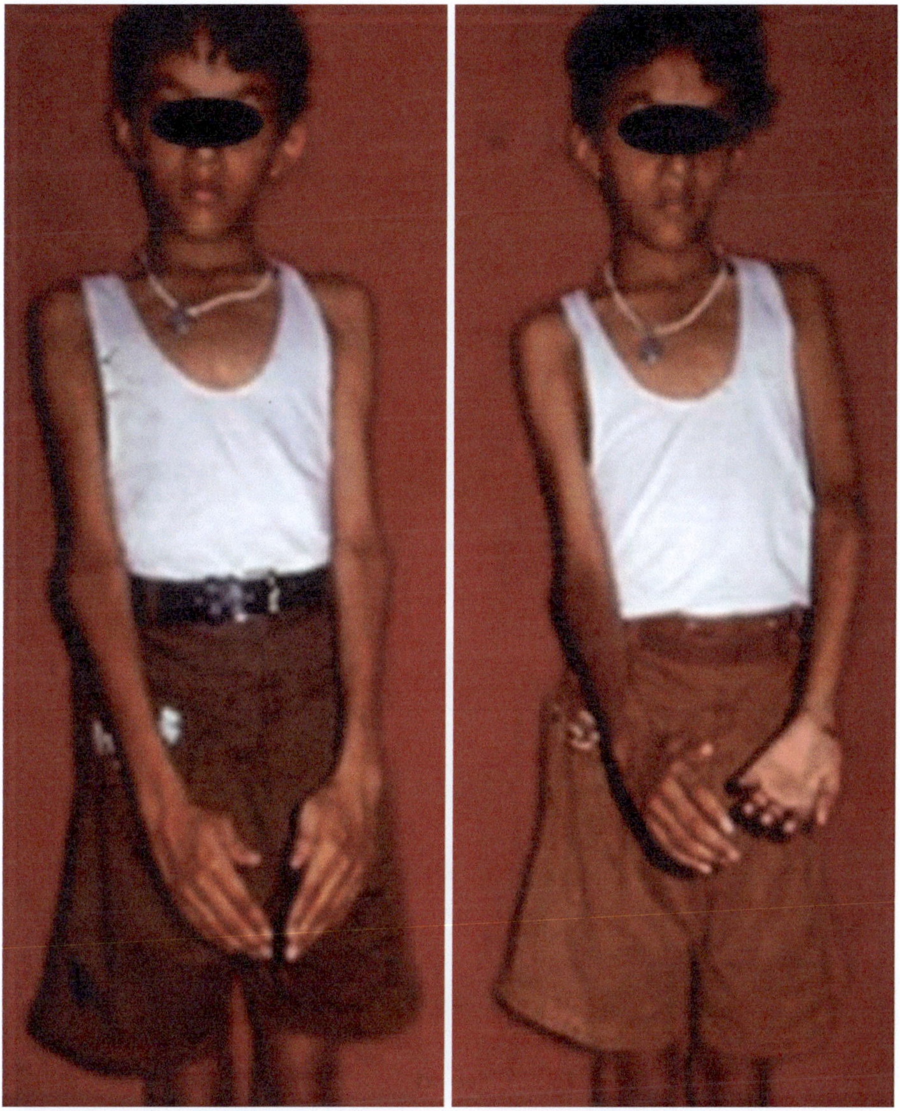

Fig. 18.10 Twin brothers showing their inability to pronate and supinate the forearm

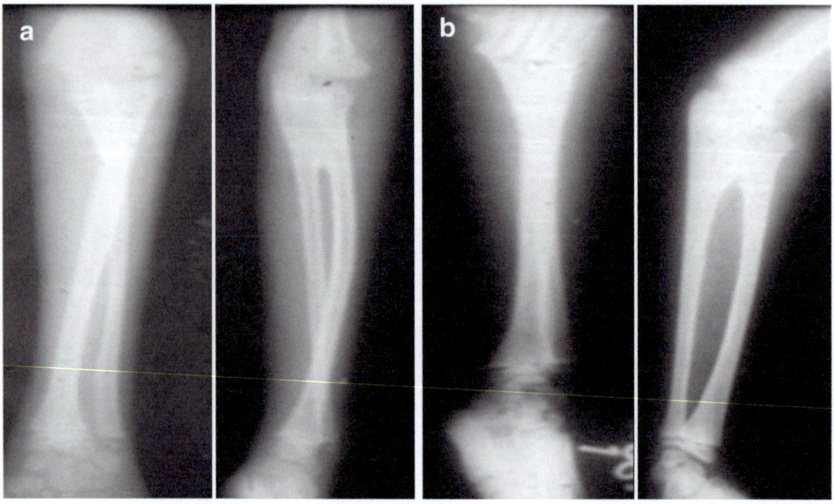

Fig. 18.11 (**a**, **b**) Roentgenogram revealing bilateral superior radio-ulnar synostosis in both patients

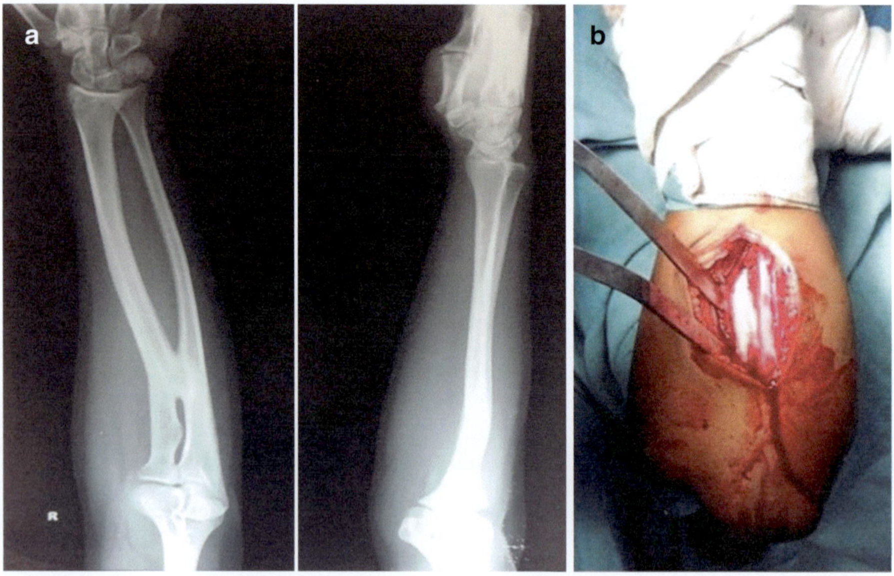

Fig. 18.12 (**a**) Roentgenogram revealing radio-ulnar synostosis-type II. (**b**) Intra operative picture after excision of the synostosis

The degree of skeletal deficiency is used to further sub classify the deficiencies into four types:

18.2.4 Type 1: Short Distal Radius

In type I the growth of the distal epiphysis of the radius is depressed which results in relatively short radius. There is mild radial deviation. The thumb and the radial carpal bones may be hypoplastic (Fig. 18.13).

18.2.5 Type 2: Hypoplastic Radius

Although the proximal and distal epiphysis of the radius are present, but the growth is defective in both areas. As a result the radius is hypoplastic and the ulna is thick and bowed to the radial side. A further shortening of the radius occurs with growth (Fig. 18.14). The function of the hand is compromised.

18.2.6 Type 3: Partial Absence of the Radius

The radius may be partially absent in the distal, middle, or proximal third. The most frequent is absence of the distal and middle third (Fig. 18.15). The hand is not supported and no stability is provided to maintain good hand function.

18.2.7 Type 4: Total Absence of the Radius (Radial Club Hand)

Type 4 defects are the most common. With total absence of the radius, the hand is completely unsupported which results in radial displacement (Fig. 18.16a–m). The soft tissue contracture is usually severe. The treatment is to release the soft tissues and provide support for the hand. The incidence of absent thumb has been reported from 17% to 86%. Pollicisation of the index finger is the treatment of the choice.

Centralization of the hand is attempted by the removal of lunate and capitate bones after adequate release of the wrist from fibrotic radial anlage. Stabilization is then obtained with Kirschner wire. The results of centralization procedures have been satisfactory. Long-term follow-up results in remodelling of the ulnar head, which becomes broad (flaring) in its distal ends, resembling a radius. The most satisfactory results are known to occur when the treatment is started during the first year of life.

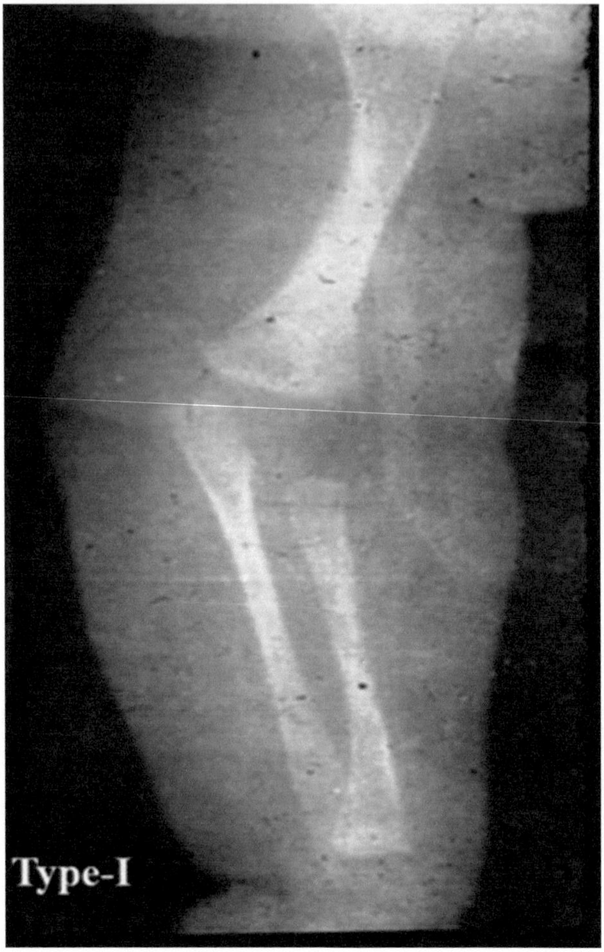

Fig. 18.13 Roentgenogram reveals short radius. The distal radial epiphysis is defective. Growth rate is decreased and is a discrepancy between the radial and ulnar growth. This results in slight radially deviated hand

18.2.8 Ulnar Deficiencies

Congenital ulnar deficiency is rare. Ulnar deficiencies have a low incidence when compared with other congenital deformities. However, the variety of malformations seen in this deficiencies are greater than those seen in radial longitudinal deficiencies (Fig. 18.17a, b). Clinically there may be ectrodactyly or short digits in various degrees, ulnar deviation of the hand, forearm shortening and bowing, diminished elbow motion and hypoplasia of humeral segment. There may be dislocation of the radial head and ankylosis of the ulnohumeral joint (Fig. 18.18a–c). Carpal anomalies and fusions are frequently presents. Syndactyly is common between the digits that are present as well as stiffness of the digits.

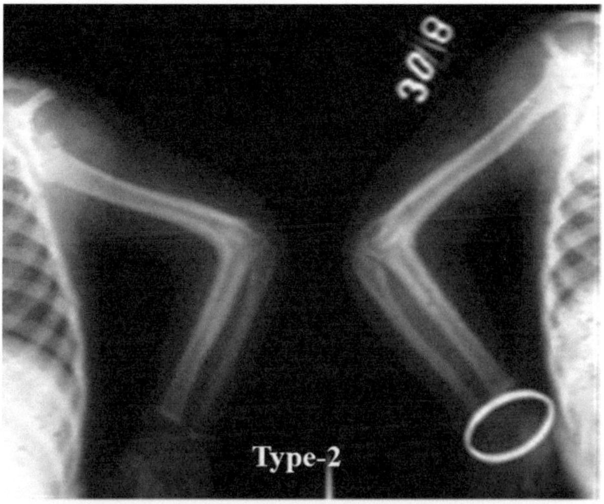

Fig. 18.14 The hypoplastic radius. Although the radius reveals proximal and distal epiphyses, but they are defective resulting in a small radius that provides no support to the hand

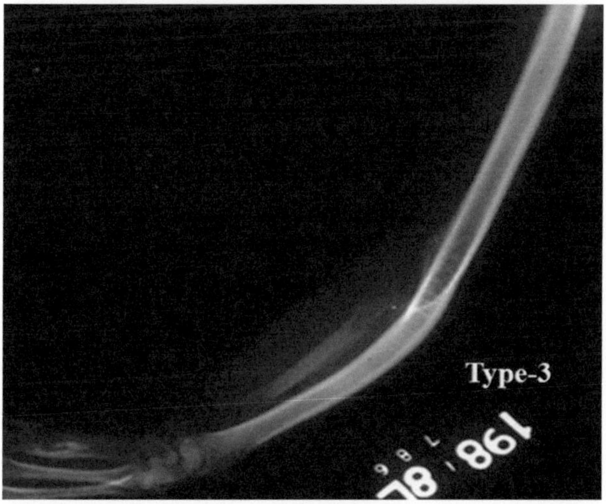

Fig. 18.15 Partial absence of radius. The radius is partially present in the proximal half, as shown. The hand is unsupported and needs centralization of the distal end of the unla

An individualized functionally oriented management programme is recommended with early resection of the fibrocartilaginous remnant to prevent shortening, radial bowing and possible malrotation. Improved motion at the elbow can be obtained by release of contractures or resection of the radial head. Wrist deformity may require resection of the ulnar remnant and a wedge osteotomy of the radius. Such patients may not report any deficiency in bimanual activities. However, they perform poorly when their congenital anomaly included radiohumeral synostosis.

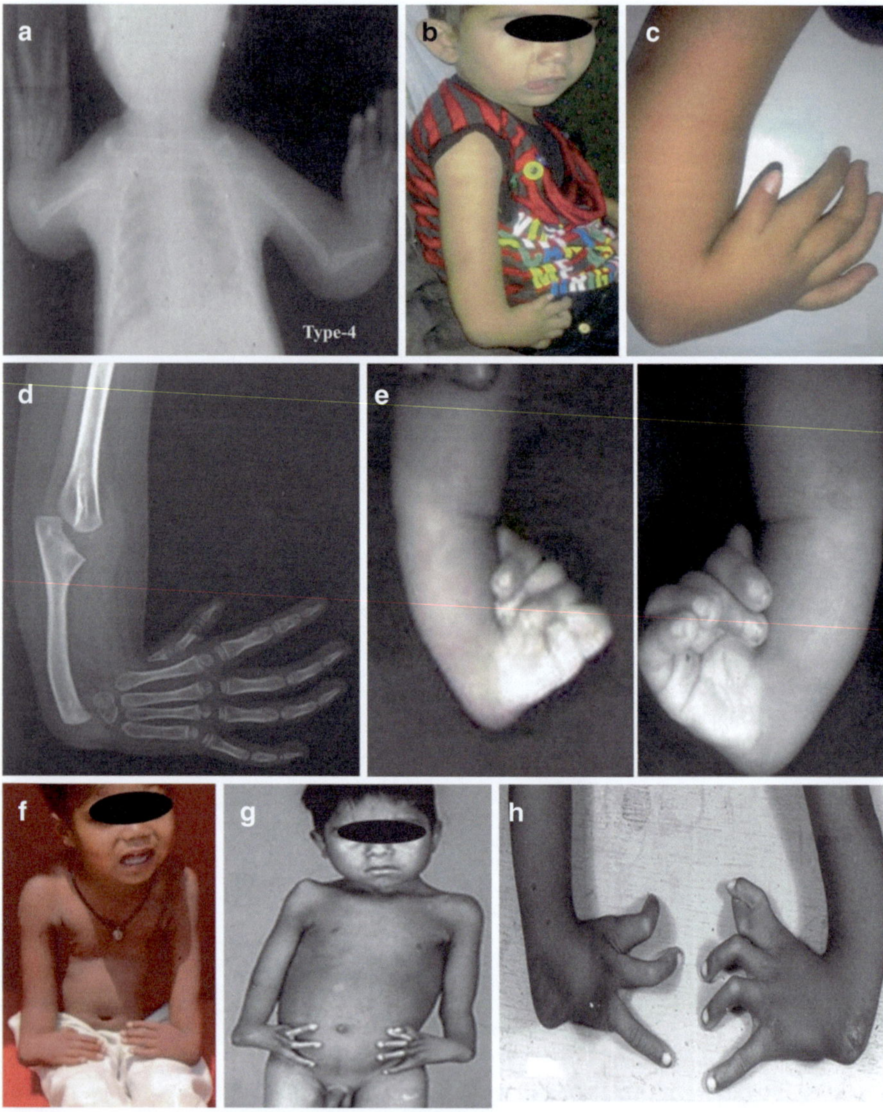

Fig. 18.16 (**a**) Radiograph reveals bilateral congenital absence of the radius with short ulnae. (**b**, **c**) Clinical presentation of a patient with grade-IV radial deficiency (Club hand). (**d**) Roentgenogram reveals complete absence of the radius with a rudimentary thumb. (**e**, **f**) Clinical picture of bilateral radial clubhands. (**g**, **h**) Bilateral radial clubhands with multiple anomalies of the hand and fingers. (**i**) Roentgenogram showing bilateral absence of radius and absence of index finger and thumb on right side. (**j**, **k**) Bilateral radial clubhands (Type 4) with absence of thumbs. (**l**) Right sided radial club hand with rudimentary thumb. (**m**) Roentgenogram revealing complete absence of the radius with an arc of the ulnar shaft and unstable wrist. Note the hypertrophy of the ulna

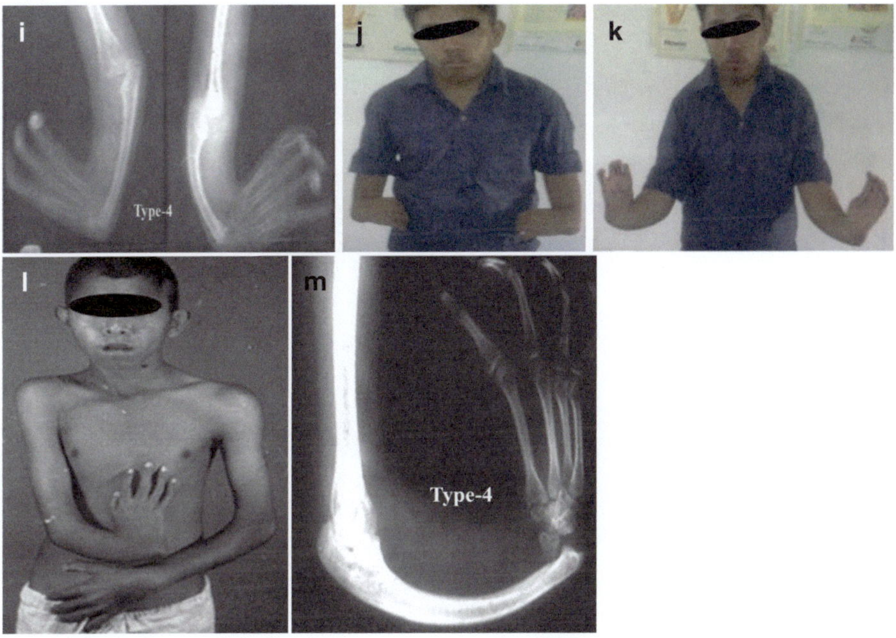

Fig. 18.16 (continued)

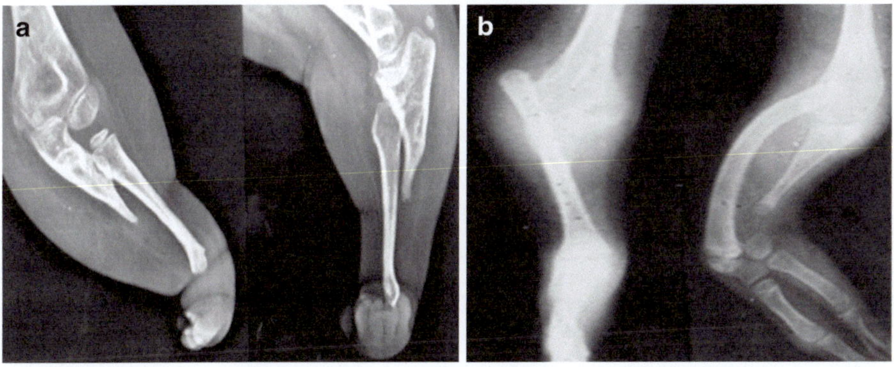

Fig. 18.17 (**a**) The roentgenogram revealed distal ulnar deficiency. The distal end of the radius is also abnormal. The wrist and hand has no bones. (**b**) Roentgenogram reveals hypertrophied radius with dislocation at both its ends. Note the oligodactyly and hypoplasia of the ulna with manus varus deformity

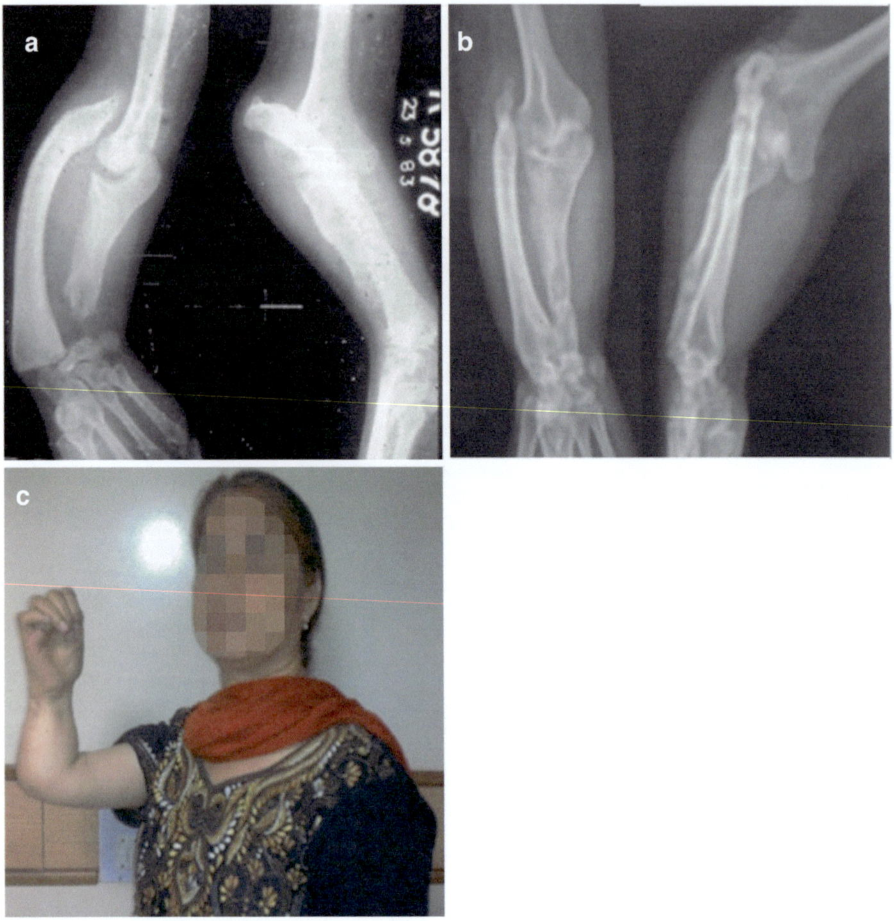

Fig. 18.18 (**a**) Radiograph shows hypoplasia of the ulna with dislocation head of the radius. (**b**) Post-operative radiograph after centralization of the bones. (**c**) Post-operative clinical picture of the straight but short forearm

18.2.9 Classification of Ulnar Deficiencies

There are number of classifications of this deficiency. Bayne's classification is as under:

Type I—There is hypoplasia of both the epiphysis of the ulna and the growth is suppressed. Since the distal ulnar epiphysis is present there is no fibrous anlage in the ulna. Radial bowing at this stage may not be appreciated. There may be mild deviation at the wrist and mild hypoplasia of the ulnar digits.

Type II—This is the most common form of ulnar deficiencies. In this type an ulnar anlage is present that replaces distal bony end of the ulna. The radius is bowed and the radial head may be dislocated. The hand is deviated in an ulnar direction with varying degree of hypoplasia involving the ulnar sided digits.

Type III—This type is rare where the ulna is completely absent and no analog is present. The radius may be straight or curved. The elbow is unstable because of the absence of the ulna. The head of the radius may be dislocated (Fig. 18.19).

Type IV—In type IV, the proximal radius is fused with humerus along with a small portion of the olecranon (Figs. 18.20 and 18.21a, b). Although the ulna is completely absent, an anlage usually attaches to the distal radial epiphysis. There is severe radial bowing with ulnar deviation of the hand. There may be other hand anomalies present.

18.2.10 Congenital Radial Head Dislocation

Congenital radial head dislocation may occur independently or in association with radio-ulnar synostosis (Fig. 18.22). Both the child and parents usually do not observe the deformity or limitation of movements until the child is 4–5 years of age. Radiologically the abnormally can be assessed and managed accordingly (Fig. 18.23a, b).

The functional limitation with congenital radial head dislocation is minimal and pain is usually uncommon. Radial head excision improves appearance and relieves pain. However, it may not increase motion.

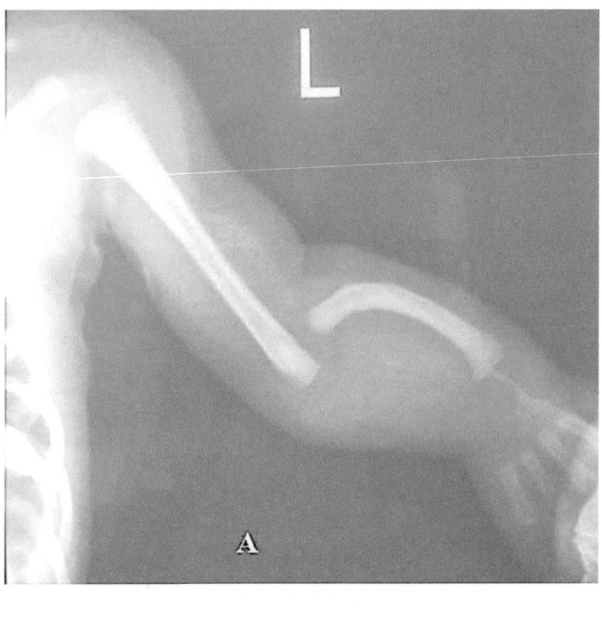

Fig. 18.19 Roentgenogram reveals complete loss of ulna with curved and deformed radius compromising the elbow and the wrist

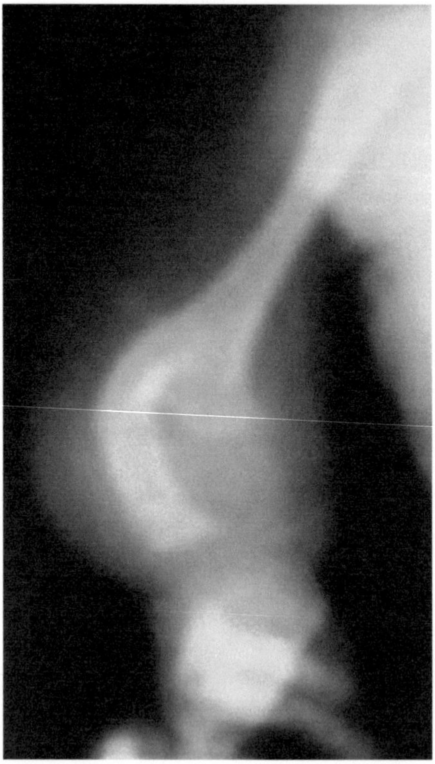

Fig. 18.20 Roentgenogram reveals pre-natal bowing of the radius, hypoplasia of the ulna associated with radio-humeral synostosis

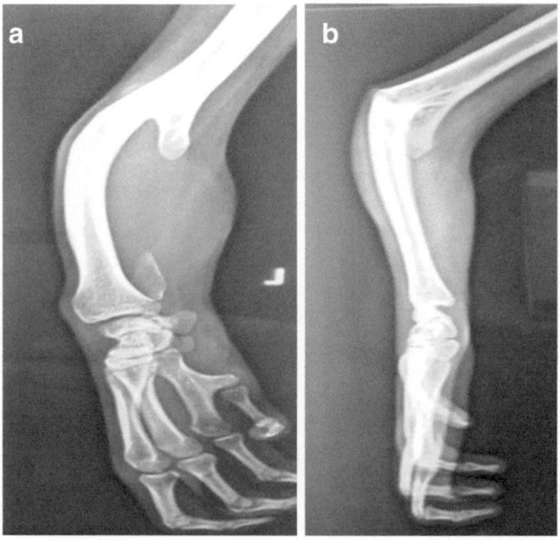

Fig. 18.21 (**a**, **b**) Radiograph reveals radio humeral synostosis. Note the rudimentary ends of the ulna with complete loss of diaphysis. The thumb and index finger are absent and the third metacarpal has a rudimentary digit

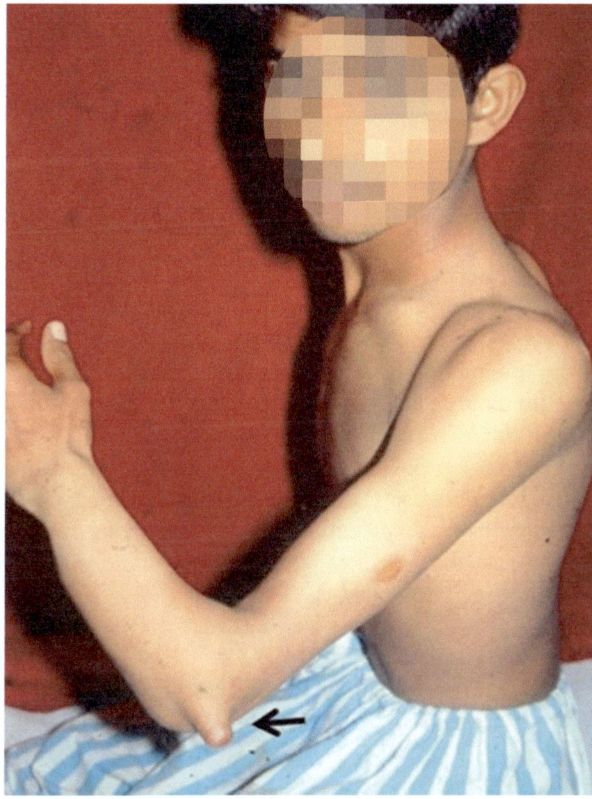

Fig. 18.22 Clinical picture showing mesomelic dysplasia and dislocation of the head of the radius (arrow)

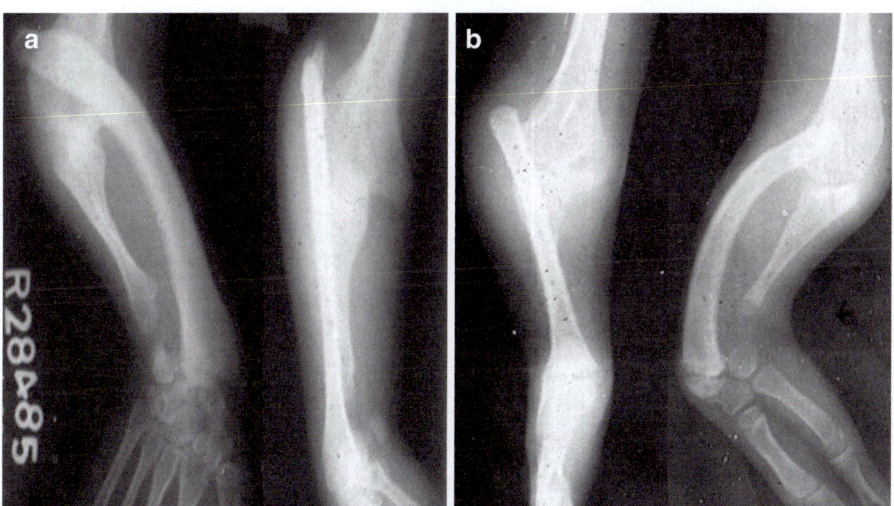

Fig. 18.23 (**a**) Radiograph revealing hypoplasia of the ulna with dislocation of the head of the radius. (**b**) Note the hyperplasia of the ulna, radial head dislocation and oligodactyly

18.2.11 Madelung Deformity

Madelung deformity is characterized by radial and palmar angulation of the distal radius caused by growth disturbance in the palmar and ulnar portion of distal radial physis (Fig. 18.24). The growth disturbance may be due to a combination of a bony lesion in the ulnar portion of the distal radius physis and an abnormal palmar ligament tethering the lunate to the radius proximal to physis. The primary made lung's deformity of the wrist may be due to dyschondrosteosis.

Madelung deformity may be congenital or acquired. There is marked radial inclination and dorsal dislocation of the ulnar head. The forearm is short and deformed. In post-traumatic cases there may be partial growth arrest at the distal radius. No treatment may be necessary for painless deformity. However, asymptomatic skeletally immature patients with progressive deformity should be considered for fibrolysis and release of the abnormal tissues.

In severe cases where ulnar head is dislocated, excision of the ulnar head may be attempted. The result is an improvement in pain and cosmetic appearance. However, the mobility may be un-altered. Spontaneous improvement in the symptoms is reported as common, so a waiting policy and observation of the patient until growth ceases, is recommended.

18.2.12 Post-Traumatic Disarticulation at the Right Shoulder and Amputation at the Left Wrist

18.2.12.1 Krukenberg Operation

An old but useful procedure for amputation at the wrist has been described. Krukenberg operation is a forearm bifurcation (splitting) procedure (Fig. 18.25a, b). It results in two long finger-like components which can close and open (Fig. 18.26a, b). The procedure has proved useful in carrying out some of the need-based functions like holding small items, papers and provides the capability to the smokers to smoke (Fig. 18.27a–c).

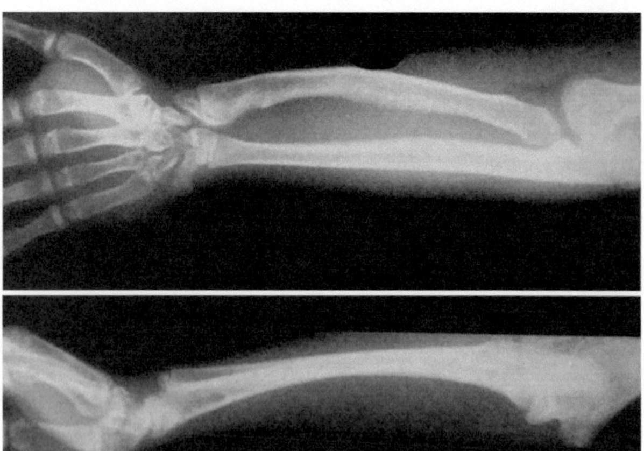

Fig. 18.24 Roentgenogram of the forearm and the wrist showing radial inclination and dorsal luxation of the head of the ulna. The carpal bones are malformed

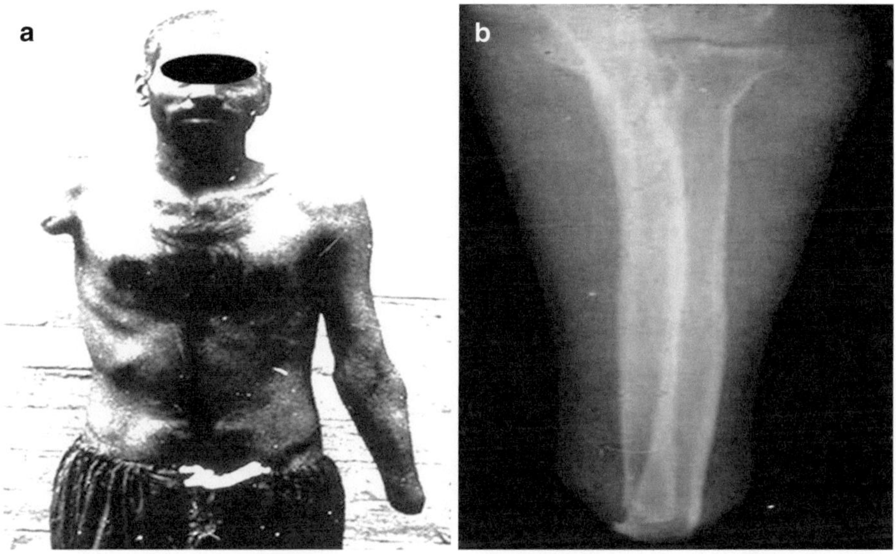

Fig. 18.25 (**a**) A 35-year-old male had disarticulation at the right shoulder and amputation at the left wrist. (**b**) Roentgenogram reveals the radius and ulna with amputation at their distal ends

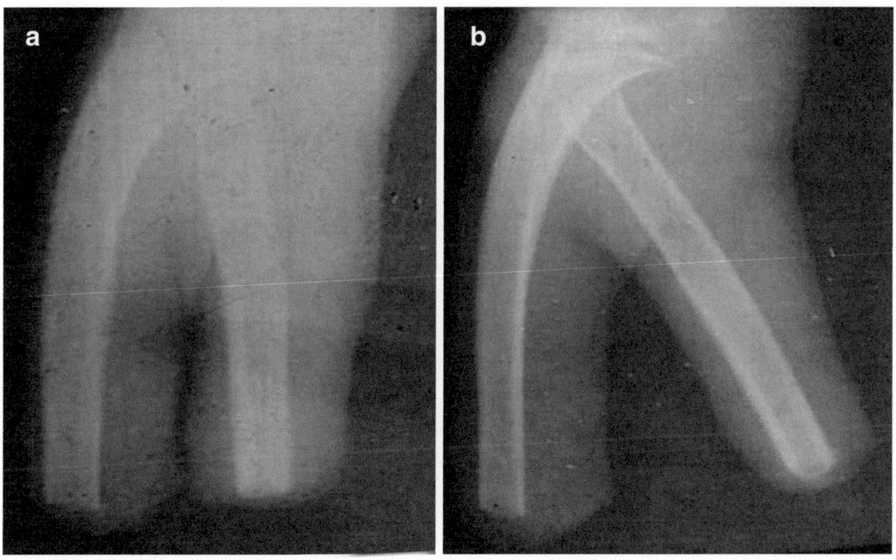

Fig. 18.26 (**a**, **b**) A forearm splitting procedure (Krukenberg operation) was performed with an objective that the split forearm will work like fingers. Roentgenograms reveal the man-made fingers, close and open

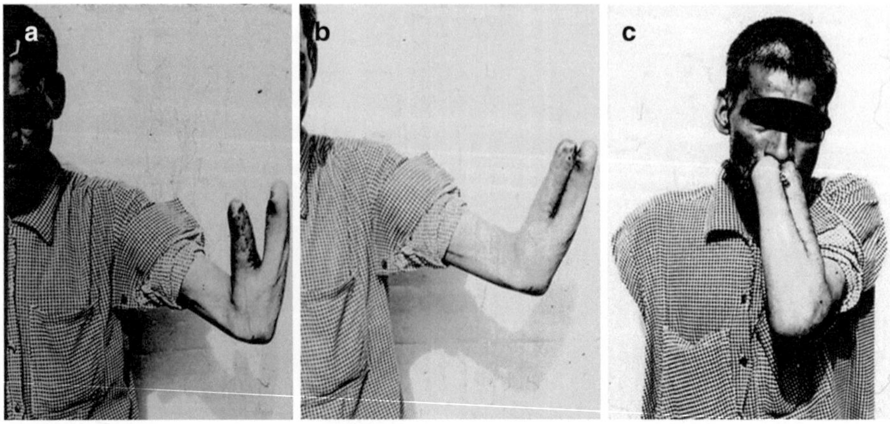

Fig. 18.27 (**a**, **b**, **c**) Clinical picture of the patient with the forearm fingers abducted and adducted. The patient being a smoker could not smoke on his own. After surgery he could smoke by holding the cigarette in the split forearm and could hold papers. He was employed as a peon

The forearm is split in two halves. The ulnar nerve and the ulnar artery are in the medial component and the median nerve and the radial artery are preserved in the lateral component. The muscles are separated according to their anatomical names. The interosseous membrane is divided.

18.3 Split Hand (Bi-Fid Hand)

Split hand, Ectrodactyly, Lobster claw hand and Central hypoplasia are the terms used to describe a cleft hand. The cleft hand is usually bilateral and of those affected, about 60% are boys (Fig. 18.28). Central deficiencies may be classified into two patterns: typical and atypical. The typical is categorized as cleft hand whereas the atypical central defect is known as symbrachydactyly (Fig. 18.29a, b). In symbrachydactyly hand often there are so many crossed bones and abnormal soft tissues that the function of the hand is already greatly compromised (Fig. 18.30).

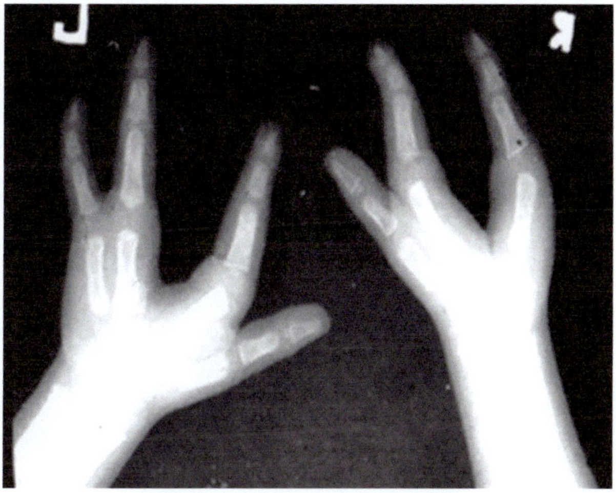

Fig. 18.28 Roentgenogram reveals typical bilateral congenital split hands. The left hand shows the absence of long finger. The right hand shows the absence of the middle and ring fingers

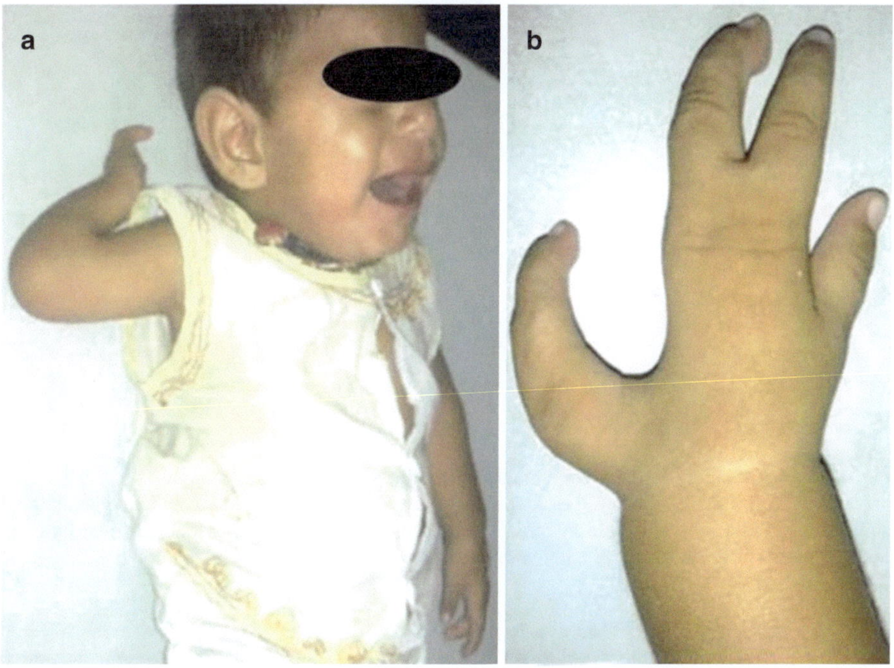

Fig. 18.29 (**a**, **b**) Clinical presentation of a child with split hand on the right side. Note the absence of ring finger resulting in a deep web

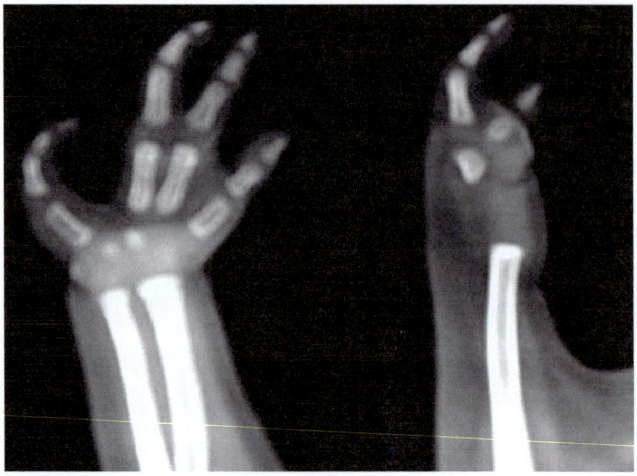

Fig. 18.30 Roentgenogram of the same patient with loss of forth ray of the hand

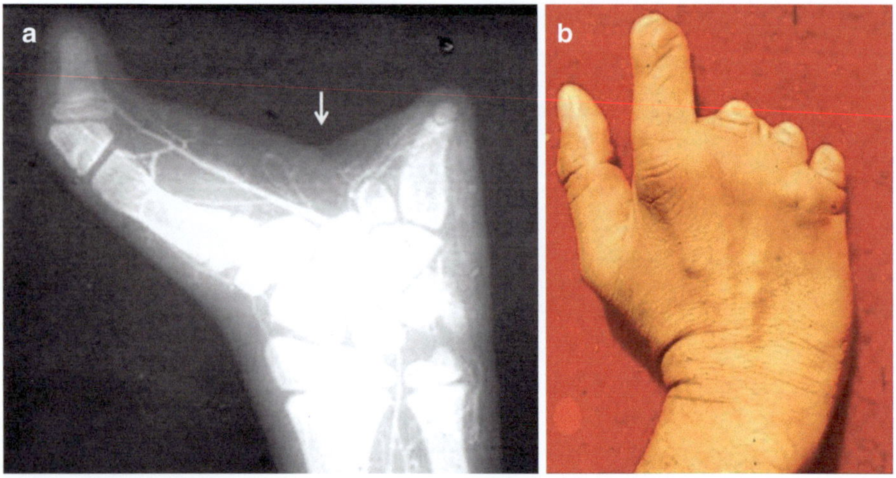

Fig. 18.31 (**a**) Angiogram showing congenital multiple errors in segmentation resulting in the present shape and size of the hand. Note the cleft (arrow). (**b**) Clinical picture revealing congenital absence of middle, ring and little fingers. However, the patient is able to carry out some function with the help of the thumb and the index finger

18.4 Manus Fissa

Manus Fissa (Lobster claw hand) is a rare congenital anomaly. This is a typical central defect also known as symbrachydactyly (Fig. 18.31a, b). In this cleft hand the central rays are absent and only ulnar and radial digits remain. An oppositional pinch between these two digits may not be possible, more so because the phalanges may be short or absent on one or both the sides. The existing digits may not be properly aligned.

18.5 Syndactylia

This disorder is characterized by marked webbing between adjacent fingers (Fig. 18.32a–e). A simple syndactyly indicates only joining of the skin whereas a complex syndactyly may indicate defect in the bony elements also (Figs. 18.33 and 18.34a, b).

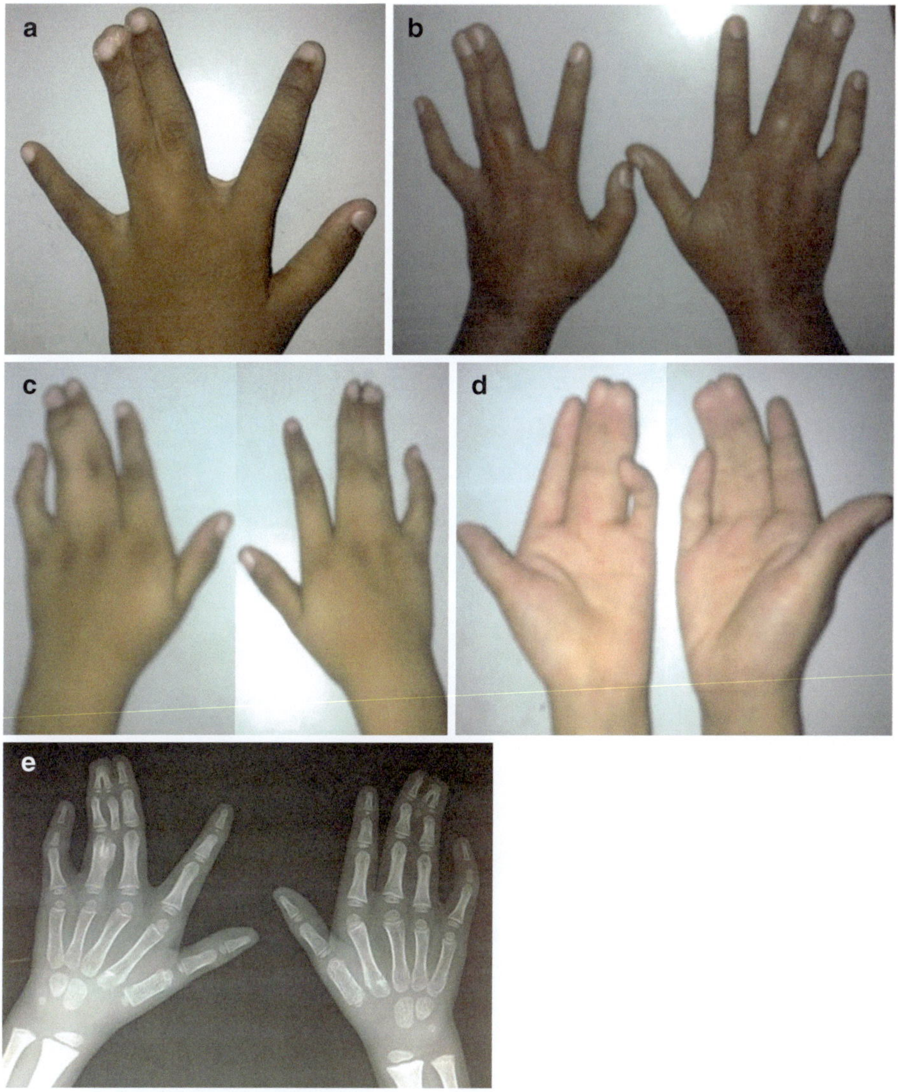

Fig. 18.32 (**a**) Clinical photograph showing syndactyly of the middle and ring fingers. Note the bilateral symmetrical syndactyly of the ring and middle fingers in a different case. (**c**, **d**) Clinical photograph showing complex syndactyly of the middle and ring fingers. (**e**) Roentgenogram of the same patient revealing a rare anomaly in the form of bifurcation of the proximal phalanx of the left ring finger and distal phalanges with fusion at the distal ends. The distal phalanx of the ring finger on the right side also reveals duplication. Note bilateral syndactyly of the middle and ring fingers

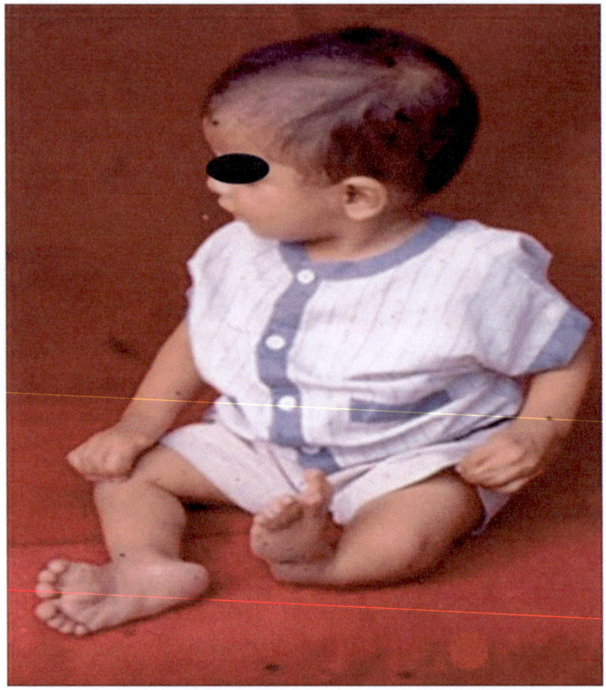

Fig. 18.33 A 2-year-old boy with polydactyly (Duplication of the toes on the right side). He also has bilateral syndactyly of the hands. Note the varus deformity of the left leg, a case of tibial hemimelia

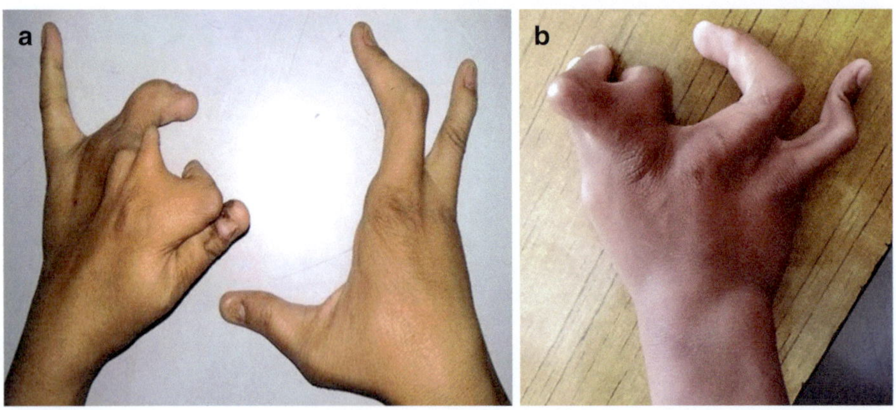

Fig. 18.34 (**a**) Clinical presentation of the left hand with complex syndactyly and deformity. Note the oligodactyly on the right side. (**b**) Clinical presentation of the right hand, almost similar to the left hand shown in (**a**)

18.6 Polydactylia Syndactilia

Polydactyly most commonly refers to the presence of more digits than normal. Polydactyly may be associated with syndactyly (Fig. 18.35a, b). The anomaly can occur by itself, or more commonly, as one feature of a syndrome of congenital anomalies (Fig. 18.36a–d). Polydactyly is one of the most common variations on the basic growth pattern of the hand.

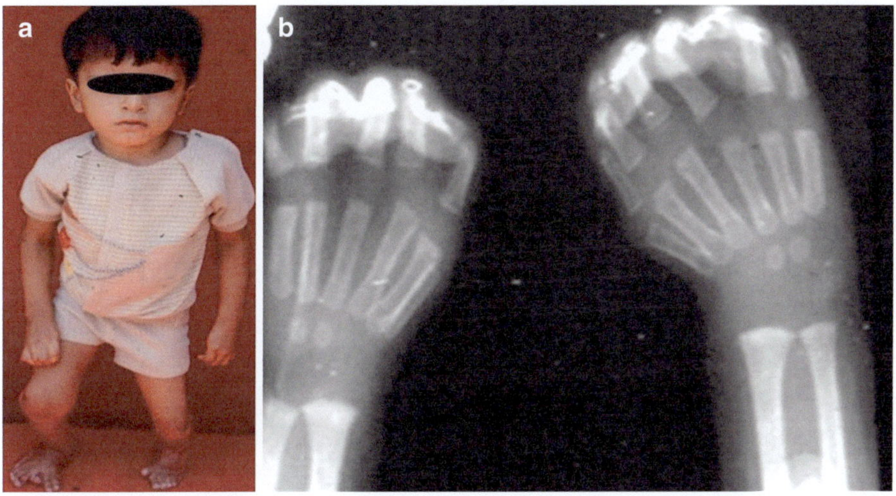

Fig. 18.35 (**a**) A 4-year-old boy with polydactyly and syndactyly of the hands and feet. There is bilateral aplasia of the thumb. (**b**) Radiograph revealing bilateral syndactyly with polydactyly. Note six metacarpals and fingers on the right side

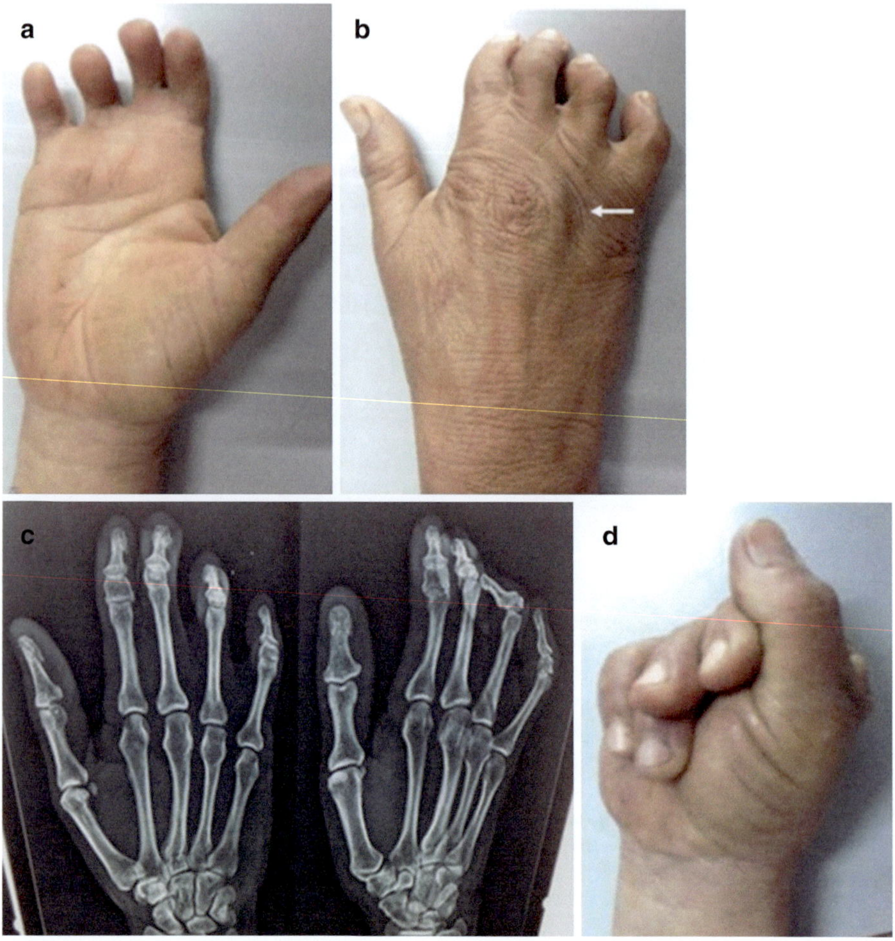

Fig. 18.36 (**a**, **b**) Clinical picture of the right hand showing small fingers due to distal extension of the palmer skin fold resulting in a long palm. However, the MP joints are in normal position (arrow). (**c**, **d**) Roentgenogram of the same hand shows the usual position of the MP joints. Note the long proximal phalanges and smaller middle and distal phalanges-a rare anomaly. (**d**) The grip/the fist making are near normal and the hand is functional

18.7 A Rare Anomaly of the Hand

18.7.1 Oligodactyly

Oligodactyly is characterized by less number of fingers and the lateral digits are relatively more involved (Fig. 18.37a, b).

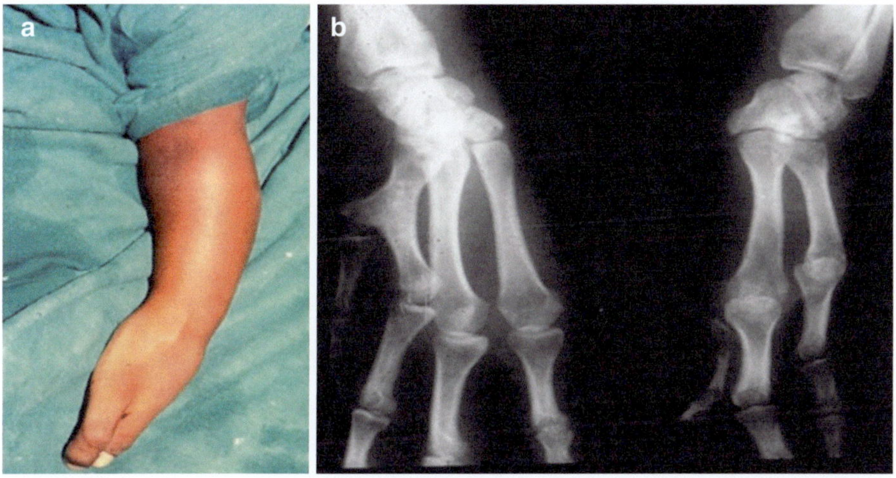

Fig. 18.37 (**a**) Congenital shortening of the forearm and absence of the three lateral rays. (**b**) Bilateral congenital anomalies of the hands. Note the absence of the thumb and the fifth ray arising from the shaft of the fourth metacarpal. There are gross anomalies of the carpal bones

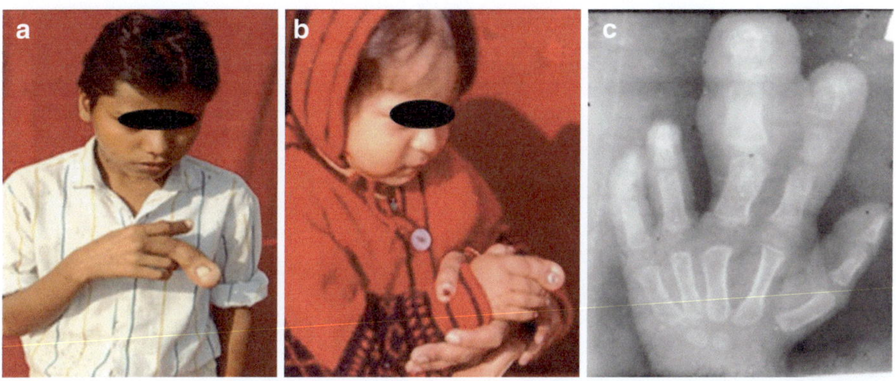

Fig. 18.38 (**a**) Macrodactyly of the right middle finger. (**b**) Another case showing the involvement of the index finger. (**c**) Roentgenogram of the hand with the involvement of the middle finger and syndactyly with the index finger

18.7.2 Macrodactylia

Macrodactylia (megalodactyly, macrodystrophia lipomatosa, dactylomegaly, local gigantism) can affect the figures. All structures are enlarged and there is no underlying pathologic condition (Fig. 18.38a–c). The soft tissue on the palmer side are diffusely enlarged and are predominantly of fibro-fat density, usually multiple adjacent digits are affected. However, a single digit may also be involved. Roentgenogram shows bony enlargement often with certain amount of deviation in the hand (Fig. 18.39a–d). The cause of this rare anomaly is not known but it may be related to fibro lipomatous tumours (Fig. 18.40a, b). Similar anomaly may also occur in the foot.

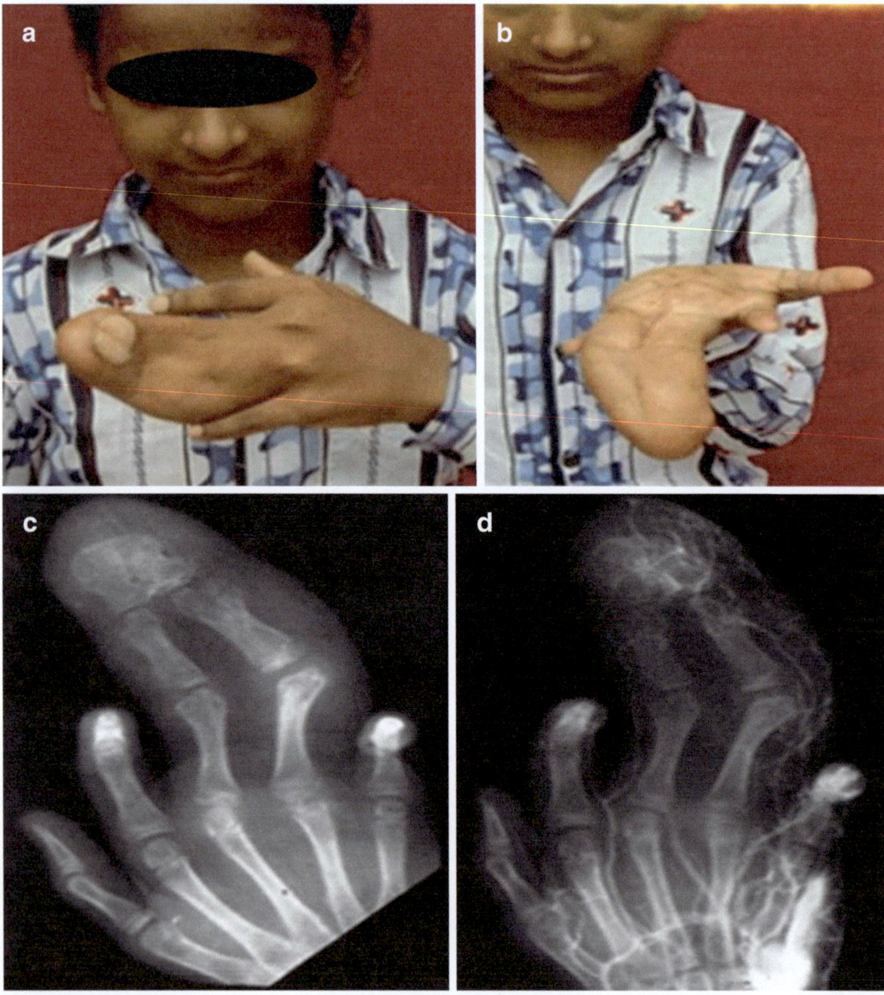

Fig. 18.39 (**a**) Macrodactylia and syndactyly of the left middle and ring fingers with fused nails. There is congenital hypertrophy of the soft tissues and bones involved of the fingers. (**b**) Note the palmer view. (**c**) Roentgenogram of the same patient showing distal inter phalangeal fusion (Acrocephalosyndactyly). (**d**) Angiogram revealing absence of digital artery on the radial side of the middle finger

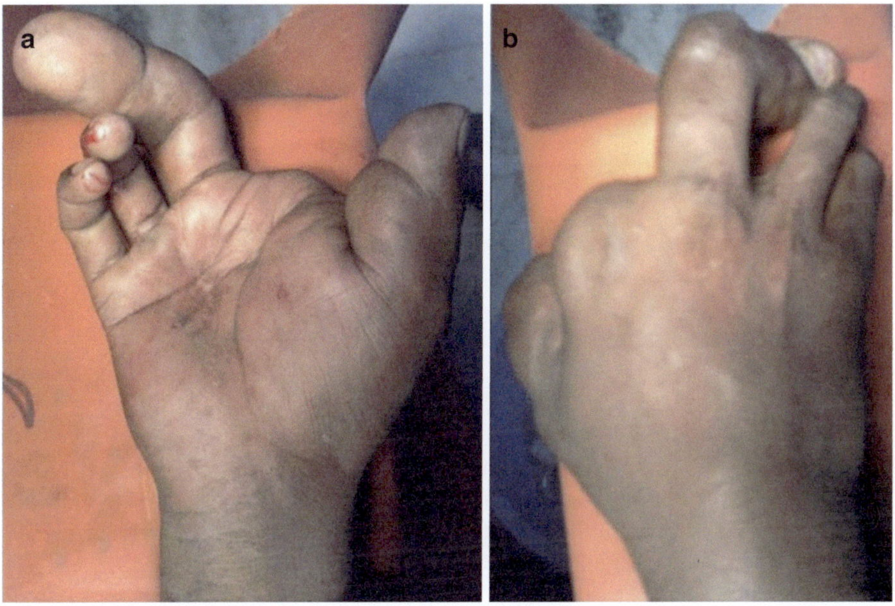

Fig. 18.40 (**a**, **b**) Clinical presentation of macrodactyly of the middle finger. The index finger was removed earlier with the same presentation

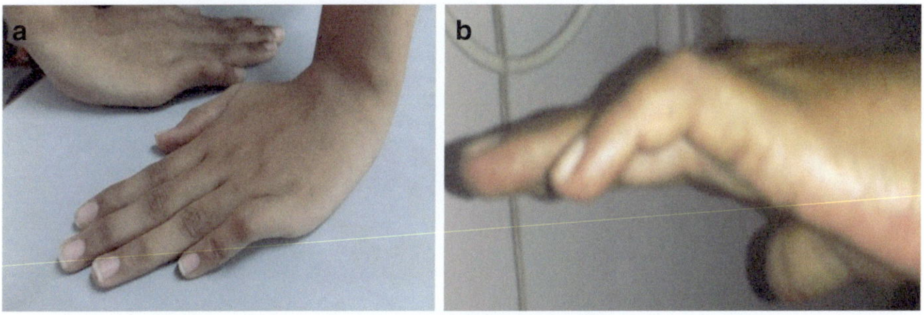

Fig. 18.41 (**a**) Clinical photograph revealing bilateral camptodactyly of the little finger. (**b**) The clinical photograph reveals camptodactyly with hyperextension at the fifth M.P. joint and flexion at both the P.I.P. joints

18.7.3 Camptodactyly

Camptodactyly means bent fingers. There is congenital developmental flexion contracture of the proximal interphalangeal joint of the finger, unrelated to any other disease (Fig. 18.41a, b). It may be unilateral or bilateral. The deformity is usually observed accidentally.

18.7.4 Clinodactyly

Clinodactyly is characterized by angular deformity of the involved finger in the radio-ulnar plane. Usually the anomaly is due to the presence of a Delta phalanx (Fig. 18.42). It may occur in any finger and is commonly bilateral. In most of the cases the problem is more cosmetic than functional. The function is seldom impaired except when the deformity is severe. When required surgical management may be carried out with certain amount of risk.

18.7.5 Thumb Anomalies

18.7.5.1 Pollex Duplex

Thumb duplication usually occurs at the metacarpo—phalangeal joint level. Four such types are known (1) Hypoplastic, (2) Ulnar deviated, (3) Convergent, (4) Divergent (Fig. 18.43a–d). Most reported cases of thumb polydactyly are sporadic. There are, however some familial cases of inheritance (Fig. 18.44). It can be associated with other congenital anomalies like syndactyly.

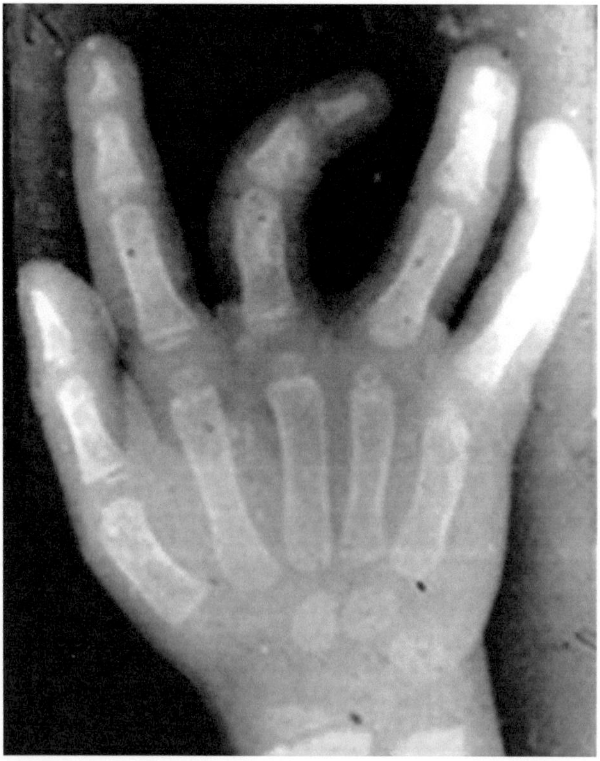

Fig. 18.42 Roentgenogram revealing clinodactyly (side bending) of the middle finger. Note the delta middle phalanx of the finger

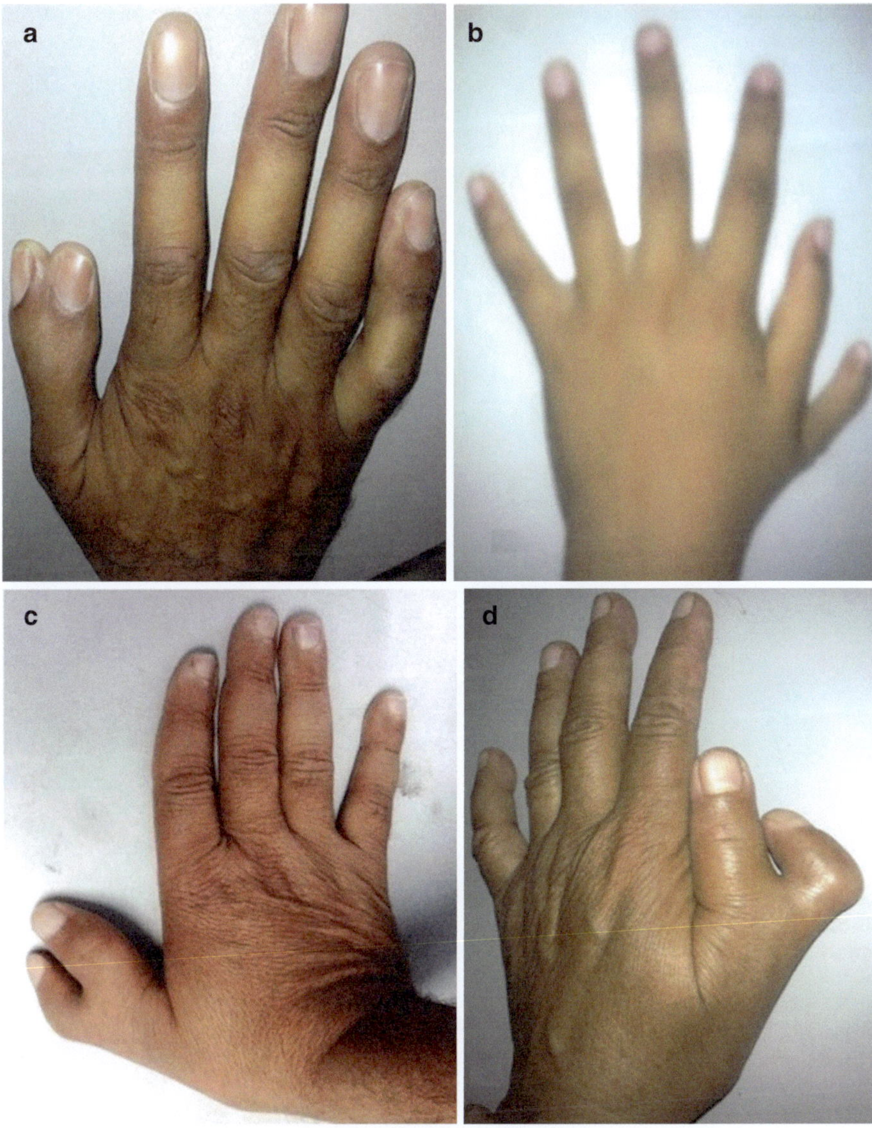

Fig. 18.43 (**a**) Divergent type of pollex duplex. (**b**) Divergent type of pollex duplex. (**c**, **d**) Hypoplastic thumb duplication

Polydactyly of the thumb covers a wide spectrum from extreme hypoplasia to complete duplication throughout the metacarpal level. Pollex duplex has also been described as a 'split' thumb (Fig. 18.45). The two thumbs that exist are never of the size of normal apposite thumb. The relations of the accessory thumbs to each other and the rest of the hand is variable.

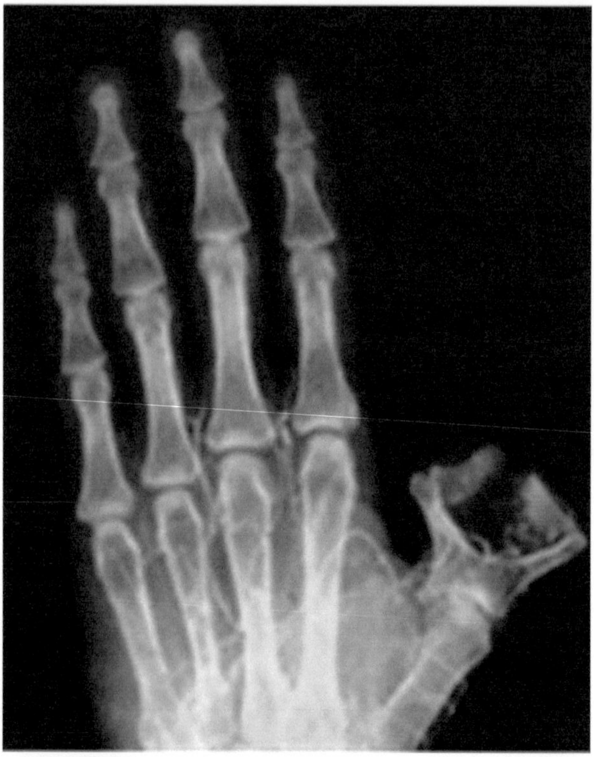

Fig. 18.44 Angiogram reveals convergent type of pollex duplex. Note the bifurcation of the proximal phalanx of the thumb with distal phalanges

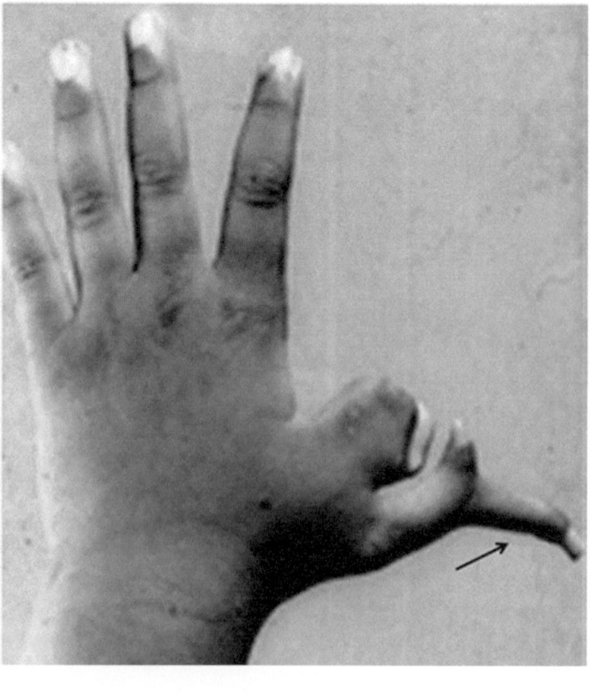

Fig. 18.45 Clinical picture of the left hand reveals bifurcation of the thumb with further bifurcation of the lateral half which we have described as a 'Crown' (arrow). Note the nail at the tip

The management of polydactyly has been designed to improve the functional use and cosmetic appearance of the thumb with the desired surgical procedure. It is commonly advocated that when the accessory thumb is a skin tag it can be tied and removed. The surgical treatment of duplicate thumb is based on the level of duplication and should remove the least functional part with reconstruction of the remaining component.

18.7.6 Pollex Duplex with a Crown

To our knowledge this anomaly has not been described earlier. The additional top is named as a 'crown'.

18.7.7 Congenital Hypoplasia of the Thumb (Floating Thumb)

This anomaly has been common and is classified in five categories—(1) Reduction in the size, (2) Hypoplasia with muscle and tendon instability (Wobbliness), (3) Small skeleton of the thumb with stiffness of the joint with an abnormal first web space, (4) A floating thumb attached by only a thread, (5) Complete absence of the thumb (Fig. 18.46a–c).

18.7.8 Congenital Absence of the Thumb

Absence of thumb is most likely to occur when there is complete absence of radius (Fig. 18.47a, b). With the total absence of the radius (Type IV), the hand is completely unsupported and marked radial displacement occurs. The incidence of thumb

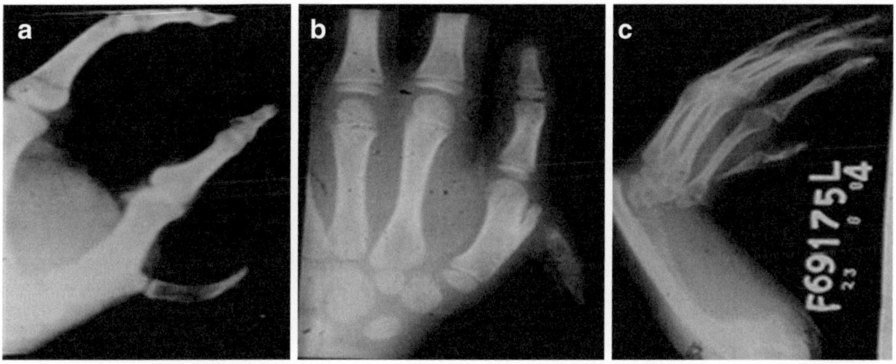

Fig. 18.46 (**a**) Roentgenogram reveals a rudimentary thumb which can be excised. (**b**) Roentgenogram reveals a tag of skin with no bones. Note the bifurcation of the distal end of the first metacarpal. (**c**) Roentgenogram reveals floating thumb which can be removed. No bony connection can be seen. Note complete absence of the radius (Type IV)

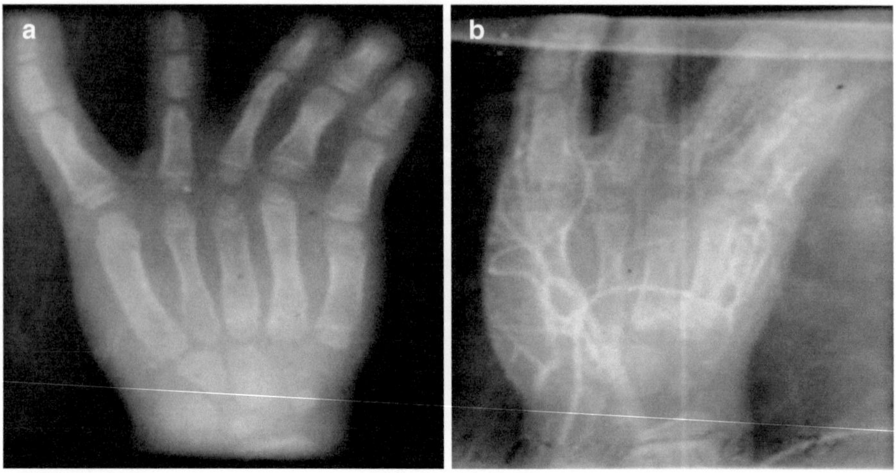

Fig. 18.47 (**a**) Roentgenogram of the hand revealing the thumb taking the shape of a finger, radial hand. Note the hypoplasia of the index and middle finger and biphangeal ring and little fingers. (**b**) Angiogram revealing the vasculature of the hand

absence has been reported to range from 17% to 86%. In the absence of thumb there may be associated deficiencies of the scaphoid and trapezium.

18.7.9 Pollicisation of the Index Finger for Loss of Thumb

Pollicisation of the index finger has proved to be a very useful procedure in absence of thumb (Fig. 18.48a–g). The index finger is surgically displaced radially and rotated to perform opponance and in a way act as a thumb.

18.7.10 Tri-Phalangeal Thumb

A tri-phalangeal thumb (TPT) is a congenital malformation where the thumb has three phalanges. The extra phalanx can vary from a very small phalanx to a triangular bone (delta phalanx) between the two phalanges of the thumb (Fig. 18.49a, b). Besides the three phalanges, there can also be other malformations. In the case of a five fingered hand it has a finger-like appearance, with the position in the plane of the four fingers, thenar muscle deficiency and additional length.

Triphalangeal thumb can be divided in three types:

1. Thumbs with a delta or abnormally shaped phalanx.
2. Thumbs with three normal phalanges in excessive length.
3. Those hands with five normal fingers and no apparent thumb.

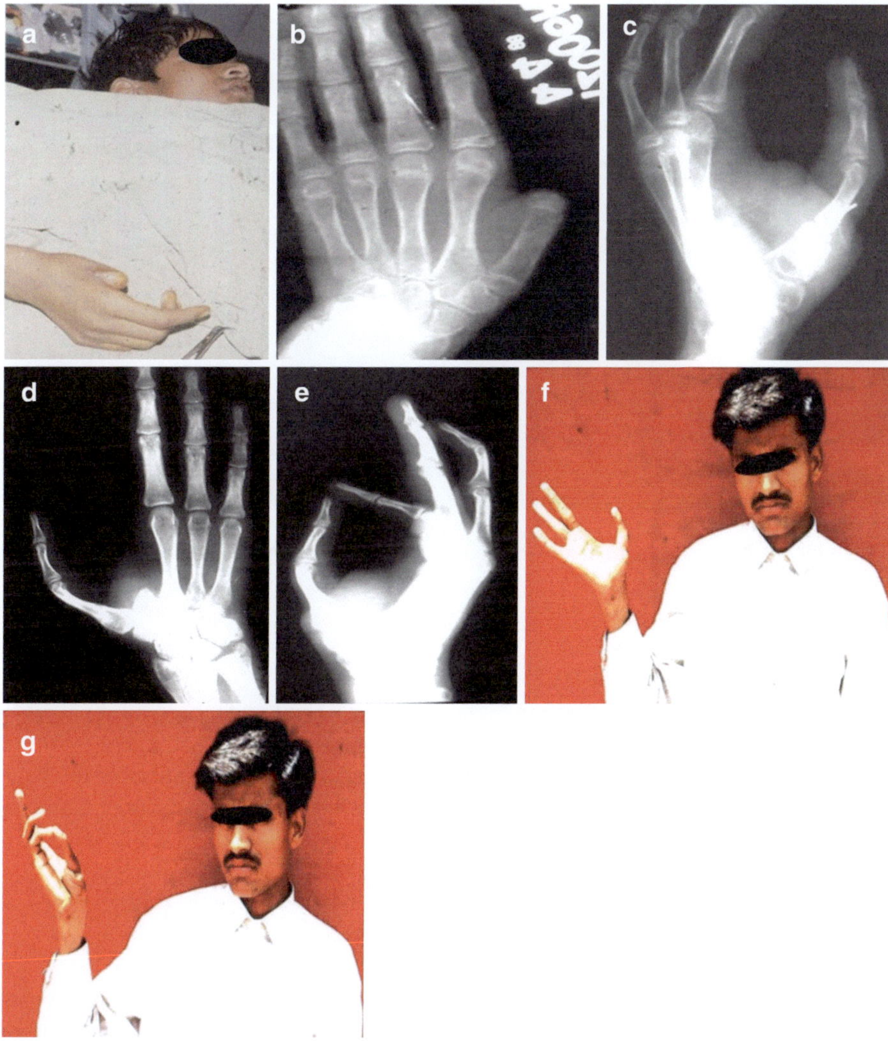

Fig. 18.48 (**a**) Clinical picture of the right hand of a 15-year-old boy showing deficient thumb. (**b**) Roentgenogram revealing deficiency distal to the first metacarpal. (**c**) Roentgenogram showing fixation of index finger with the first metacarpal by cross Kirschner wires. (**d**, **e**) Roentgenogram revealing final function after pollicisation showing abduction and oppnance of the pollicised finger. (**f**, **g**) Patient showing the pollicised finger in abduction and opponance with the little finger after a follow-up of 5 years

Many anomalies could be associated with triphalangeal thumb like polydactyly and congenital tibial defects.

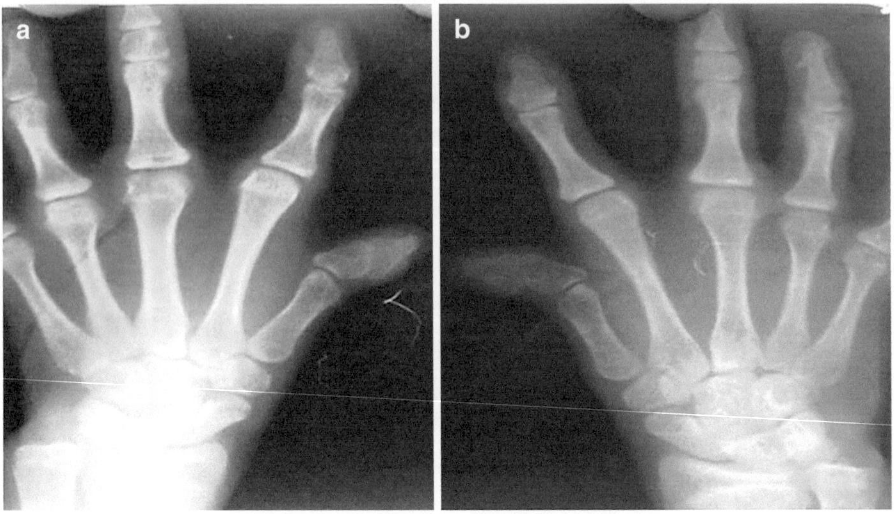

Fig. 18.49 (**a**, **b**) Roentgenogram revealing bilateral triphalangeal thumb and bi-phalangeal index and the ring fingers

18.7.11 Pollex Flexus

Flexion anomaly of the thumb in the interphalangeal joint and a hard knot palpable along the long flexor tendon of the thumb over the metacarpophalangeal joint are the diagnostic steps (Fig. 18.50). Constriction of the synovial sheath over the base of the thumb is the key factor of such a presentation. The anomaly may occur in fingers also. The anomaly may be unilateral or bilateral (Fig. 18.51a, b).

18.7.12 Short Metacarpal

In this rare condition one or more metacarpals are short resulting in short appearance of the corresponding finger (Fig. 18.52a, b). It may be associated with other congenital mal developments. It may be unilateral or bilateral (Fig. 18.53a, b). There may be associated shortening of the metatarsals.

18.7.13 Dupitrins Contracture

Dupitrins contracture (Palmar fibromatosis) results due to gradual idiopathic contracture of the middle and the ring fingers (Fig. 18.54). It is commonly seen in adults or the elderly which may be bilateral in 50% of cases. It may be because of subcutaneous contracture that the patient is unable to extend the fingers fully. A complete need-based surgical release/excision of the contracted tissue may result in relief.

Fig. 18.50 Photograph shows flexion attitude of the interphalangeal joint of the thumb. In a case of pollex flexus

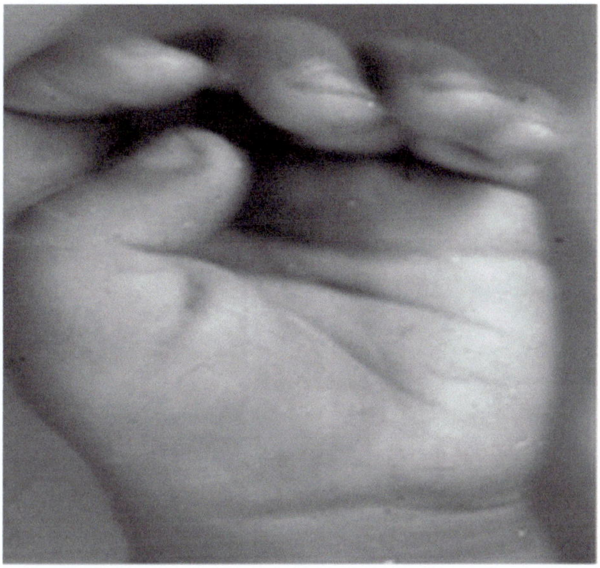

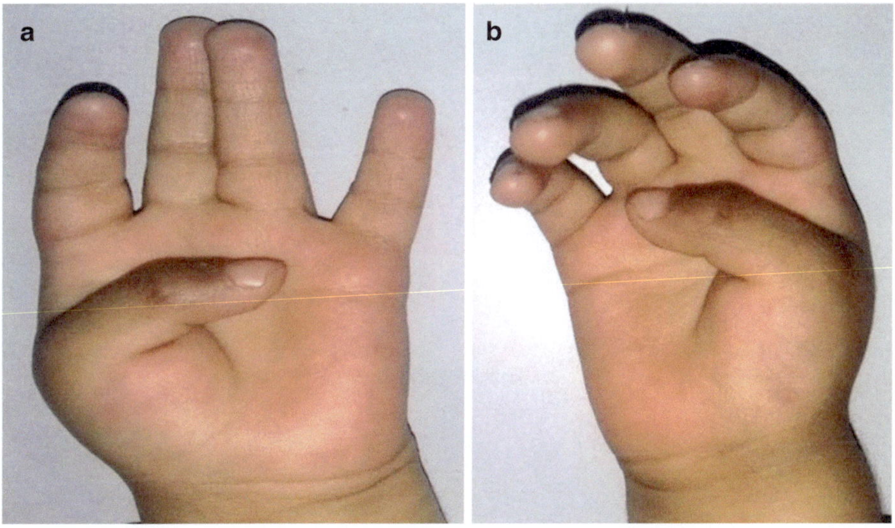

Fig. 18.51 (**a**, **b**) A 2-year-old boy showing bilateral pollex flexus deformity

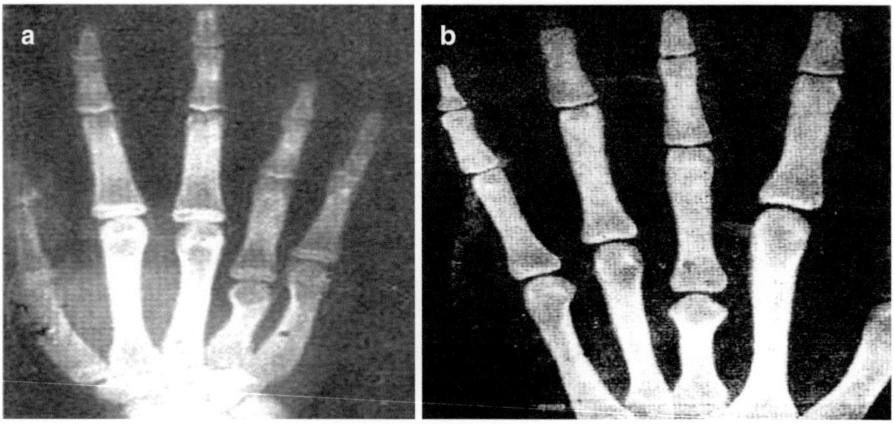

Fig. 18.52 (**a**) Roentgenogram shows congenital shortening of the fourth metacarpal. Note the unusually long proximal phalanges. (**b**) Roentgenogram revealing severe congenital shortening of the third metacarpal with a longer proximal phalanx. Note the bi-phalangeal index finger on the right side

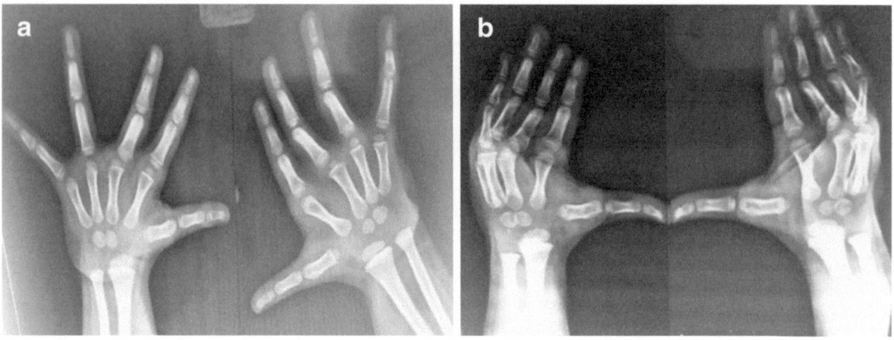

Fig. 18.53 (**a**, **b**) Roentgenograms reveal short metacarpals and long phalanges. Note the luxations at the metacarpophalangeal and interphangeal joints

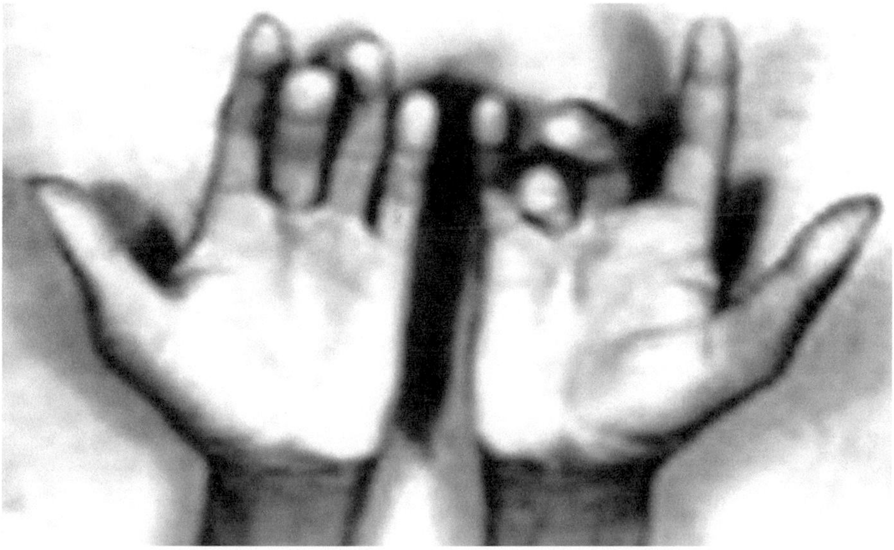

Fig. 18.54 Patient showing bilateral Dupitrins contracture of the hands involving the middle finger on the left side and middle and the ring fingers on the right side

18.8 Phalangeal Variations

Acrosyndactyly is an adjacent fusion of the digits at their distal ends (Fig. 18.55a). There is usually proximal fenestration between the digits distal to the normal web space. Brachysyndactyly refers to shortened fingers that are joined on adjacent sides.

An additional category called complicated complex syndactyly has features that not only include syndactyly but also clinodactyly, camptodactyly, symphalangism, or associated syndromes, such as Poland, Apert and amniotic band syndrome. In this category there is more than the simple sharing of a soft tissue envelope with bony connection at the distal portion (Fig. 18.55b–e).

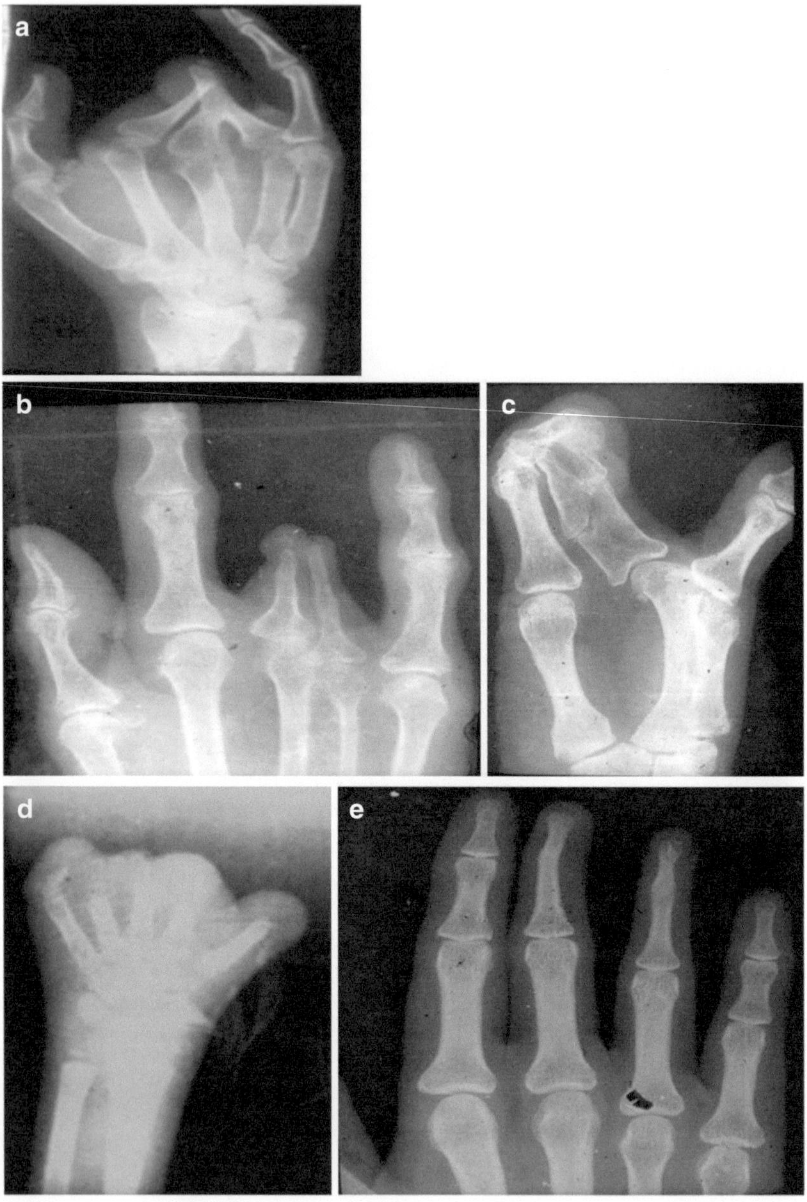

Fig. 18.55 (**a**) Roentgenogram showing Acrocephalosyndactyly, a variation of acro syndactyly manifested by a severe form of complex syndactyly. Note the fusion of distal phalanges of the second, third and fourth rays (Apert hand)—a rare presentation. (**b**) Roentgenogram reveals irregular segmentation and hypoplasia of the third and fourth metacarpals and phalanges—symphalangism. (**c**) Roentgenogram reveals congenital errors in the segmentations with multiple defects in size and shape of the metacarpals and phalanges in a case of oligodactyly. (**d**) Roentgenogram shows congenital absence of fingers and the thumb on the left side (aphalangia). (**e**) Roentgenogram reveals congenital symphalangism of the middle and distal phalanges of the middle and ring fingers

Suggested Readings

Dinham JM, Meggit DF. Trigger thumbs in children. J Bone Joint Surg Br. 1974;56:153.
Hudson HW. Snapping thumb in childhood reported eight cases. N Engl J Med. 1935;210:854.
Sprecher EE. Trigger fingers in infants. J Bone Joint Surg Am. 1949;672
Hansen OH, Anderson NO. Congenital radial ulnar synostosis: report of 37 cases. Acta Orthop Scand. 1970;41:225.
Miura T, Nakamura R, Suzuki M, Kanie J. Congenital radial ulnar synostosis. J Hand Surg Br. 1984;9:153.
Simmons BP, Southmayd WW, Riseborough EJ. Congenital radial ulnar synostosis. J Hand Surg. 1983;8:829.
Almquist EE, Gordon LH, Blue AI. Congenital dislocation of the head of the radius. J Bone Joint Surg Am. 1966;51:118.
Hirayama T, Takeimitsu Y, Yagihara K, Mikita A. Operation for chronic dislocation of the radial head in children: reduction by osteotomy of the ulna. J Bone Joint Surg Br. 1987;9:639.
Kelikian H. Congenital deformities of the hand and forearm. Philadelphia: WB Saunders; 1974. p. 902.
Mardam-Bey T, Ger G. Congenital radial head dislocation. J Hand Surg. 1979;4:316.
Neilsen JB. Madelung's deformity: a follow-up study of 26 cases and a review of the literature. Acta Orthop Scand. 1977;48:379.
Beals RK, Lovrien EW. Dyschondrosteosis and Madelung's deformity. Report three kindreds and review of the literature. Clin Orthop. 1976;116:24.
Golding JSR, Blackburne JS. Madelung's disease of the wrist and dyschondrosteosis. J Bone Joint Surg Br. 1976;58:350.
Blair WF, Shurr DG, Buckwalter JA. Functional status in the ulnar deficiency. J Pediatr Orthop. 1983;3:37.
Broudy AS, Smith RJ. Deformities of the hand and wrist with ulnar deficiency. J Hand Surg. 1979;4:304.
Carroll RE, Bowers WH. Congenital deficiency of the ulna. J Hand Surg. 1977;2:169.
Laurin CA, Farmer AW. Congenital absence of the ulna. Can J Surg. 1959;2:204.
Riordan DC. Congenital absence of the ulna. In: Lovell WW, Winter RB, editors. Pediatric orthopaedics. 2nd ed. JB Lippincott: Philadephia; 1978. p. 714.
Riordan DC. Congenital absence of the radius or ulna. Abstr J Bone Joint Surg Br. 1972;54:381.
Swanson AB, Tada K, Yonenobu K. Ulnar ray deficiency: its various manifestations. J Hand Surg Am. 1984;9:658.
Bora FW Jr, Nicholson JT, Cheema HM. Radial meromelia: the deformity and its treatment. J Bone Joint Surg Am. 1970;52:966.
Bora W Jr, Osterman AL, Kaneda RP, Esterhai J. Radial club hand deformity. J Bone Joint Surg Am. 1981;63:741.
Carrol RE, Louis DS. Anomalies associated with radial dysplasia. J Pediatr. 1974;84(3):409.
Carrol RE. Use of the fibula for reconstruction in congenital absence of the radius. Proceedings of the American Academy of Orthopaedic surgeons. J Bone Joint Surg Am. 1966;48:1012.
Lamb DW. Radial club hand, a continuing study of sixty-eight patients with one hundred and seventeen club hands. J Bone Joint Surg Am. 1977;59:7.
Sherik S, Flatt A. The anatomy of congenital radial dysplasia. Clin Orthop. 1969;66:125.
Urban MA, Osterman AL. Management of radial dysplasia. Hand Clin. 1990;6(4):589.
Watson HK, Beebe RD, Cruz NL. A centeralization procedure for radial club hand. J Hand Surg Am. 1984;9(4):541.
Urban MA, Lee Osterman A. Management of radial dysplasia. The pediatric upper extremity. Hand Clinics. 1990;6(4):589.
Conway H, Bowe J. Congenital deformities of the hands. Plast Reconstr Surg. 1986;18:2.
Maisel SDO. Lobster claw deformity of the hands and feet. Br J Plast Surg. 1970;23:269.

Phillips RS. Congenital split foot (lobster claw) and triphalangal thumbs. J Bone Joint Surg Br. 1971;53:247.

Barsky AJ. Cleft hand classification: incidence and treatment. J Bone Joint Surg Am. 1964;46:1707.

Buck-Gramcko D. Congenital malformations: syndactyly and related deformities. In: Nigst H, Buck-Gramcko D, Millesi H, Lister G, editors. Hand surgery. New York: Thieme; 1988. p. 12.

Ruby L, Goldberg MJ. Syndactyly and polydactyly. Orthop Clin North Am. 1976;7:361.

Wood VE. Treatment of central polydactyly. Clin Orthop. 1971;74:196.

Marks TW, Bayne LG. Polydactyly of the thumb: abnormal anatomy and treatment. J Hand Surg Am. 1978;3(2):107–16.

Barsky AJ. Marcrodactyly. J Bone Joint Surg Am. 1967;49:1255.

Edgerton MT, Turek DB. Macrodactyly (digital gigantism): its nature and treatment. In: Littler JW, Carmer LM, Smith JW, editors. Symposium on reconstructive hand surgery, vol. 9. CV Mosby: St. Louis; 1974. p. 157.

Jones KG. Megalodactylism: case report of child treated by epiphyseal resection. J Bone Joint Surg Am. 1963;45:1704.

Kalen V, Burwell DS, Omer GE. Macrodactyly of the hands and feet. Am J Hand Surg. 1988;8:311.

Ackland MK, Uthoff HK. Idiopathic localized gigantism: a 26 year follow-up. J Pediatr Orthop. 1988;8:618.

Ben-Bassat M, Casper J, Laron Z. Congenital macrodactyly. J Bone Joint Surg Br. 1966;48:359.

Dennyson WG, Bear JN, Bhoola KD. Macrodactyly in the foot. J Bone Joint Surg Br. 1977;59:355.

Herring JA, Tolo VT. Instructional case: macrodactyly. J Pediatr Orthop. 1984;4:503.

Khanna N, Gupta S, Hanna S, Tripathi F. Macrodactyly. Hand. 1975;7:212.

Tsuge K. Treatment of macrodactyly. J Hand Surg Am. 1985;10:968.

Tibbles SA, Rogers JG. Macrodactyly hemihypertrophy and connective tissue mevi: report of a new syndrome and review of the literature. J Pediatr. 1976;89:924.

Currariano G, Waldman I. Camptodactyly. AJR. 1964;92:1312.

Kourtemanche AD. Camptodactyly: etiology and management. Plast Reconstr Surg. 1969;44:451.

Oldfield MC. Camptodactyly flexion contracture of the fingers in young girls. Br J Plast Surg. 1956;8:312.

Goldberg MD, Bartoshesky LE. Congenital hand anomaly: etiology and associated malformations. Hand Clin. 1985;1(3):405–15.

Burke F, Flatt AE. Clinodactyly. A review of a series of cases. Hand. 1979;3:269.

Andrew JG, Sykes PA. Duplicate thumbs: a survey results of 20 patients. J Hand Surg Br. 1988;13:50.

Ezaki M. Radial polydactyly. Hand Clin. 1990;6(4):577.

Lister G. Pollex adductus in hypoplasia and duplication of the thumb. J Hand Surg Am. 1991;16:626.

Marks TW, Bayne LGP. Polydactyly of the thumb: abnormal anatomy and treatment. J Hand Surg. 1978;3:107.

Cotta N, Jaeger M. The familial triphalangia of the thumb and its operative treatment. Arch Orthop Unfallchir. 1965;58:282.

Poznanski AK, Garn SM, Holt JF. The thumb in the congenital malformation syndrome. Radiology. 1971;100:115.

Buck-Gramcko D. Pollicization of the index finger. J Bone Joint Surg Am. 1971;15:1605.

Barsky AJ. Congenital anomalies of the hand. J Bone Joint Surg Am. 1951;33:35.

Jaeger M, Refior HJ. The congenital triangular deformity of the tubular bones of the hand and foot. Clin Orthop. 1971;81:139.

Wood VE, Flatt AE. Congenital triangular bones in the hand. J Hand Surg. 1977;2:1979.

Goldberg MJ, et al. Congenital hand anomaly: etiology and associated malformations. Hand Clin. 1985;1(3):405.

Nutt JN, Flatt AE. Congenital central hand defect. J Hand Surg. 1981;6:48.

Lower Extremity Malformations 19

19.1 Proximal Femoral Focal Deficiency (PFFD)

The most frequent partial aplasia of the femur is the proximal femoral focal deficiency (PFFD), when the femur is shortened, flexed, abducted and externally rotated. Flexion contractures of the hip and knee are present. PFFD is one of several terms applied to a condition in which the femur is short, and there may be an apparent absence of continuity between neck and the shaft of the femur. In many instances, the radiographically delineated defect in the upper femur of the infant ossifies as the child matures. The overall frequency of PFFD is 1/100,000.

PFFD has been classified in four types-Aitken's classification for PFFD:

Type A, has a shortened femur is present proximally, ending at or slightly above the level of the acetabulum. The femoral head is often absent but later ossifies; the acetabulum is well-developed. After ossification, there is subtrochanteric varus deformity of the proximal femur (Fig. 19.1a–d). There may be pseudoarthrosis in the subtrochanteric area usually heals by skeletal maturity (Fig. 19.2).

Type B, has a more severe defect, the femoral segment is shorter than in type A. There is often a tuft at the proximal end of the femur, and the acetabulum is well formed. The femoral head may not be ossified at birth, but it appears as the child gets older (Fig. 19.3a–d). The upper portion of the femur usually lies proximal to the acetabulum, and at maturity, there is no ossific connection of the shaft to the head segment of the femur.

Type C, has an absent femoral head that does not ossify and has a markedly dysplastic acetabulum. The femoral shaft is shorter than type B. The entire proximal femur, including the trochanters, does not develop (Fig. 19.4a–c). There is a proximal tuft, the femoral head is absent, and the acetabulum is poorly formed with a flat lateral wall of the pelvis in the place of the acetabulum.

S. S. Yadav, *Skeletal Dysplasias & Congenital Malformations*,
https://doi.org/10.1007/978-981-99-5679-1_19

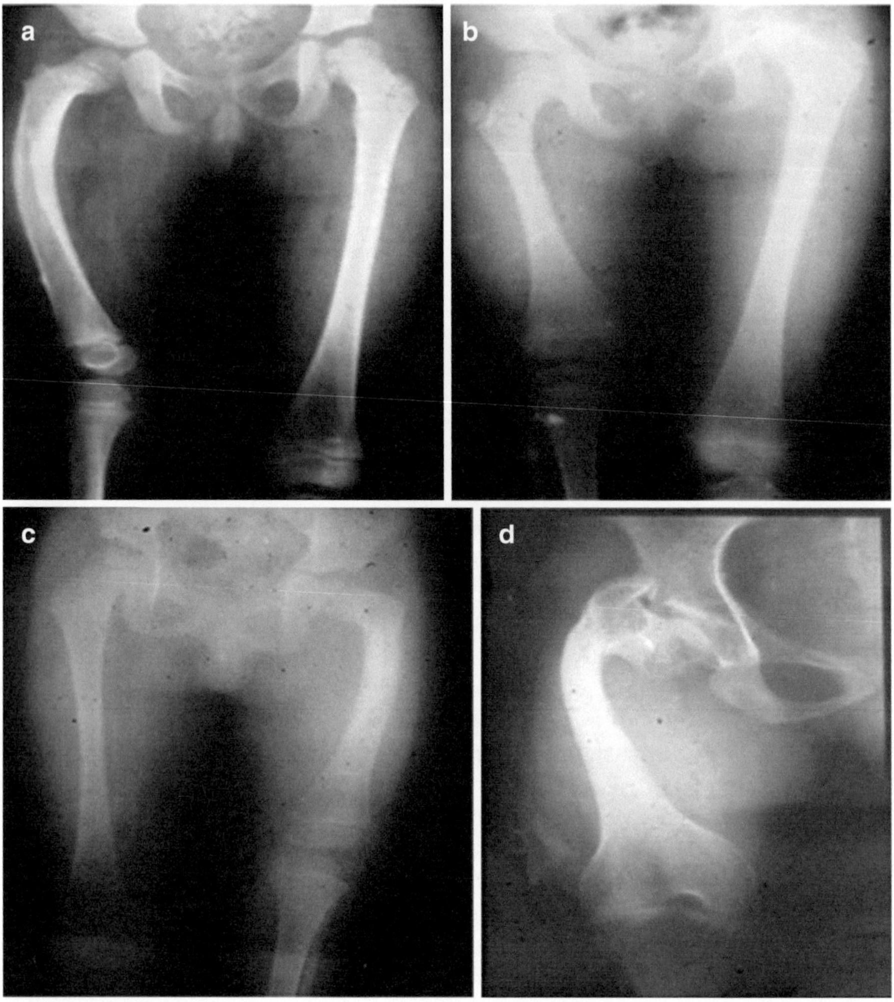

Fig. 19.1 (**a**) Roentgenogram reveals congenital short and curved femur on the right side with external rotation deformity during growth. Note the lateral picture of the knee with anteroposterior view of the hip. (**b**) Roentgenogram reveals congenital short femur. (**c**) Roentgenogram showing congenital short femur of the left side. (**d**) Roentgenogram reveals short, stout and curved right femur with severe coxa-vara. Note the proximal femur lies above the level of acetabulum

Type D, the femoral head is absent and the acetabulum poorly formed; in addition, the femoral shaft is extremely short or even absent (Fig. 19.5a–g).

The clinical appearance of a PFFD is distinctive. The thigh is very short; the hip is flexed, abducted and externally rotated; and there is usually a flexion contracture of the knee. The foot usually lies at the level of the contra-lateral knee.

The goal of treatment is equalization of the limb lengths (Fig. 19.6a–g). In some cases, a single femoral lengthening combined with contra-lateral epiphysiodesis can correct the leg length discrepancy, but in other cases, two femoral lengthening may be necessary. Lengthening procedures for congenital femoral shortening are arduous and prone to many complications, and they should not be undertaken lightly.

19.1.1 Type A

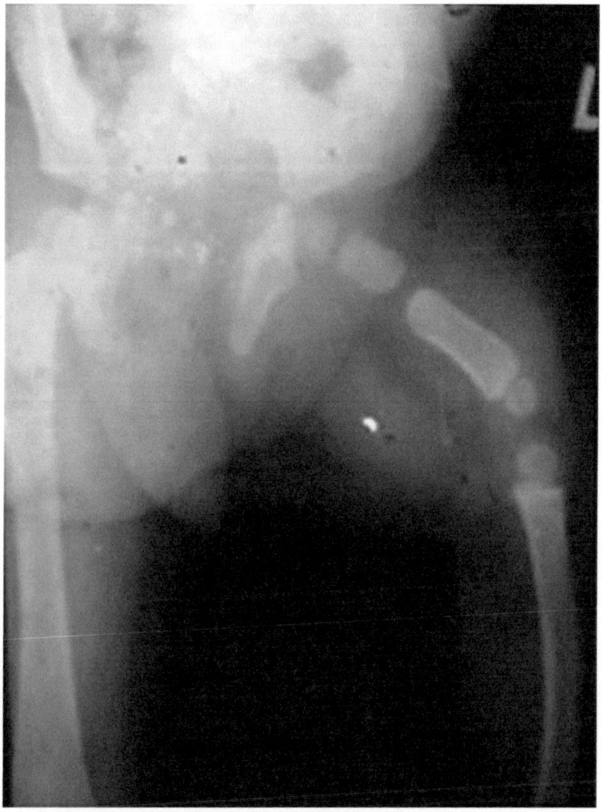

Fig. 19.2 Roentgenogram reveals shortening of the left femur with pseudoarthrosis of the proximal shaft of the femur which unites at maturity

19.1.2 Type B

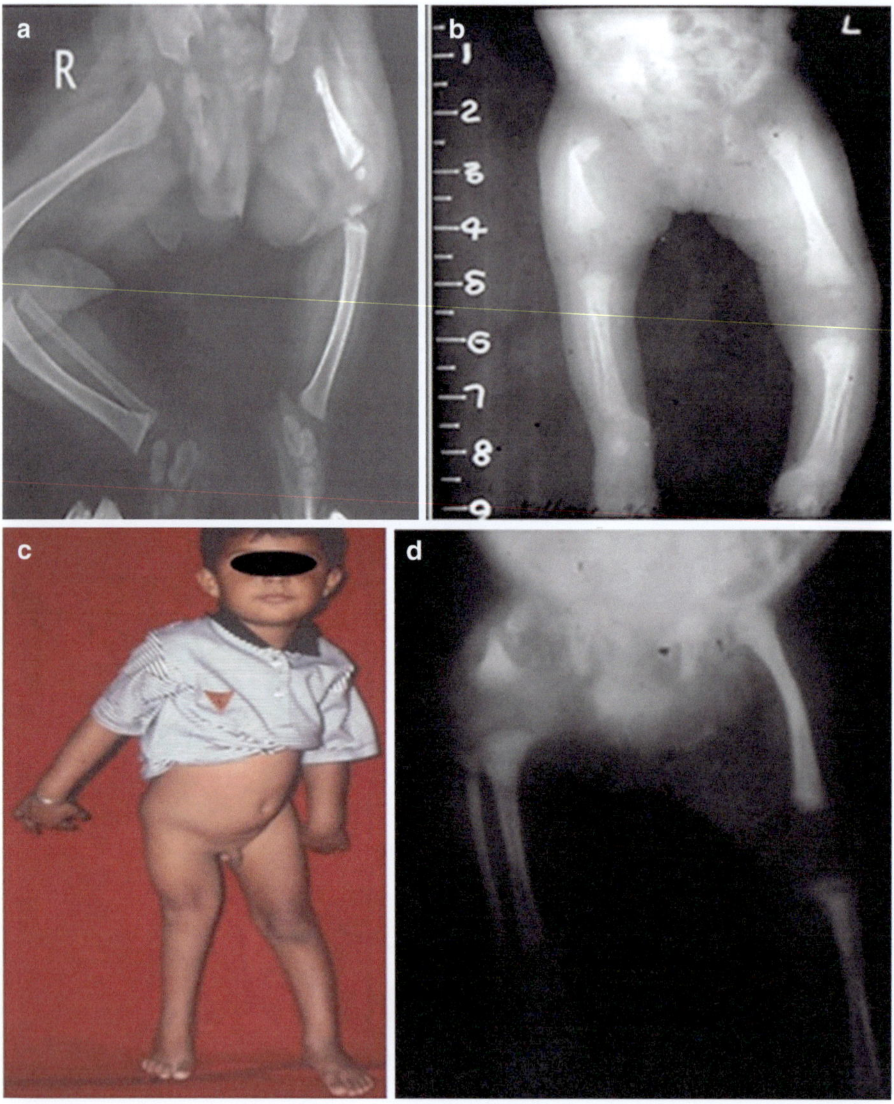

Fig. 19.3 (**a**) Roentgenogram shows PFFD (type-B) on the left side with an unstable hip. (**b**) Roentgenogram reveals PFFD on the right side. (**c**) PFFD on the right side with severe shortening of the thigh. (**d**) Roentgenogram shows PFFD (type-B) with pseudoarthrosis of the proximal femoral shaft with shortening. Note the bilateral dislocation of the hips

19.1.3 Type C

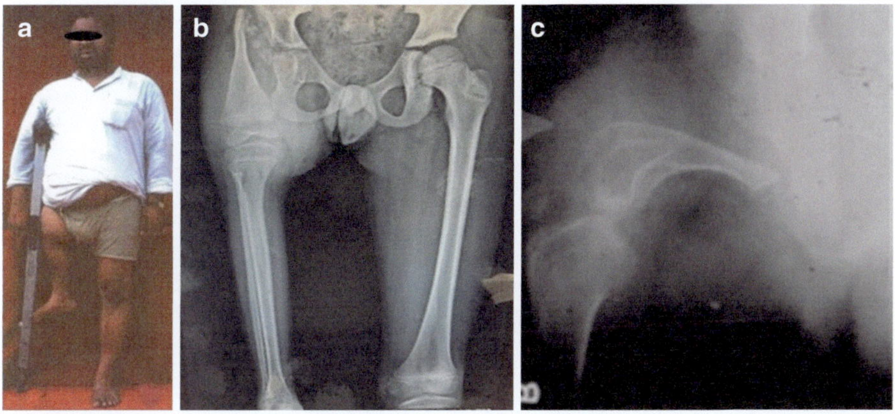

Fig. 19.4 (**a**) Clinical picture reveals severe shortening at the right thigh. (**b**) Roentgenogram shows PFFD (type-C) on the right side with the absence of proximal femur. Note the proximal migration of the remaining femur and a flat acetabulum. (**c**) Roentgenogram reveals PFFD (type-C) in an adult the acetabulam is not well formed

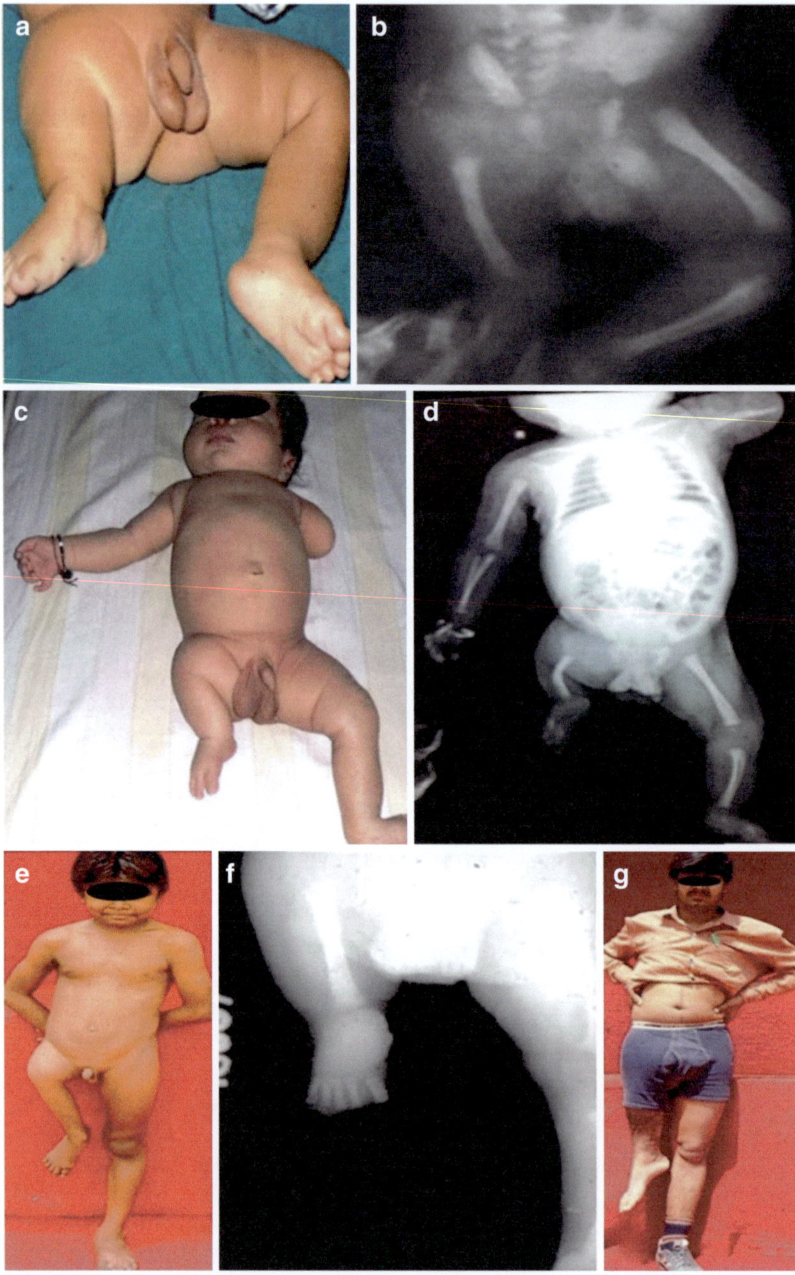

Fig. 19.5 (**a**) Congenital severe shortening of the right thigh. The thigh does not show any crease. (**b**) Roentgenogram reveals absence of the femur on the right side. (**c**) Clinical picture showing congenital shortening of the right lower limb and congenital amputation at the left arm. (**d**) Roentgenogram reveals PFFD (Type-D) on the right side. The proximal end of the tibia is facing the acetabulum. (**e**) An 8-year-old boy showing congenital severe shortening of the right thigh. The right leg is also wasted and short. The right ankle is at the level of left knee. (**f**) Roentgenogram reveals congenital absence of the right femur. (**g**) Clinical picture of a 25-year-old patient with PFFD on the right side (type-D). Note the external rotation of the extremity involved

19.1.4 Type D

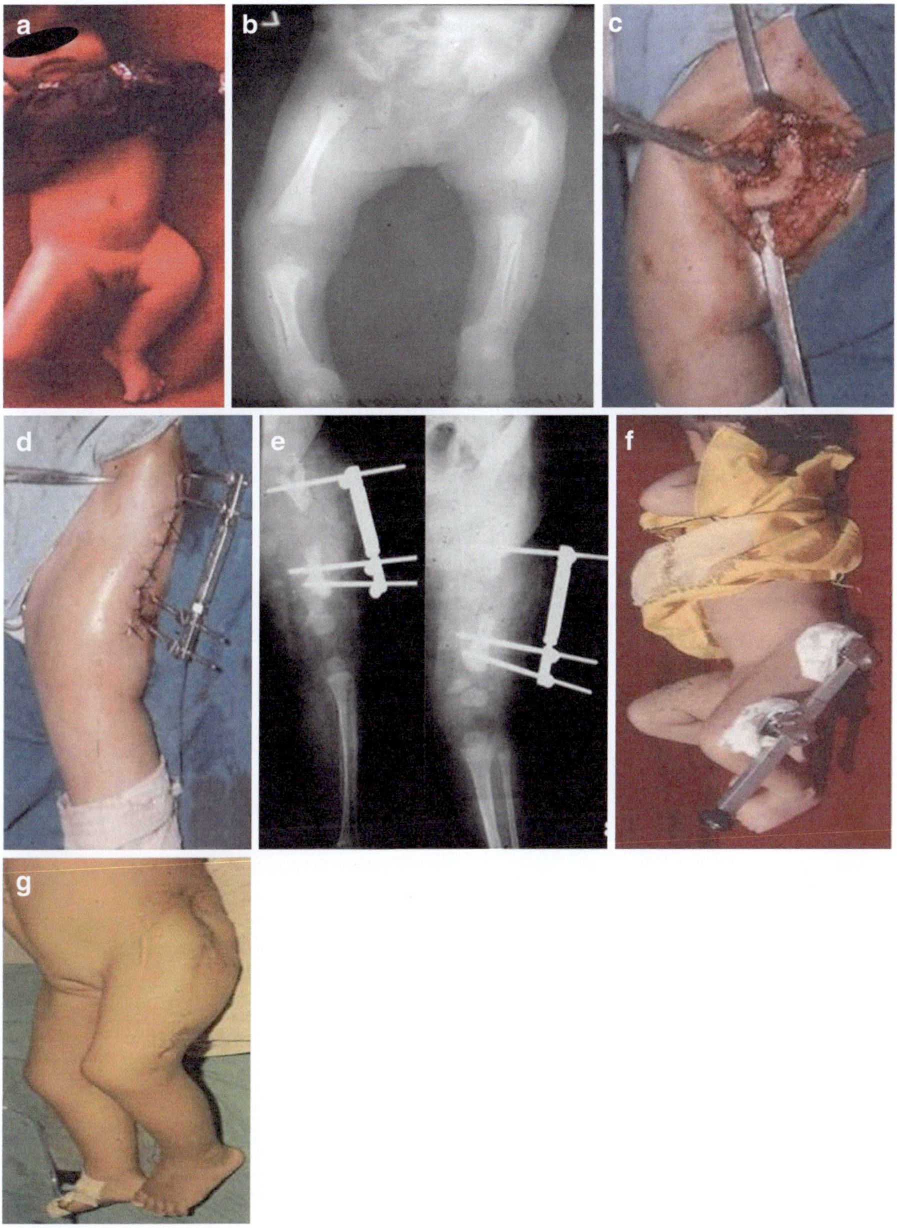

Fig. 19.6 (**a**) Clinical picture showing shortening of the left thigh. (**b**) Roentgenogram reveals (Type-B) PFFD. (**c**) Intra-operative picture showing short and curved left femur. (**d**) Picture showing application of a fixator for lengthening of the femur after osteotomy. (**e**) Roentgenogram revealing the gap after osteotomy during the process of lengthening. (**f**) Clinical picture of the left thigh. The fixator was changed. (**g**) Clinical picture of the left thigh after lengthening

19.2 Rotationplasty

Van Nes described a rotationplasty of the tibia in which the foot is rotated 180° and the toes face posteriorly to enable the ankle and foot to control a prosthetic knee. The motor control of the ankle is primarily through the gastrocsoleus muscle, which becomes a 'knee' extensor in the new configuration. In addition to motor control, the sensory feedback from the ankle allows better proprioceptive control of the knee. Gillespie reported good initial function after the Van Nes procedures, but the feet are gradually derotated as the children grew, with 10 of 21 cases requiring repeated rotationplasty procedures.

19.3 Congenital Dislocation of the Hip

Congenital Dislocation of the Hip (CDH) is due to spontaneous displacement of femoral head from its normal position. Usually bilateral, it may be unilateral where the skin crease is asymmetrical (Fig. 19.7a, b). At birth, the acetabulum and femoral head are normal in size and shape. The Ortolani test and Barlow's manoeuvre are useful in making the diagnosis. Its effects girls more than the boys (Figs. 19.8a–c and 19.9a, b). The abnormity is usually observed when the child starts walking. Thigh skin crease asymmetry and perceptible clicking of the hip may be noted in infancy. There is waddling gait in bilateral cases. In subluxated hip, there may be varying degree of upward and lateral displacement of the head from its normal position.

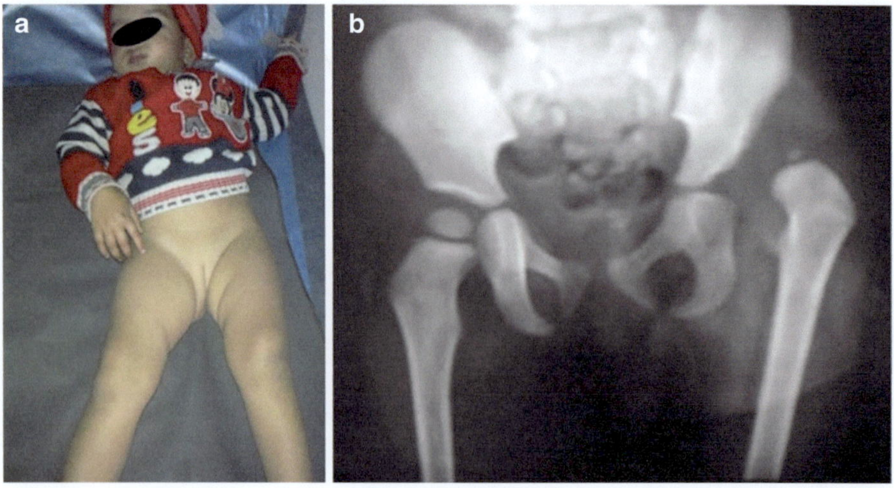

Fig. 19.7 (**a**) A child with CDH on the left side. Note the shortening of the thigh and the skin folds. (**b**) Roentgenogram showing left sided CDH. The femoral epiphysis is poorly developed on the left side

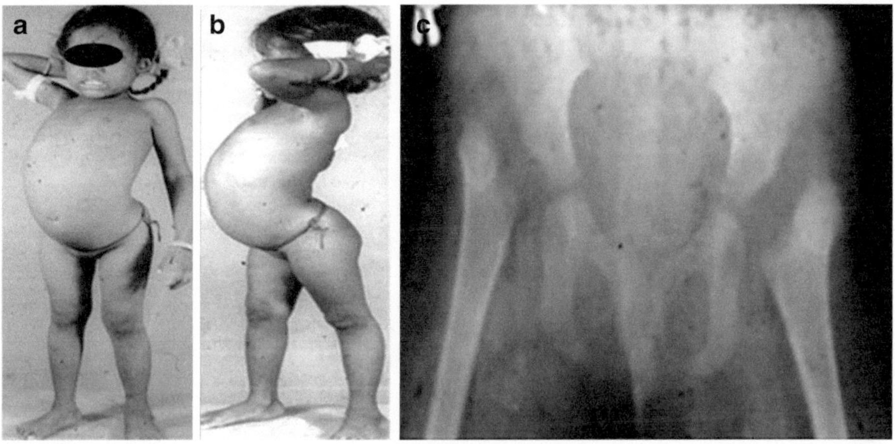

Fig. 19.8 (**a, b**) A 6-year-old girl revealing broad pelvis. Note the severe lumber lordosis with secondary protuberant abdomen. She had waddling gait. (**c**) Roentgenogram of the same patient reveals bilateral congenital dislocation of hip with external rotation

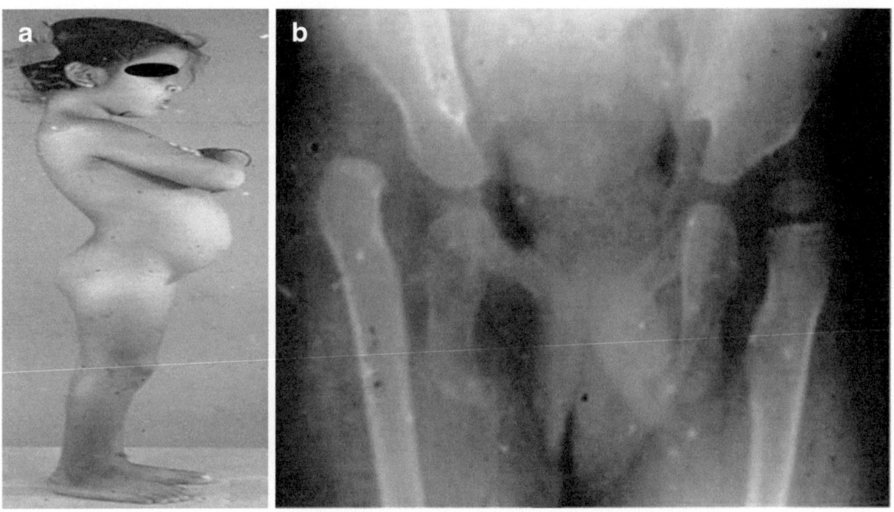

Fig. 19.9 (**a**) A 7-year-old girl with bilateral congenital dislocation of hip. Note the enhanced lumber lordosis with protuberant abdomen. She also had waddling gait. (**b**) Roentgenogram showing bilateral neglected CDH. Note the coxa-valga

Radiologically there is retarded development and delay in appearance of the ossific nucleus of the femoral head. The bony acetabular roof may have an excessive upward slope. Varying degrees of upward and lateral displacement of the head of the femur may be noted.

In mild cases a close reduction of the femoral head in the acetabulum stabilized with POP cast has been recommended (Figs. 19.10a–c and 19.11a, b). In severe

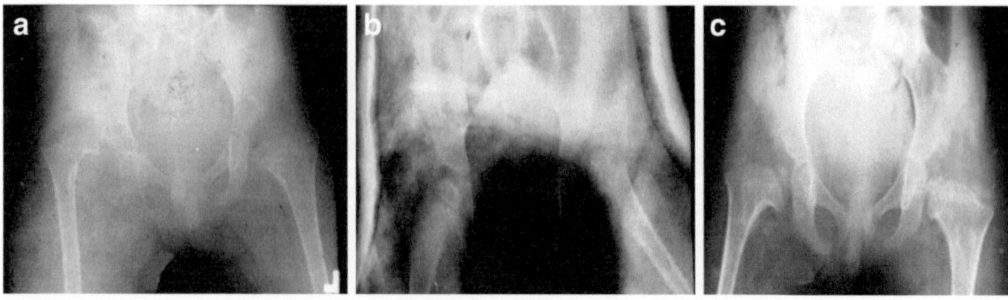

Fig. 19.10 (**a**) The roentgenogram reveals dislocation of the hip on right side (the left hip was managed earlier by close reduction. (**b**) Roentgenogram reveals close reduction of the right hip (in a POP spica cast). (**c**) Roentgenogram showing treated bilateral CDH with close reduction

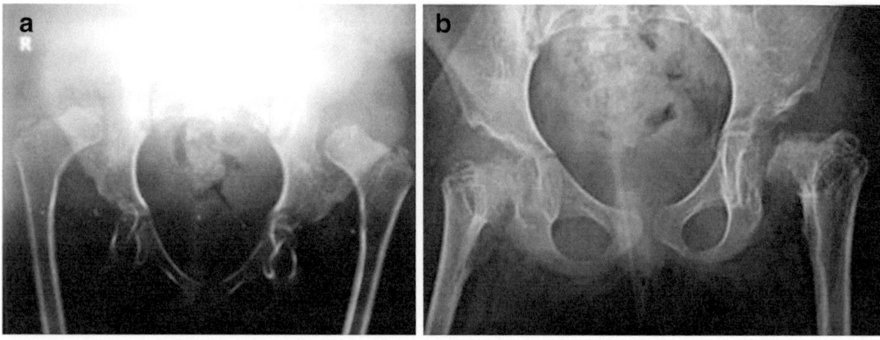

Fig. 19.11 (**a**) Roentgenogram reveals very high dislocated femoral heads on either side in a case of Ehlers-Danlos syndrome. (**b**) Roentgenogram of the same patient reveal reduced hips after open reduction. The high femoral heads could be reduced without any additional procedure

cases operative reduction of the head of the femur is necessary. To locate and retain the head in an acetabulam the femur goes in for internal rotation. The head is stabilized in the acetabulam with twin Kirschner wires (Fig. 19.12a). To bring the extremity in neutral position, a derotation osteotomy at the subtrochantric level is necessary (Fig. 19.12b). The extremity is encased in a POP spica cast. Neglected CDH may be observed in different abnormal positions (Fig. 19.13a, b).

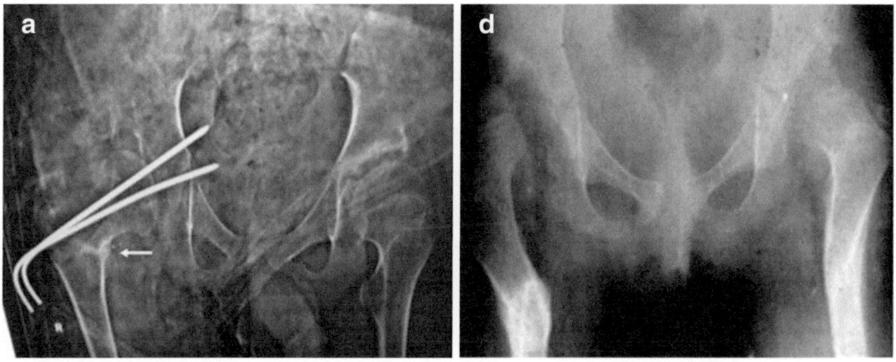

Fig. 19.12 (**a**) Roentgenogram showing open reduction and stabilization of the head of the femur with two Kirschner wires. Note the sub-trochanteric derotation osteotomy (arrow). (**b**) Roentgenogram reveals reduced CDH and late signs of sub-trochanteric derotation osteotomy on both sides

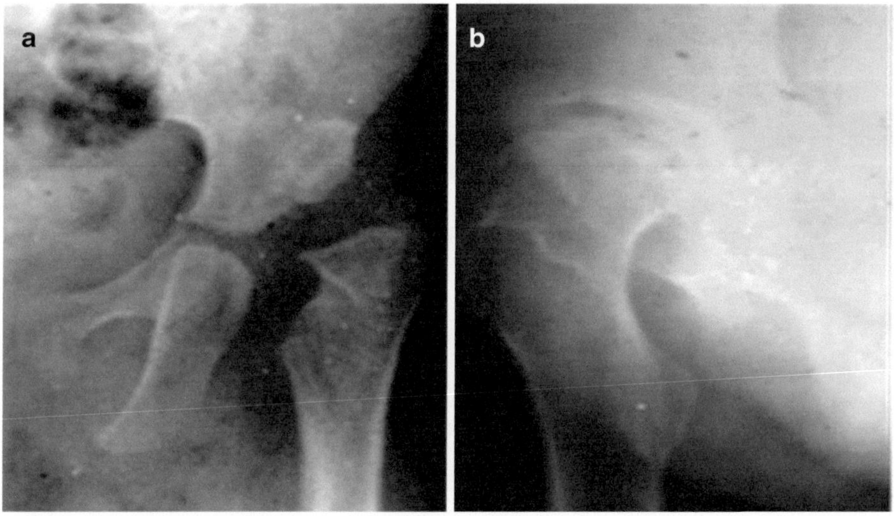

Fig. 19.13 (**a**) Roentgenogram reveals a neglected CDH with hypoplastic epiphysis. The acetabulum is also not well developed. (**b**) Roentgenogram reveals a neglected case of right sided CDH. Note the secondary acetabulum at the dorsum iliac and coxa-valga

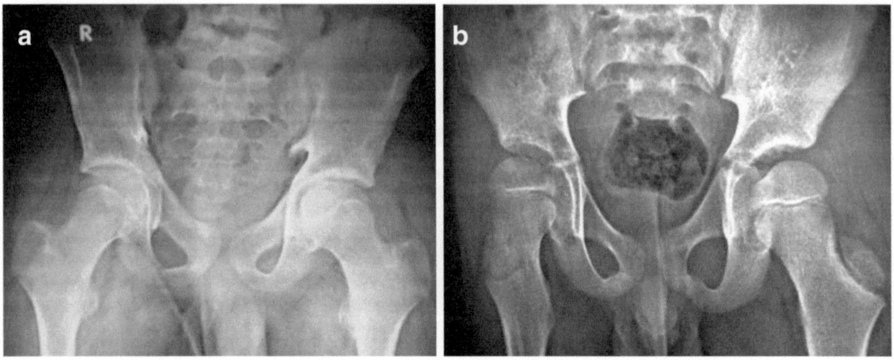

Fig. 19.14 (**a**) Roentgenogram revealing coxa-valga in a case of cerebral palsy. (**b**) Radiograph showing bilateral coxa-valga in a case of PPRP

19.4 Coxa-Valga

Coxa-valga is a developmental deformity of the proximal femur. A characteristic feature of this is the enhanced collodiaphyseal angle (135°) associated in some cases with shallow, dysplastic acetabulum and subluxation of the femoral head. It can be a source of pain following development of secondary osteoarthritis in the hip joint. Coxa-valga is also often observed in spastic patients in association with muscle imbalance around the hip joint (Fig. 19.14a, b). Coxa valga is also associated with some of the skeletal dysplasias. The condition is more often detectable in female than male patients.

19.5 Congenital Coxa-Vara

Congenital coxa-vara is a rare developmental disorder and is characterized by decreased femoral neck shaft angle less than 120 which present at birth and the deformation progress with growth. There may be painless gait abnormalities. There is limping in unilateral cases. Patients with bilateral involvement present with a waddling gait patron associated with increased lumbar lordosis (Fig. 19.15a–d). On physical examination the greater trochanter is elevated with abductor muscle weakness and positive Trendelenburg gait. There may be associated hip flexion contracture.

Radiological observations of developmental coxa-vara have certain classic findings. These radiographic features include a decreased femoral neck-shaft angle (often below 90°) a more vertical position of the physeal plate, a tri-angular metaphyseal fragment in the inferior femoral neck surrounded by an inverted radiolucent Y pattern, a decreased angle of anteversion and there may be acetabular dysplasia

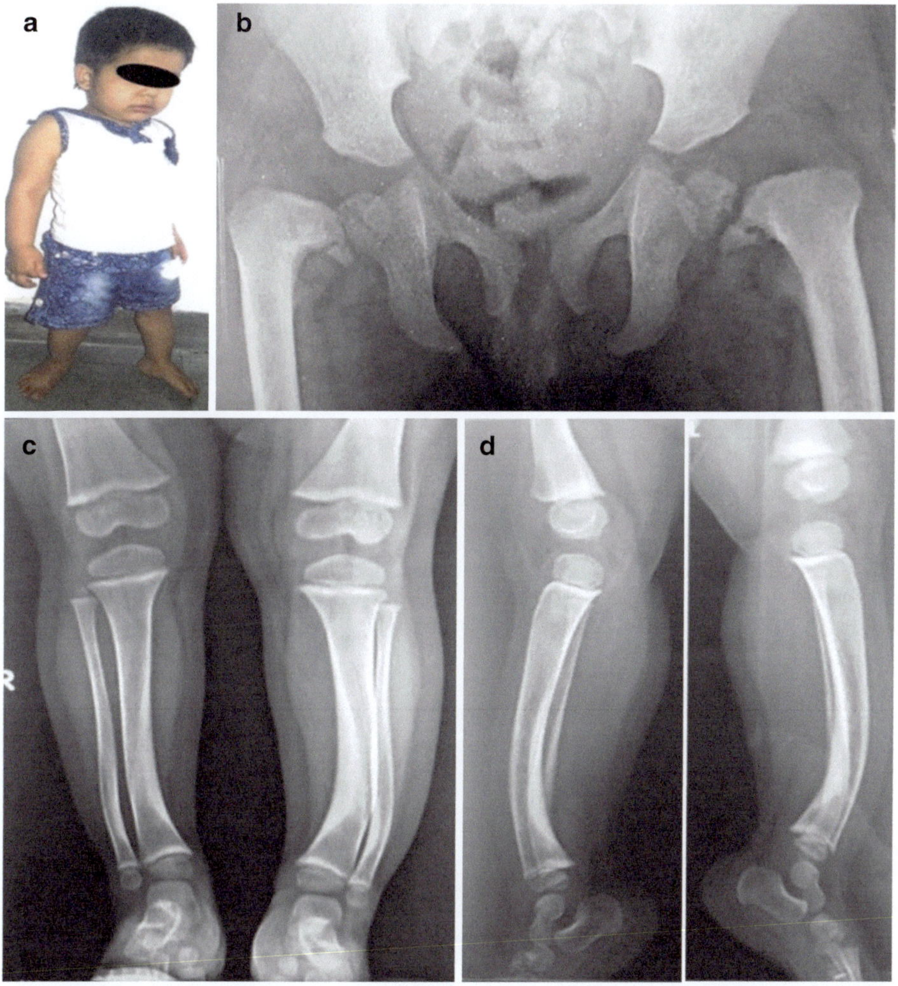

Fig. 19.15 (**a**) Clinical picture of a 2-year-old child with bilateral congenital coxa-vara. (**b**) Roentgenogram showing bilateral congenital coxa-vara. Note the triangular bone piece on either side. (**c, d**) The same child also had bilateral anterolateral bowing of the tibia and fibula

(Fig. 19.16a, b). The radiolucent area actually represents a zone of widening of inferior portion of the physeal plate, with associated abnormal ossification and an interposed triangular segment of dystrophic bone (Fig. 19.17a, b).

Surgical correction with a sub-trochanteric derotation proximal femoral osteotomy has been reported to be quite successful in the management of congenital coxa-vara.

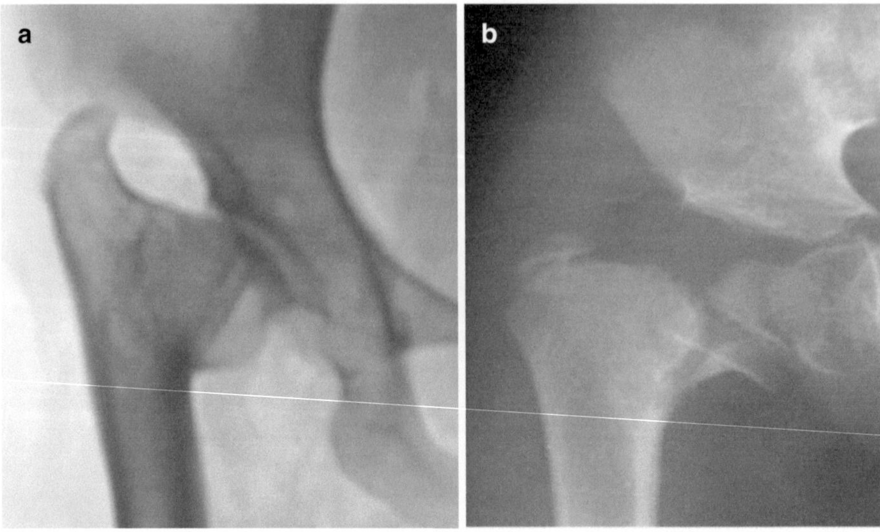

Fig. 19.16 (**a**) Roentgenogram shows congenital coxa-vara with more vertical physeal plate and inferior triangular metaphyseal surrounded by an inverted Y. (**b**) Roentgenogram reveals a hip affected with developmental coxa-vara. Note the projecting tip of greater trochanter and coxa-vara less than 90°

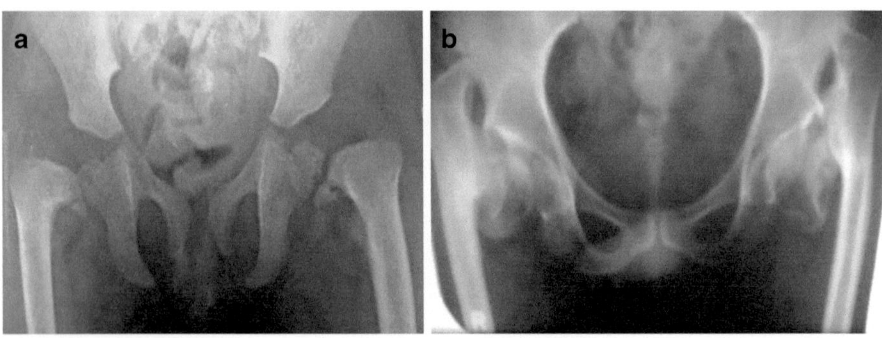

Fig. 19.17 (**a**) Roentgenogram showing a case of bilateral congenital coxa-vara. The triangular piece situated inferiorly is diagnostic. (**b**) Roentgenogram reveals a rare case of bilateral neglected congenital coxa-vara. Classically both the trochanters have migrated proximally. Note the secondary degenerative changes in both the hips

19.6 Slipped Capital Femoral Epiphysis (SCFE)

The condition is characterized by displacement of Capital Femoral Epiphysis from the femoral neck through the physeal plate. The femoral head stays in the acetabulum and the neck usually displaces anteriorly resulting in a varus deformity of the proximal femur. Rarely the neck may get displaced posteriorly creating a proximal femoral valgus deformity. Obesity is known to be associated with SCFE which may be unilateral or bilateral. The etiology of SCFE is not known. Local trauma is known to be a contributed factor in SCFE. Endocrine imbalance is known to change physeal plate physiology and has been considered as an important causative factor, particularly in bilateral cases.

Clinically the patient may complain of pain in the hip or groin, limitation of movements and difficulty in walking. The pain may be referred in the distribution of obturator nerve. The pain usually becomes severe and persistent. The effected limb may lie in external rotation and there is limited flexion, abduction and medial rotation.

Radiologically, in pre-slip stage, the radiographs may reveal a minor widening and fuzziness of the physeal plate which at times may not be noted. Gradually the physeal plate gets widened, irregular and decrease in epiphyseal height in the centre of the acetabulum (Figs. 19.18a, b and 19.19a, b). A crescent-shaped area of increased density may be identified in the proximal portion of the femoral neck and is known as the 'blanch sign of Steel'.

An AP and a true lateral radiograph help in diagnosis. The physis may also be displaced in valgus (Fig. 19.20). A frog-leg lateral radiograph demonstrates the posterior displacement of the femoral neck. Ultrasonography, computed tomography and MRI are helpful in early as well as late diagnosis.

The treatment includes prevention of further slipping by stabilizing the diseased physis. The overall management is usually divided in three stages, treatment to prevent further slippage, treatment to reduce the degree of slippage and salvage treatment. In earlier stages nonoperative treatment in the form of traction may be of great help. Operative treatment includes reduction and internal fixation. However, the choice depends on the treating surgeon.

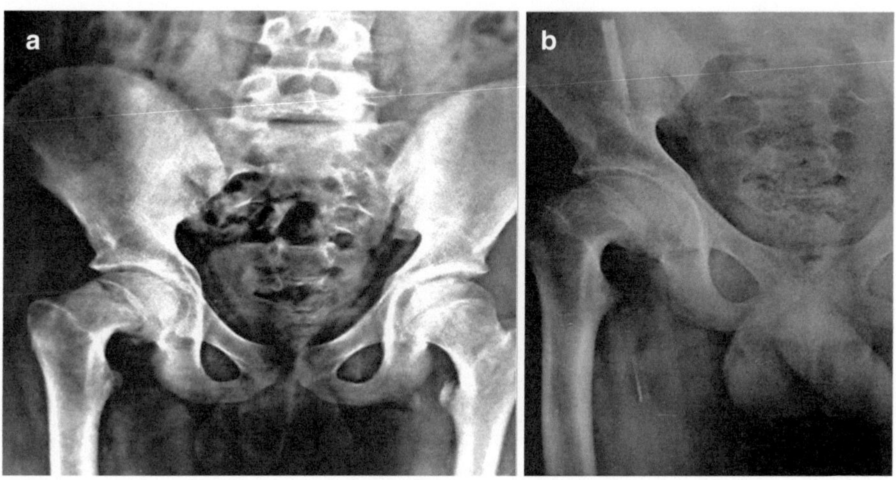

Fig. 19.18 (**a**) A patient with 2-months-old right sided slipped capital femoral epiphysis in varus. The patient had no complains except minimal occasional pain in the hip. (**b**) The lateral roentgenogram reveals minimal radiological changes

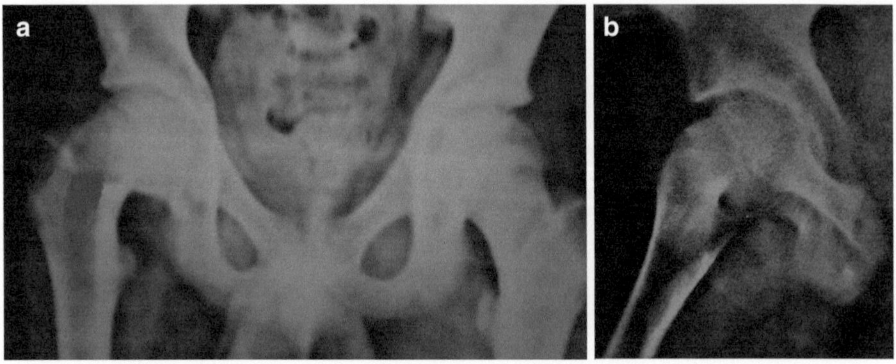

Fig. 19.19 (**a**) Roentgenogram of the same patient after 2 years. The patient had pain in the right hip. (**b**) The lateral radiograph reveals signs of early osteoarthrosis

Fig. 19.20 An old healed case of bilateral slipped capital femoral epiphysis with valgus displacement. The patient had occasional bilateral hip pain

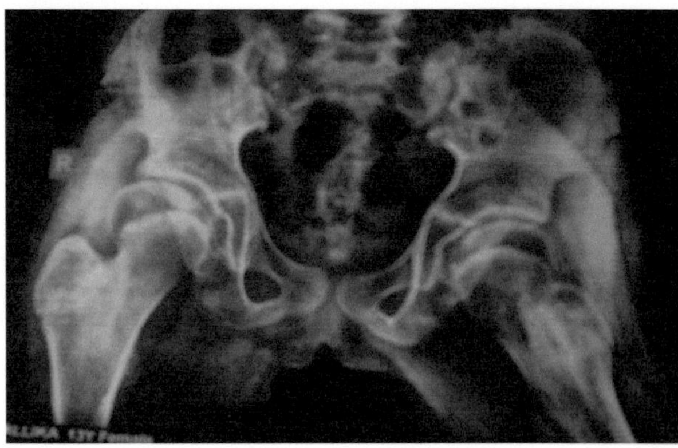

19.7 Congenital Gluteus Maximus Contracture

Congenital gluteus maximus contracture is a rare anomaly where the child is not able to flex the hip in the frontal plane but flexion is possible after external rotation of the hip. Although the defect is congenital, the symptoms appear during childhood. The buttock is not well developed because of the fibrosis and contracture in the central fibres of gluteus maximus. Clinically the diagnosis can be made in lying down position when the child can flex the hip only after external rotation (Fig. 19.21a–c). Radiologically the hip appears normal (Fig. 19.22). A surgical release/lengthening (Z-Plasty) proves useful (Fig. 19.23).

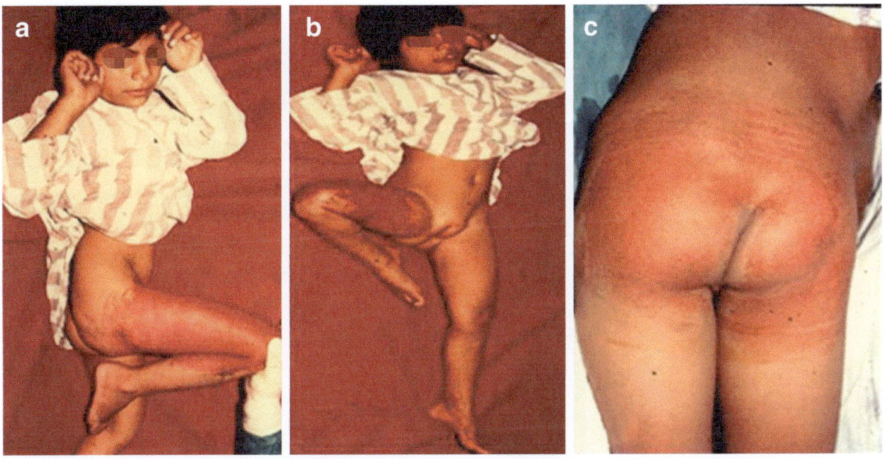

Fig. 19.21 (**a**) A 6-year-old girl with Congenital Gluteus Maximus Contracture who is not able to flex the right hip in the frontal plane. (**b**) The same hip can be flexed further with external rotation. (**c**) Note the wasting of the right buttock

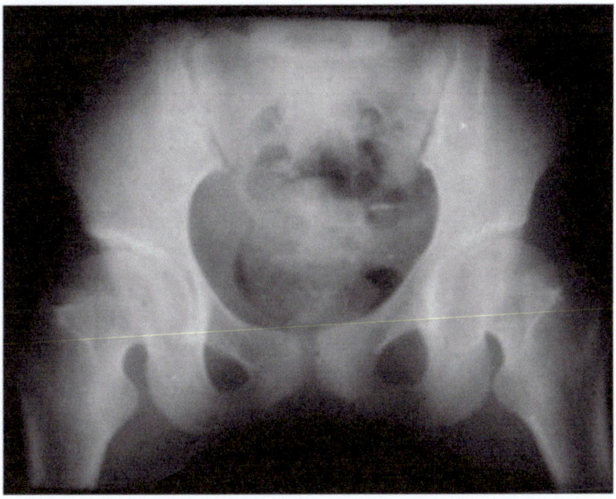

Fig. 19.22 Roentgenogram reveals normal right hip joint

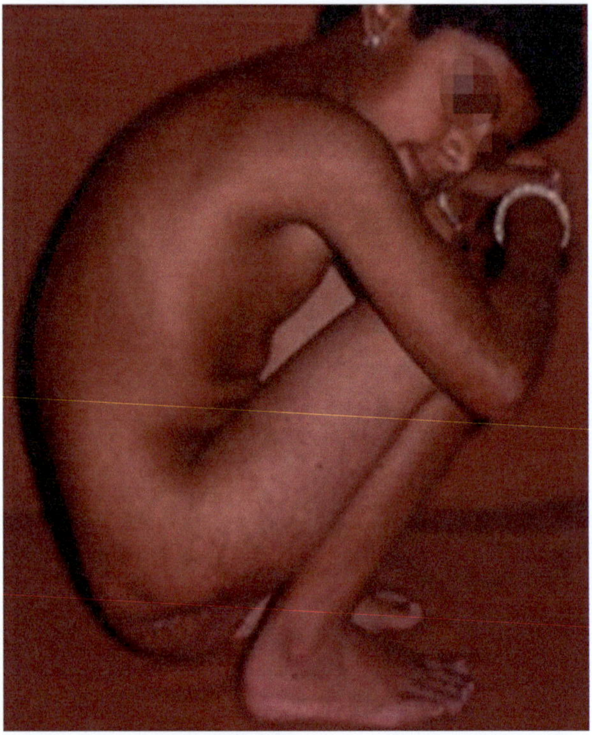

Fig. 19.23 The same patient after Z-plasty of the gluteus maximus. Note the scar of the right buttock

19.8 Congenital Dislocation of the Knee

Congenital dislocation of the knee is a rare anomaly, occurring 40–80 times more rarely than congenital dislocation of the hip. Muscular abnormality has always been observed in such cases. May be due to shorting and fibrosis of the quadriceps muscle, resulting in muscular imbalance during pregnancy (Fig. 19.24a–e). Nearly 50% cases are bilateral. The defect is obvious at birth in the form of hyperextension of varying degree. Complete dislocation (Grade-III) has been reported to be most frequent (Fig. 19.25a–c). Family history has been traced in some cases (Fig. 19.26). It may be associated with other congenital anomalies. Although conservative treatment has been recommended, almost all the children require operative treatment in the form of lengthening of the quadriceps tendon. Early management in the form of lengthening of the quadriceps may be helpful in restoring the knee function (Fig. 19.27a–f). In severe case there may be subluxation or dislocation of tibia on the femur (Fig. 19.28a, b). Untreated/neglected cases have been reported which result in gross disability and loss of function (Figs. 19.29a–d and 19.30a, b).

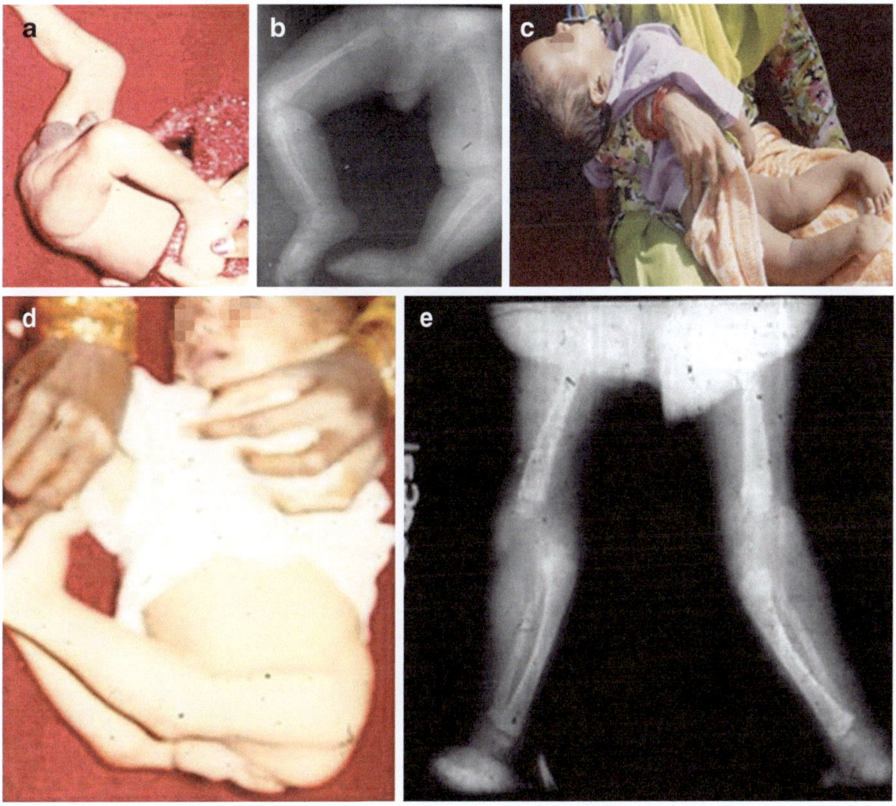

Fig. 19.24 (**a**) An 18-month-old boy showing gross congenital anterior dislocation of the left knee. (**b**) Roentgenogram of the same knee showing severe luxation. (**c**) 1-year-old child with bilateral dislocation of both knees. Child also had bilateral CTEV. (**d**) A child with bilateral congenital dislocation of knees. (**e**) Roentgenograms reveal congenital anterior dislocation of the knees

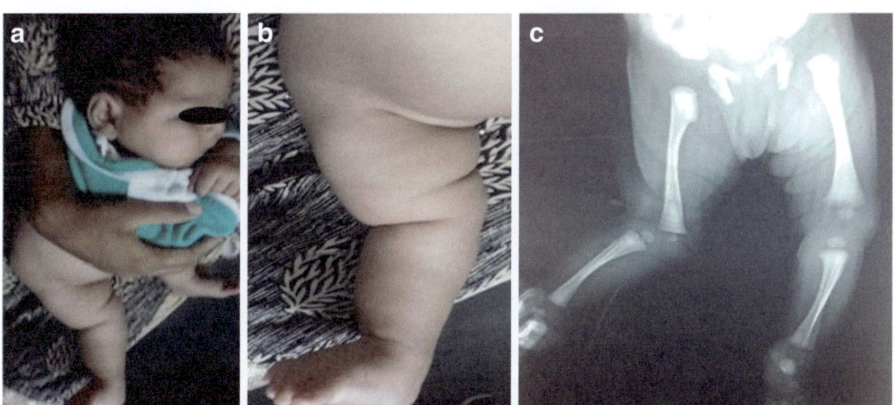

Fig. 19.25 (**a, b**) Clinical picture showing a recurvatum deformity at the right knee. (**c**) Roentgenogram shows anterior dislocation of the right knee

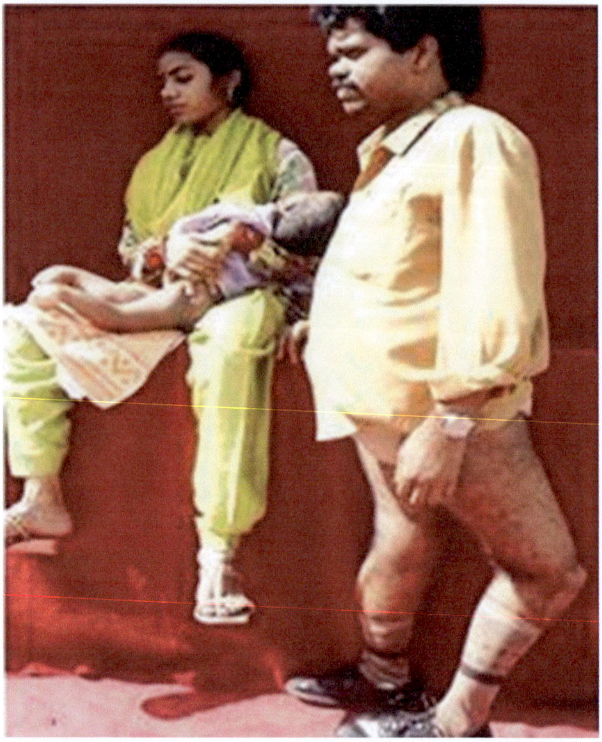

Fig. 19.26 Clinical picture of the father and the son revealing dislocation of the knees

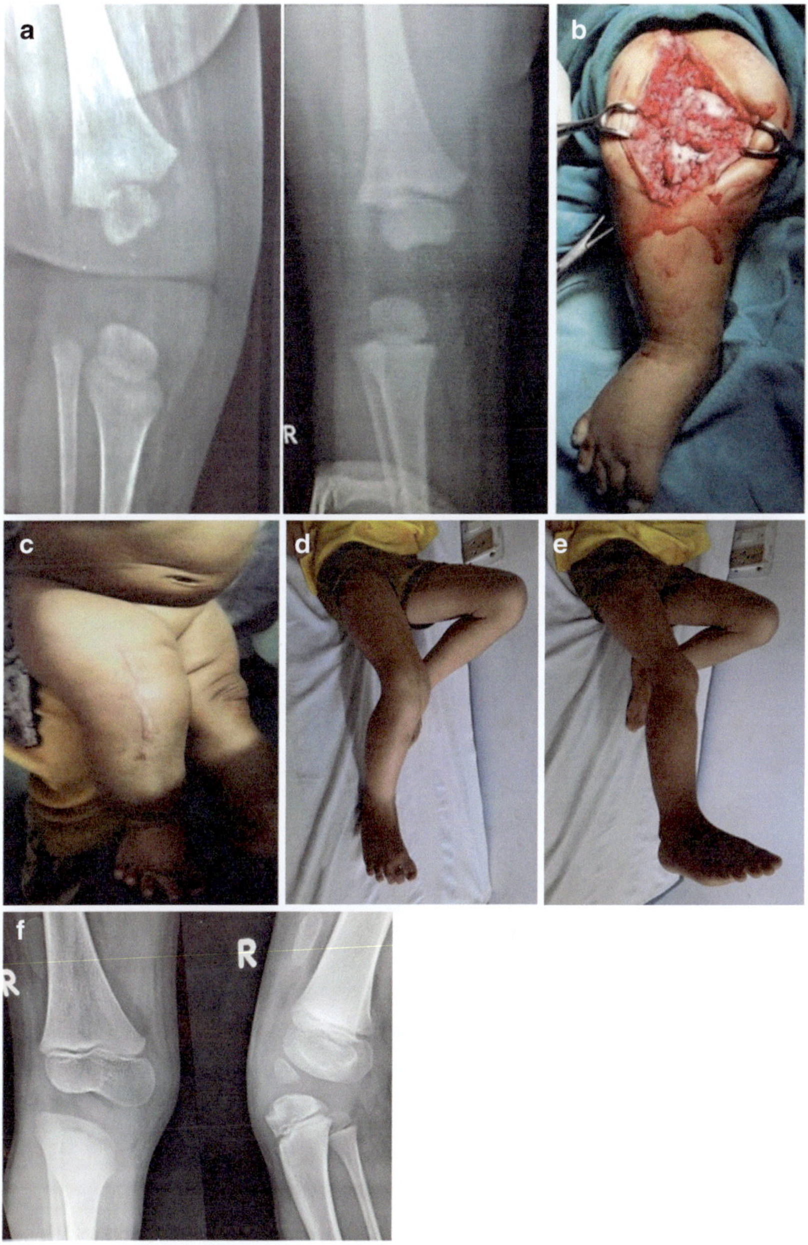

Fig. 19.27 (**a**) Intra-operative picture after Z-Plasty of the quadriceps. (**b**) Note the operation scar line on the front of the right knee. (**c**) Post-operative roentgenogram revealing relocation of the knee joint after Z-Plasty of the quadriceps tendon. (**d, e**) A 3-year follow-up of the same patient showing active flexion and extension with exercises she has been gradually gaining some flexion at the knee. (**f**) Roentgenogram of the patient after 3 years

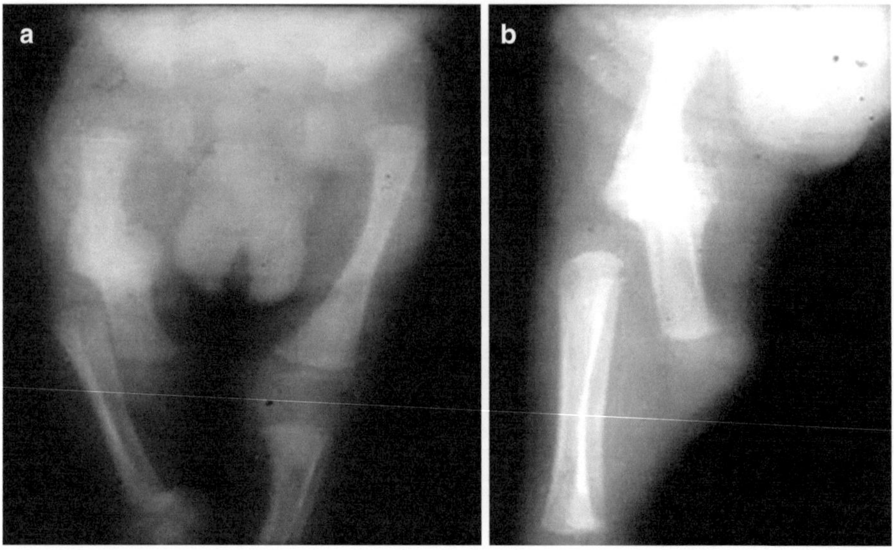

Fig. 19.28 (**a, b**) Roentgenogram reveals neglected gross dislocation of the right knee. The tibia is markedly displaced anteriorly. Note the healed fracture of the femoral shaft

19.8.1 Neglected Cases with Dislocation of the Knee

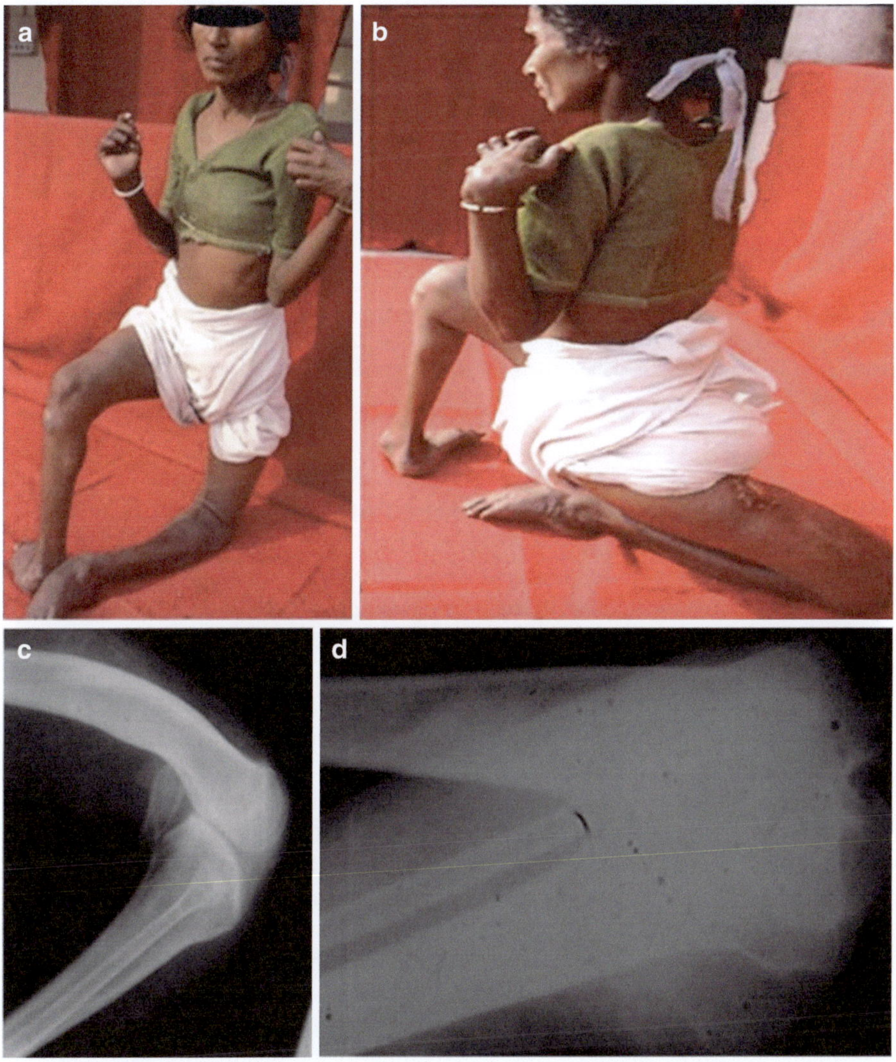

Fig. 19.29 (**a, b**) Clinical picture of a lady-bagger revealing grossly neglected congenital dislocation of the left knee—a case of PPRP. The patella is caught between the thigh and the leg. Note multiple deformities of both feet. (**c**) Roentgenogram reveals anterior dislocation of the left knee in the same patient. (**d**) Roentgenogram shows complete dislocation of the knee in the above patient. The patella is caught between abnormally placed femoral and tibial condyles-a very rare presentation

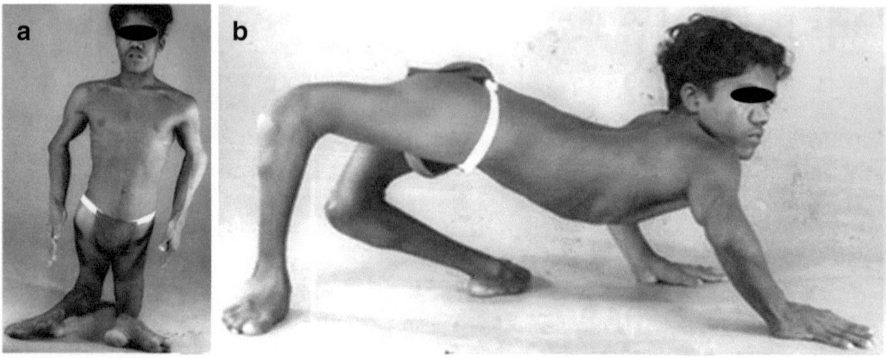

Fig. 19.30 (**a, b**) A 26-year-old patient with rhizomelic shortening of the limbs with neglected congenital dislocation of both the knees. Because of the severe contracture of the quadriceps on either side, the patient stands at the lower ends of the femora and propels on fours. Note the callosities under the heels

19.9 Congenital Dislocation of the Patella

Congenital dislocation of the patella is a rare disorder of the knee joint in which the patella is permanently displaced, even in extension it may be fixed on the lateral aspect of the femoral condyle. The patella often shows abnormal appearances (Fig. 19.31). In infancy, radiographs may show the angular deformity, but the patella has not begun to ossify in the infant. Ultrasonography has been useful in diagnosing a lateral congenital patellar dislocation when it is unossified. MRI can demonstrate the position of the cartilaginous patella at this age. The child has a knee flexion contracture and holds the lower leg externally rotated, with genu-valgum noticed later. A radiograph with a skyline view is necessary to confirm the diagnosis (Fig. 19.32a–e). The condition must be differentiated from recurrent and habitual patellar dislocations. If left untreated, flexion, valgus and external rotation deformity may develop.

Surgical treatment is necessary to realign the entire extensor mechanism of the knee which allows good range of knee motion. The results of surgery are better in young children. Surgical treatment usually involves a combination of lateral release and medial plication, together with lengthening of the quadriceps, especially the vastus intermedius and rectus femoris (Fig. 19.33a–c).

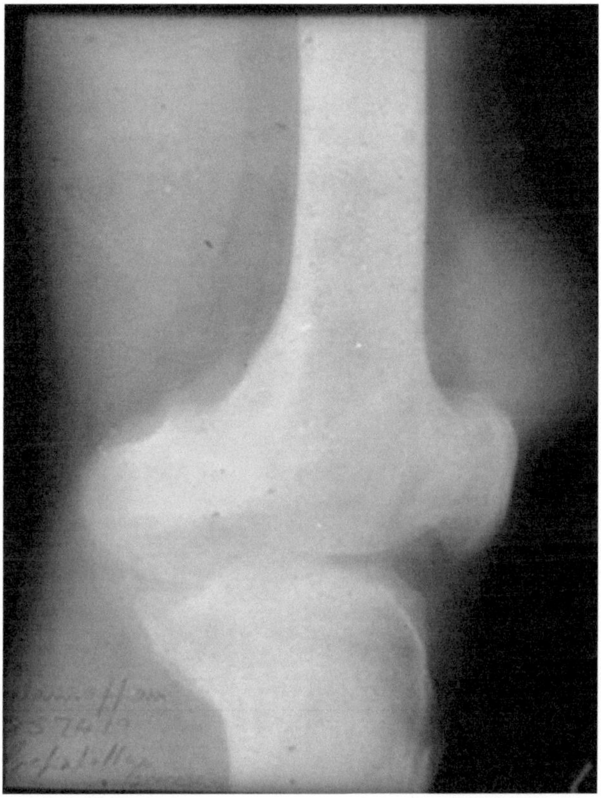

Fig. 19.31 Roentgenogram shows supero-lateral dislocation of the hypoplastic patella on the right side

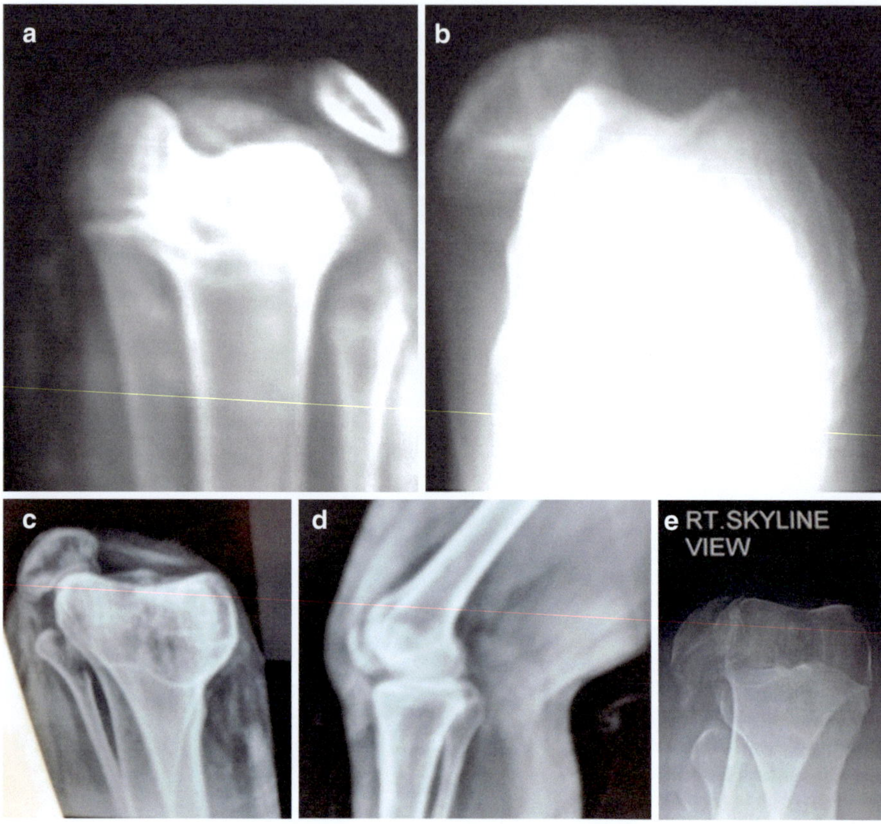

Fig. 19.32 (**a**) Roentgenogram (Skyline view) reveals lateral dislocation of the hypoplastic patella on the left side. (**b**) Roentgenogram shows supero-lateral dislocation of the hypoplastic patella on the right side. (**c**) A skyline radiograph reveals supero-lateral dislocation of the patella. (**d**) A lateral radiograph shows a hypoplastic patella. (**e**) Roentgenogram reveals complete supero-lateral dislocation of the right patella

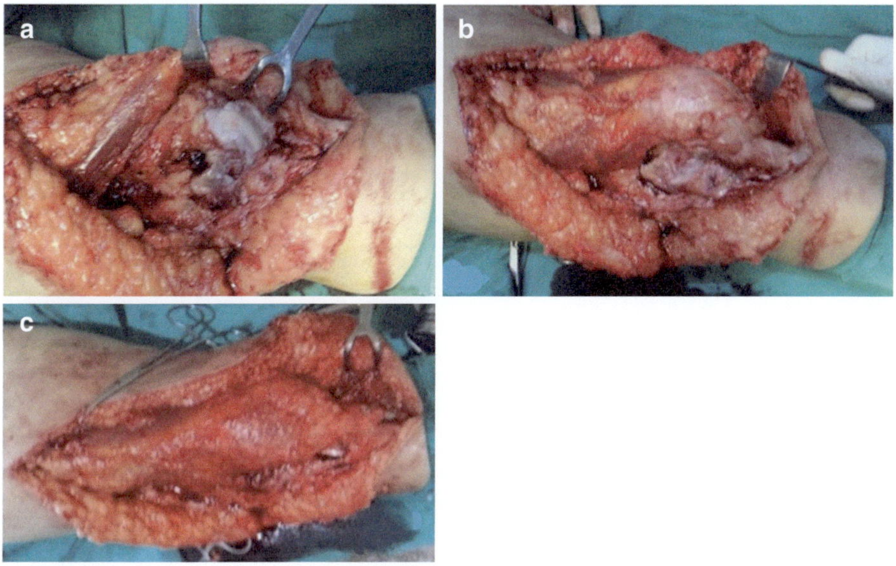

Fig. 19.33 (**a**) Intra-operative photograph showing lateral release and medial plication of the quadriceps muscle (Campbell operation). (**b**) Medial and distal positioning of tibial tubercle (Hauser operation). (**c**) The final position of the quadriceps after complete repair

19.10 Tibial Hemimelia

Congenital absence of the tibia (complete paraxial tibial hemimelia, terminal or intercalary) is a rare deformity of unknown etiology with resultant eqinovarus deformity of the foot and variable limb shortening. Tibial hememilia is a rare anomaly and often bilateral. Other congenital anomalies may also be associated with the absence of tibia. The quadriceps function is impaired and the patella is usually absent. The foot is in a severe varus position relative to the distal fibula and has little functional movement.

19.10.1 Classification: John et al.

1. Complete absence of Tibia—Type-I.
2. A proximal tibia of varying size is present—Type-II.
3. Distal tibia evidence—proximal absent—Type-III.
4. Short tibia with proximal and distal migration of fibular ends. (Distal Tibia-Fibular diastasis)—Type-IV.

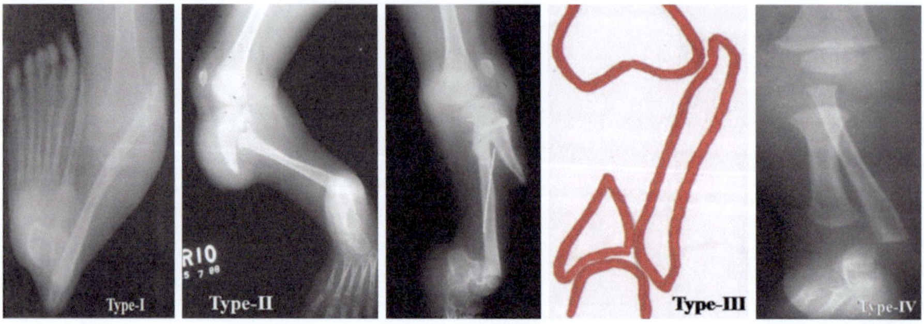

Type-I Type-II Type-III Type-IV (Diastasis)

Jone has classified tibial hemimelia into four categories. In type I, the tibia is not seen radiographically at birth (Fig. 19.34). In subtype IA, there is an upper tibia, and in IB, the upper tibia is late to ossify. If there is no ossification of the distal femoral epiphysis or it is small, it is likely that no upper tibia is present. In type II, the proximal tibia is ossified, and the distal tibia is absent. In type III, the distal tibia is ossified, and the proximal tibia is absent, an exceedingly rare variation. Type IV consists of the entity formerly called congenital diastasis of the ankle. In these cases, the tibia and fibula diverge at the ankle, the tibia is short, the talus is proximally displaced and the distal tibial articular surface is absent.

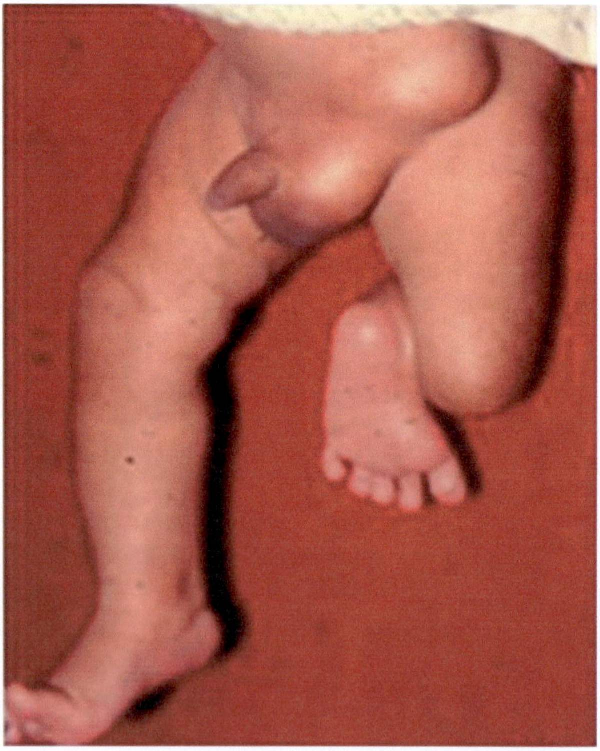

Fig. 19.34 A 1-year-old boy with tibial hemimelia type-I on the left side with short and deformed leg. Note the inguinal hernia on the left side

Treatment options are limited and are primarily through knee amputation, knee arthrodesis with through ankle amputation and fibular centralization. Centralization of the fibula was popularized by Brown has been commonly attempted to reconstruct the knee. It is more useful in a grade II deficiency where the quadriceps power is reported to be is near normal. However, in grade I deficiency quadriceps power is deficient and is not considered necessary. Considerable flexion contracture at the knee has been reported to occur in due course of time.

The recommended treatment of complete tibial hemimelia (type-I) is knee disarticulation, centralization of the fibula combined with Syme's amputation and application of lifelong prosthetic appliances has been recommended. However, the results after an amputation have not been very appreciative.

To avoid an amputation at any level, the author has recommended a one time stabilizing procedure called 'Femoro-fibulo-calcaneal arthrodesis' (FFCA), even in bilateral cases (Figs. 19.35a–g, 19.36a–l, and 19.37a–d). In this way the extremity is preserved with its sensation and proprioception. The author had ten patients with 18 legs having tibial hemimelia (type-I) and has recommended three stages of management to achieve the suggested femoro-fibulo-calcaneal arthrodesis (Figs. 19.38a–d, 19.39a–d, 19.40a–e, and 19.41a–e). The stages are as follows:

1. **Stage of Loosening—**The soft tissues, particularly around both the ends of the fibula are widely released. In older children skeletal traction is applied proximally through the lower end of the femur and at the distal end through the talus/calcaneus. In younger children suitable external fixator is applied.
2. **Stage of Lengthening—**Traction is applied proximally (Head side) and distally over the horizontal part of Bohler-Brown splint. Weight (depending on the age of the patient) is increased daily on either side till the fibular ends are placed under the femoral condyles and over the body of the talus (Centralization of the fibula). A suitable external fixator can also be used, more so in a young patient. The major aim of the treatment in such cases is leg lengthening and centralization of the fibula. With gradual traction the leg is lengthened and the fibula is automatically centralized after fibrinolysis at both its ends.
3. **Stage of Stabilization—**Once the fibular ends are centralized, they are fixed to the inter condylar notch of the femur and the body of the talus/calcaneus with cross Kirschner wires. The straight and elongated leg is now encased in above knee P.O.P. cast. The objective is to stabilize the construct. This has been achieved proximally by an arthrodesis between the proximal fibula and the intercondylar notch and distally between fibula and talus/calcaneus in plantigrade position of the foot. The physes at the fibular ends is not disturbed to allow future growth.

The reconstruction by talo-fibular arthrodesis avoids a Syme's amputation. The technique has been proved to be very useful in almost all the cases described in the series. If considered necessary, the child can use a suitable orthotic appliance till consolidation occurs at the knee and the ankle.

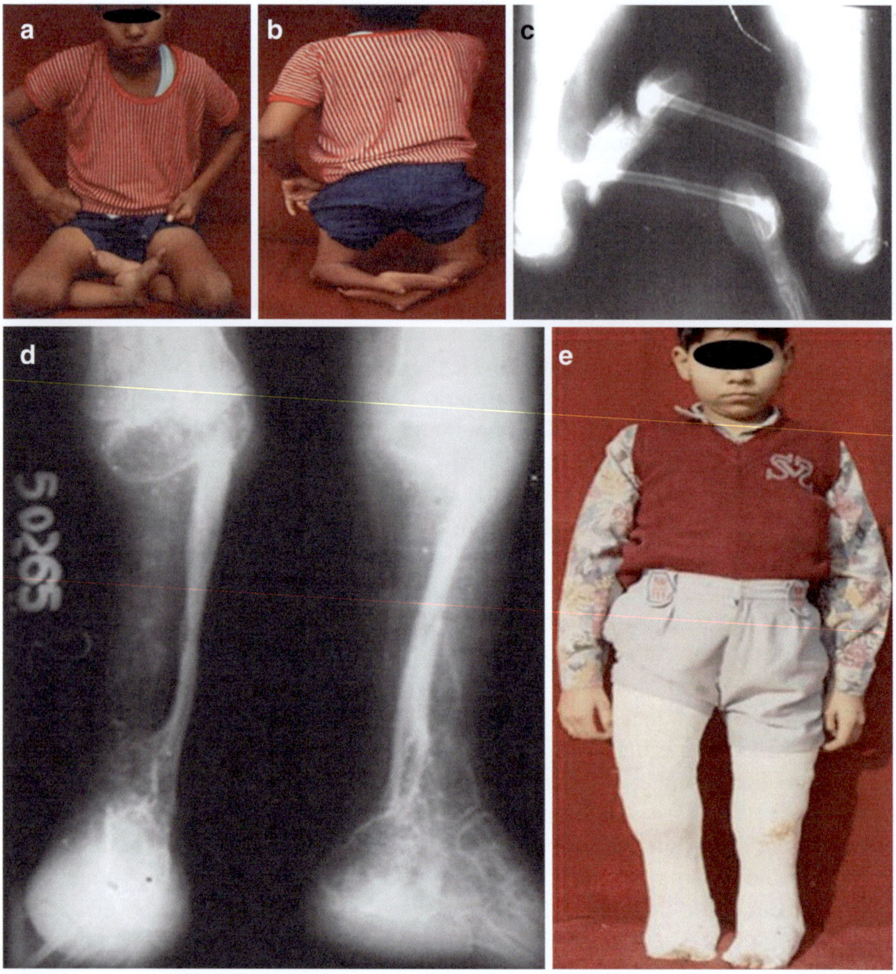

Fig. 19.35 (**a, b**) Clinical picture of a 12-year-old boy with tibial hemimelia (Type-I). (**c**) Roentgenogram of the patient showing complete absence of both tibiae(type-I) deficiency. (**d**) Roentgenogram revealing centralisation of the fibula in an attempt to achieve femoro-fibulo-calcaneal-arthrodeses. (**e**) Patient in bilateral long leg cast during the course of treatment. (**f**) Roentgenogram showing complete healing at the knee and ankle. Note the hypertrophy of the fibula. (**g**) Clinical picture of the patient after treatment. Note three medial rays absent on the right side

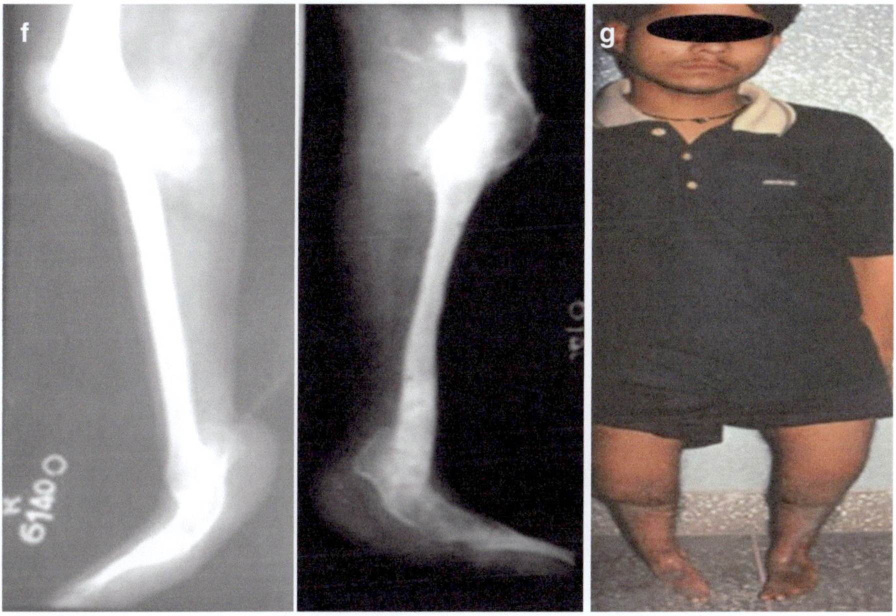

Fig. 19.35 (continued)

The main advantage of preserving the foot is that the limb is preserved and the patient has sensations and proprioception at the sole. Moreover, the expenses required for periodic change of prosthetic appliances can be avoided. The results of this technique have been encouraging. A follow-up of 9 and 35 years have been reported in the series which has been published as an original article **(JBJS Am. Open Access. 2019:e0029. http://doi.org/10:2106/JBJS.OA.18.00029).**

19.10.2 Tibial Hemimelia (Type-II)

In type II, the proximal tibia is ossified, and the distal tibia is absent. For partial tibial hemimelias, the outlook is much better. When the proximal portion of the tibia is present (Fig. 19.42a, b), fusion of the proximal fibular end to the upper tibia provides an excellent functional result. The fibula may be fused to the tibia in a side-to-side position with screw fixation or an end-to-end position-Huntington Procedure (Fig. 19.43a–h).

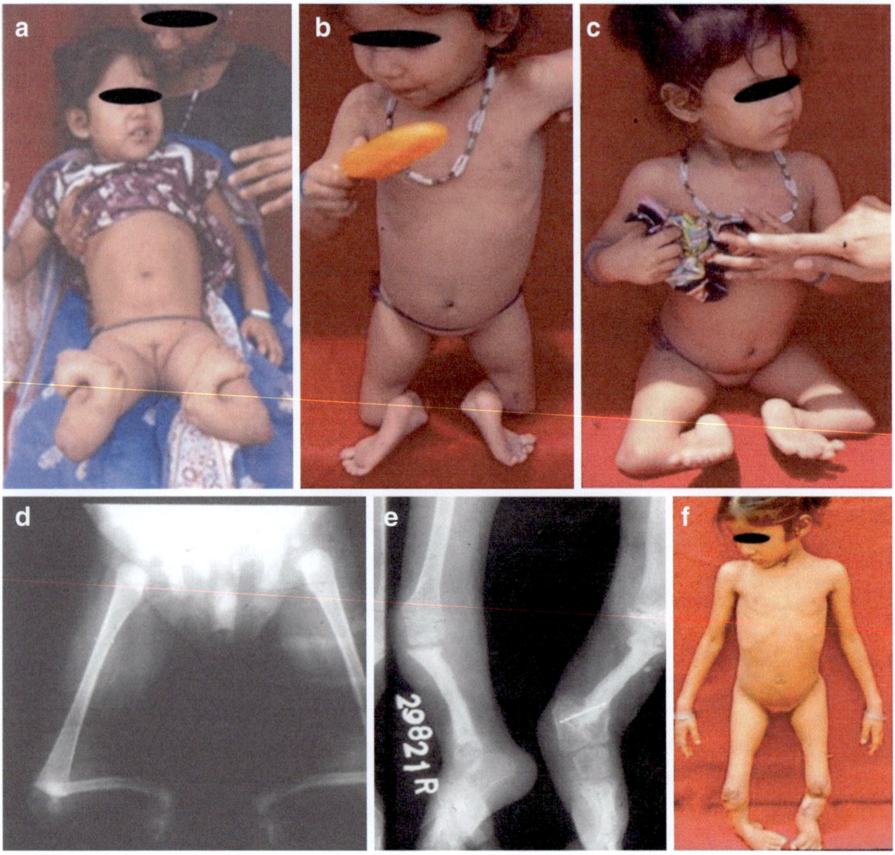

Fig. 19.36 (**a, b, c**) Clinical picture of a 2-year-old girl with bilateral tibial hemimelia (Type-I) revealing multiple deformities. (**d**) Roentgenogram revealing total absence of tibiae on both sides. (**e**) Roentgenogram showing centralized hypertrophied fibulae. (**f**) The patient stands on her own feet. The right foot was corrected in POP. (**g**) The final roentgenogram reveals femoro-fibulo-calcaneal arthrodesis. Note the hypertrophied fibula and planti-grade feet with a follow-up of 36 years. (**h, i, j**) The patient can walk on her own planti-grade feet. She has been an active teacher in a school. She is married and has two sons. The elder son has type-II tibial hemimelia on right side. The second son does not reveal any abnormality. (**k**) The patient with the second son who does not reveal any abnormality. (**l**) Clinical picture of both the sons of the patient

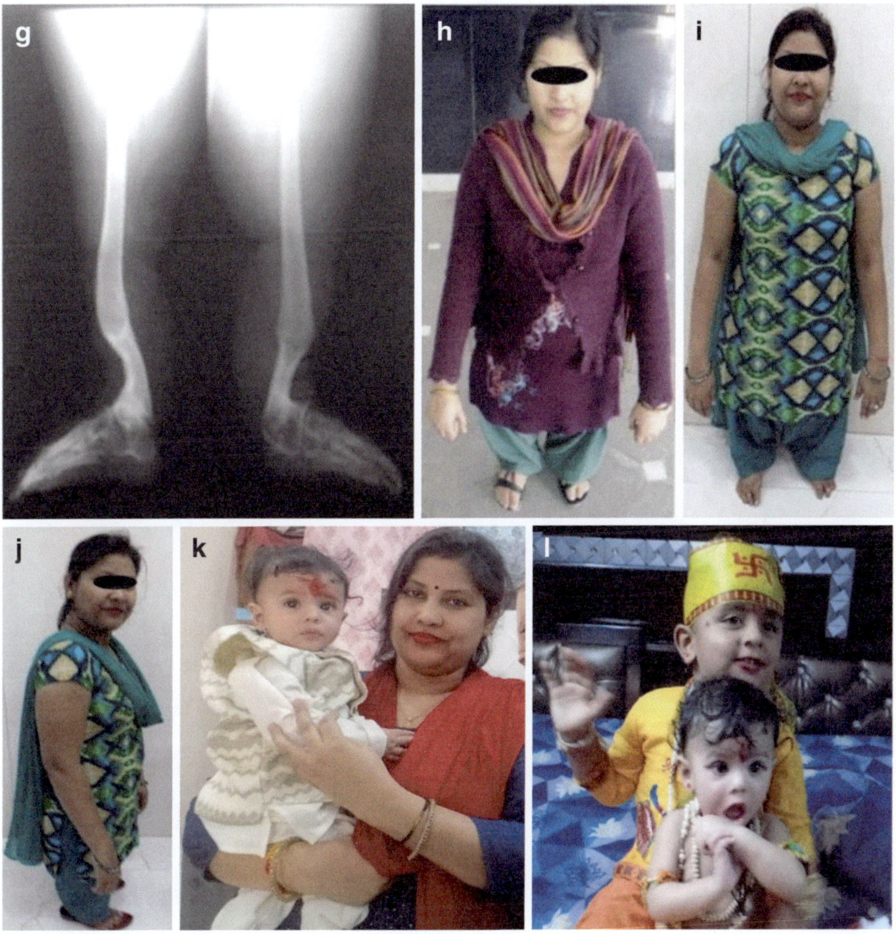

Fig. 19.36 (continued)

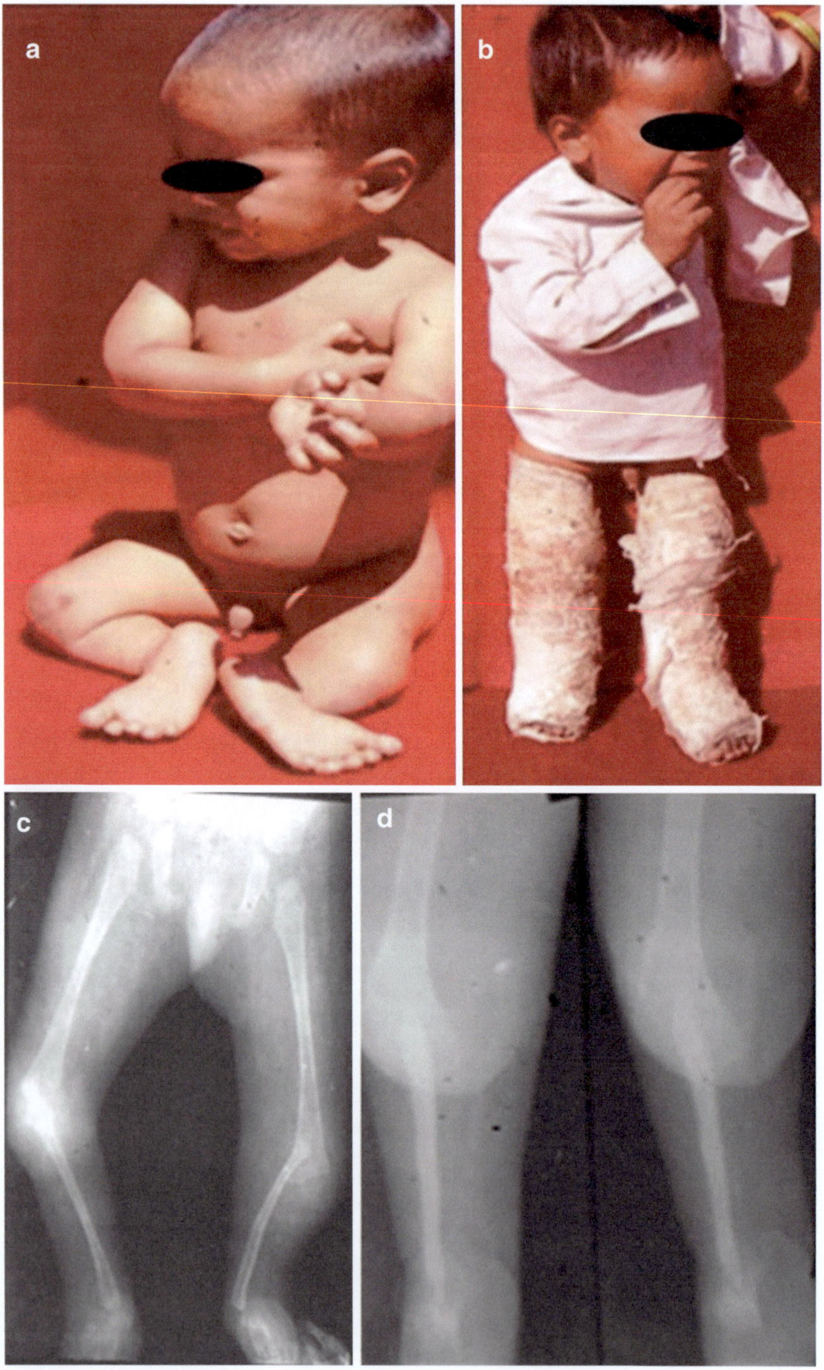

Fig. 19.37 (**a**) A case of bilateral tibial hemimelia type-I. (**b**) The patient stands with a POP cast. (**c**) Roentgenogram reveals centralized fibulae in the process of achieving femoro-fibulo-calcaneal arthrodesis. (**d**) Roentgenogram showing final appearance of femoro-fibulo-calcaneal arthrodesis. Note the hypertrophy of the fibulae

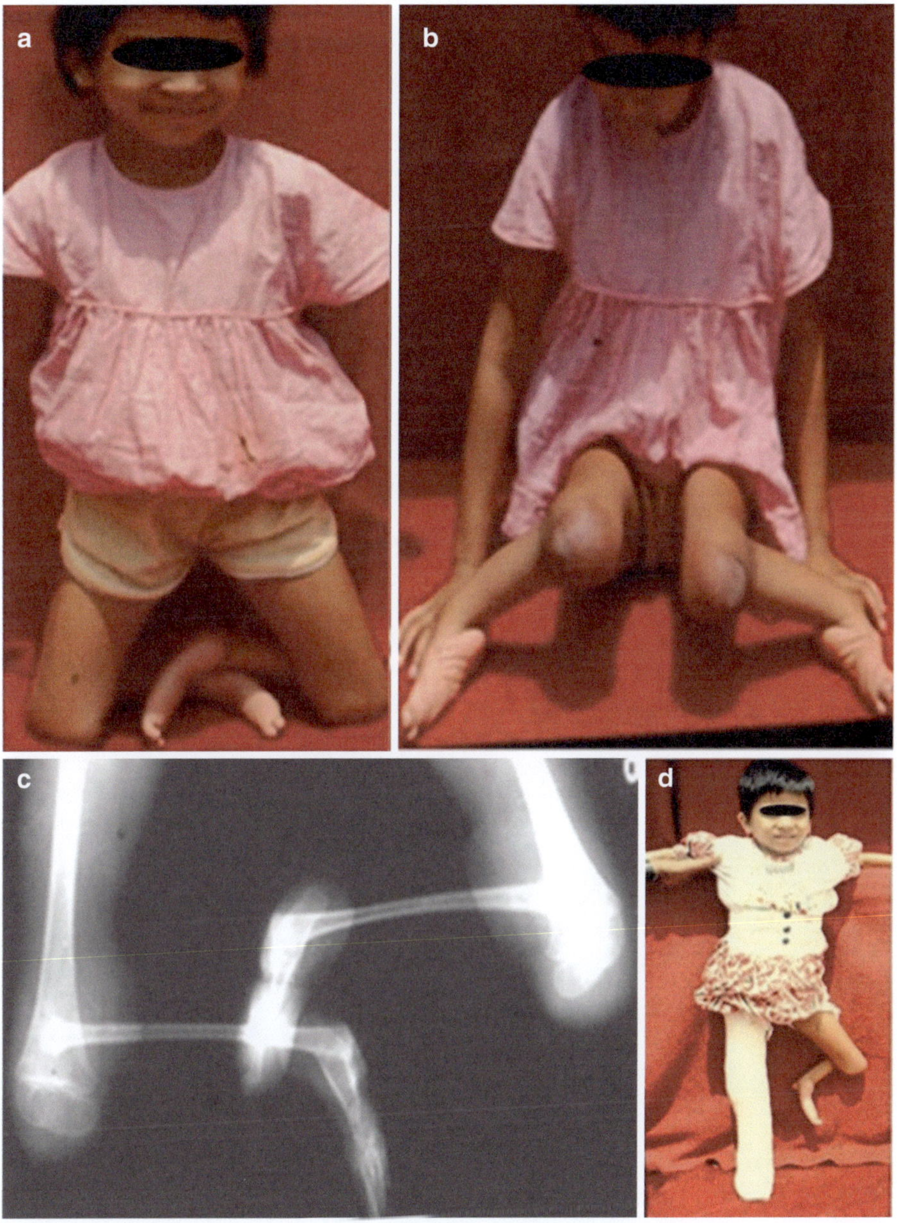

Fig. 19.38 (**a**, **b**) A 10-year-old girl from Bangladesh shows bilateral tibial hemimelia type-I. She had two digits on either side. (**c**) Roentgenogram confirms the diagnosis. (**d**) The patient stands in the POP cast on the right side. Note the lengthening of the leg achieved (Left side was to be corrected). The patient could not come for further treatment

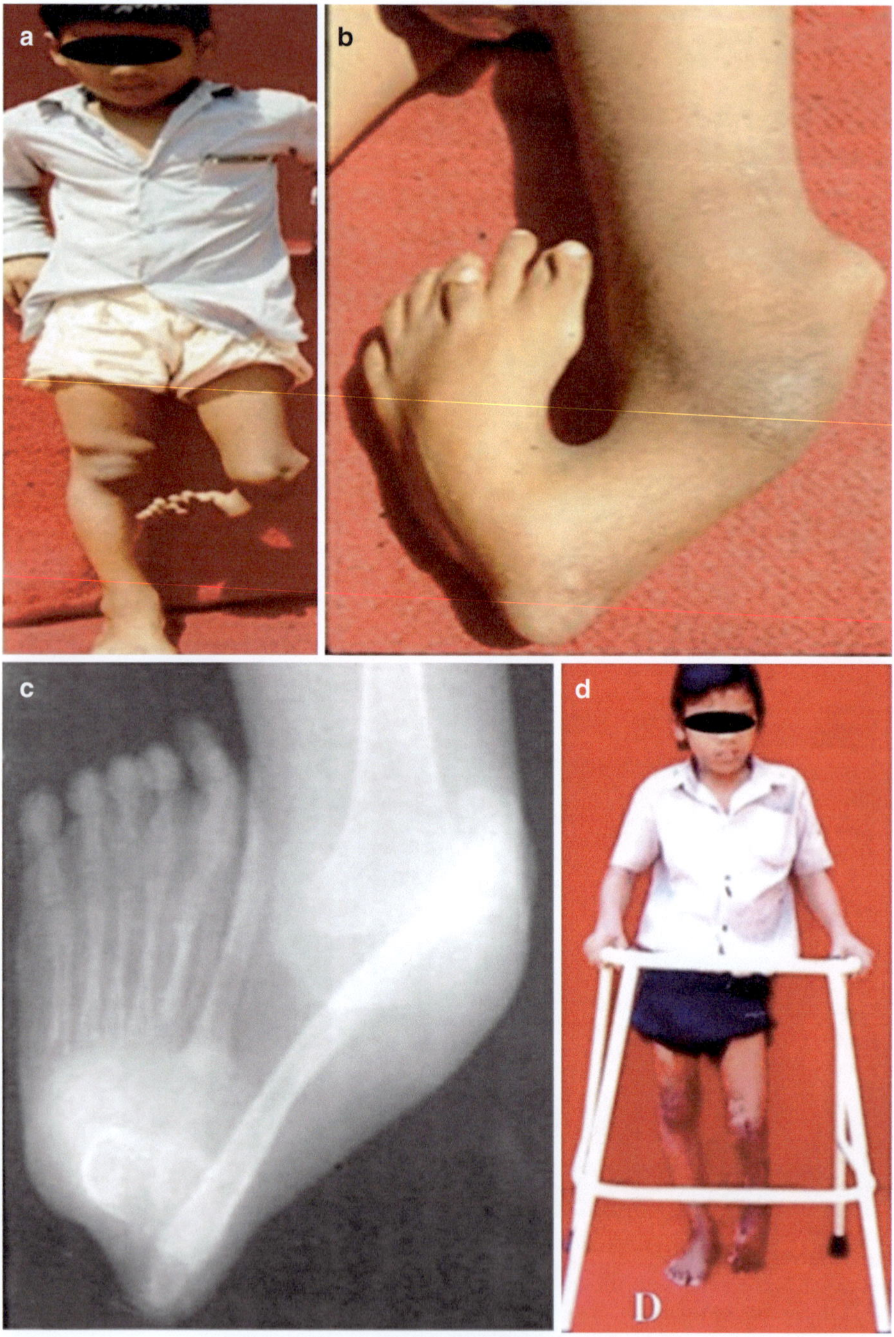

Fig. 19.39 (**a**) A 12-year-old patient with severe tibial hemimelia on the left side with polydactyly. (**b**) A close-up view of the deformity. Note associated polydactyly. (**c**) Roentgenogram reveals the position of the fibula with polydactyly. (**d**) The patient walks with a walker after the fixator is removed. Now needs stabilization of the fibular ends

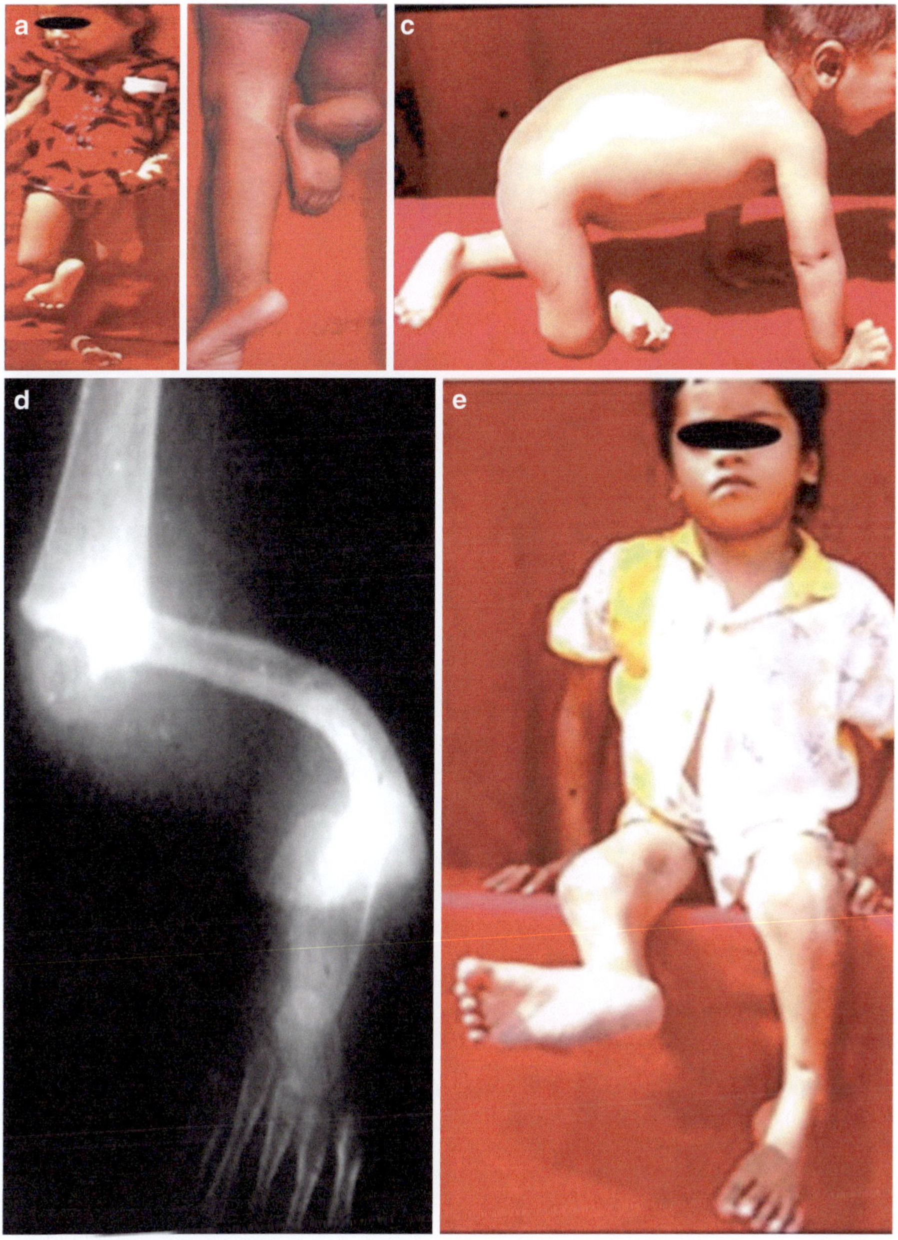

Fig. 19.40 (**a, b, c**) A 2-year-old boy with tibial hemimelia type-I on the right side with severe deformity of the leg. (**d**) Roentgenogram of the patient with curved fibula. (**e**) The patient after the treatment. The leg is straight and long

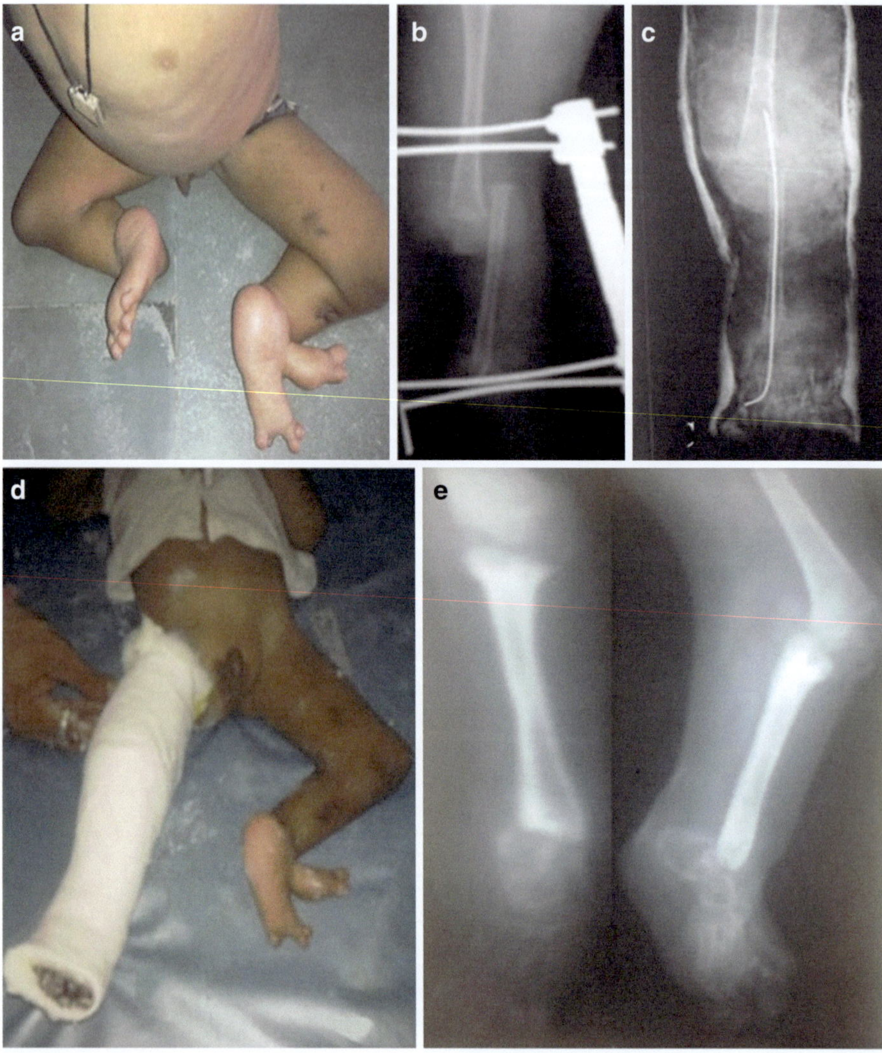

Fig. 19.41 (**a**) A 2-year-old boy with bilateral tibial hemimelia type-I. Note the bi-fid foot on the left side. (**b**) Roentgenogram shows the lengthening process in a fixator. (**c**) The AP roentgenogram reveals a long Kirschner wire passed through the calcaneus fibula and to the femur. The construct is further supported by a long leg POP cast. (**d**) The patient in a long leg cast. The right extremity is straight and long. (**e**) Roentgenogram reveals centralized and hypertrophied fibula. Arthrodesis at the knee is yet to take place. The patient died at home—unknown cause

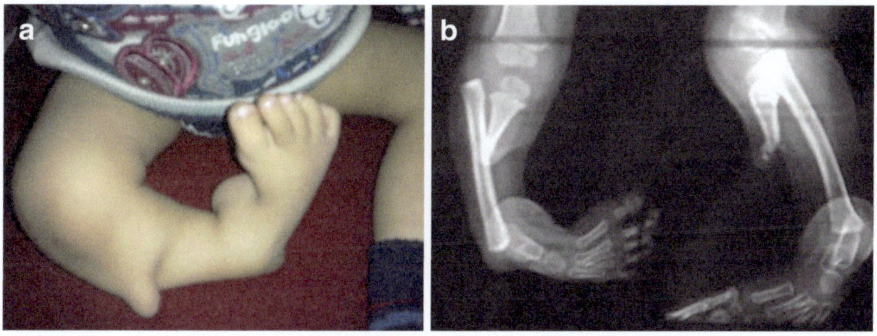

Fig. 19.42 (**a**) Clinical picture of the elder son of the patient (Fig. 19.36a–l) with type-II tibial hemimelia and CTEV on right side. (**b**) Roentgenogram confirms the diagnosis of type-II anomaly

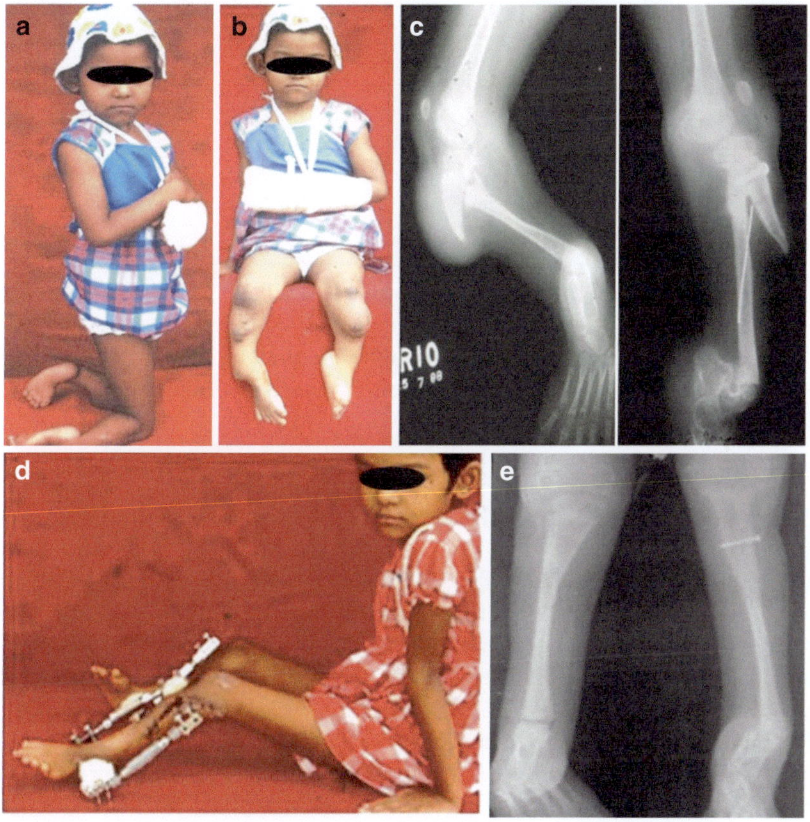

Fig. 19.43 (**a**) A 9-year-old girl had tibial hemimelia type-II came walking on the knees. Note the callosities below the knees on both the sides with deformed feet. She was operated outside for the loss of middle finger of the left hand and on the left leg (In POP). (**b**) Roentgenogram showing type-II tibial hemimelia with callosities. (**c, d**) Clinical picture during the course of the treatment. (**e**) Final roentgenogram shows healing at both the ends and hypertrophy of the fibula after Huntington procedure. The knees are functional and the ankles stable. (**f**) The patient stands in POP cast. (**g**) The legs are straight and lengthened. (**h**) The final picture of the girl

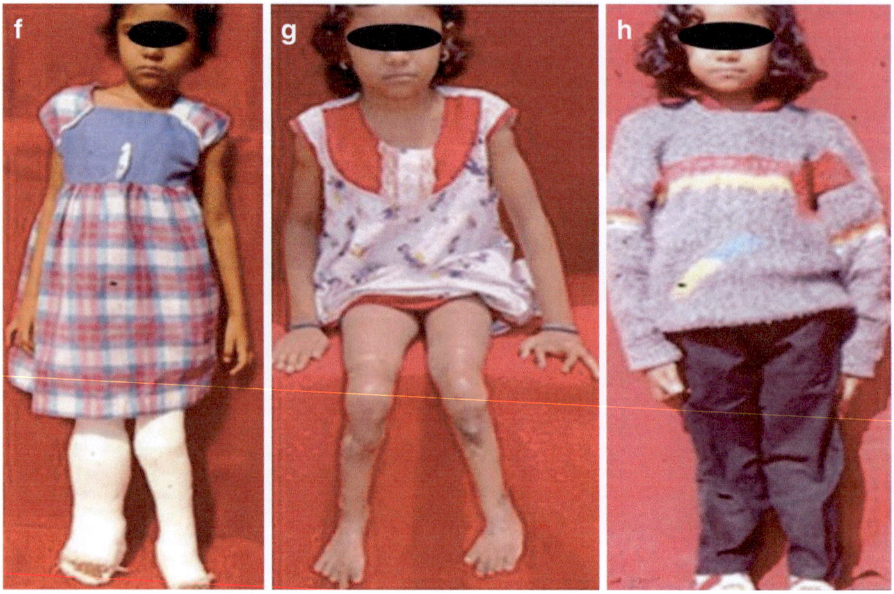

Fig. 19.43 (continued)

19.11 Neglected Tibial Hemimelia Type-I

It is advised to get the children born with tibial hemimelia type-I treated in a good reconstructive centre. Neglected cases result in severe deformity and gross disability (Fig. 19.44a–d).

19.11.1 Congenital Fibular Hemimelia

Congenital absence of the fibula, also called fibular hemimelia, has several manifestations. These range from complete absence of the fibula with missing lateral rays of the foot (i.e., terminal longitudinal deficiency) to absence of only a portion of the fibula without foot involvement, i.e., intercalary longitudinal deficiency (Figs. 19.45a, b and 19.46a–c).

The entity has been classified according to the degree of fibular deficiency. Type I includes all cases with a portion of the fibula present. In type IA, the proximal fibular epiphysis is distal to the level of the tibial growth plate, and the distal fibular physis is proximal to the dome of the talus. In type IB, there is 30–50% decrease in the length of the proximal fibula and distally, the fibula did does not support the ankle. In type II, the fibula is completely absent.

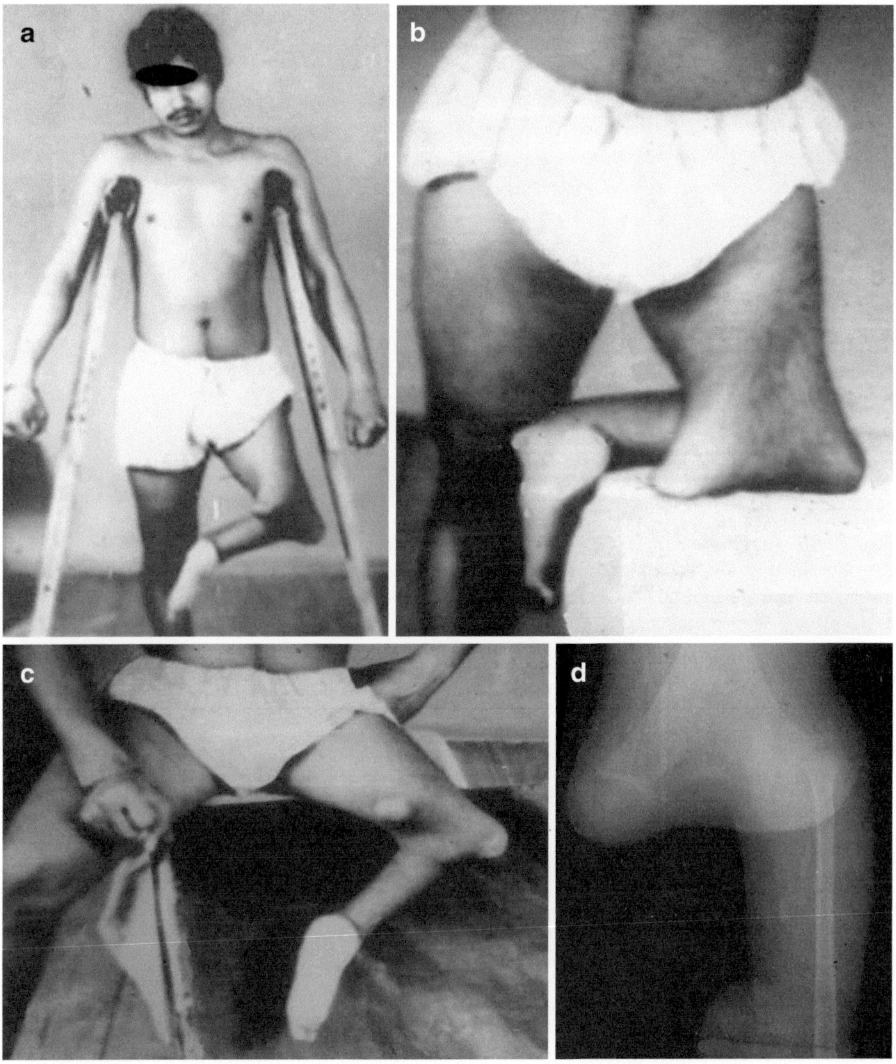

Fig. 19.44 (**a, b**) A neglected/untreated case of left sided tibial hemimelia type-I with bi-fid femur. The leg and the foot are under developed. (**c**) There is 90° flexion contracture at the knee and equinus at the ankle. (**d**) Radiograph of the knee and the leg reveals bi-fid femur and eccentric position of the fibula. The patient is totally disabled and walks with support

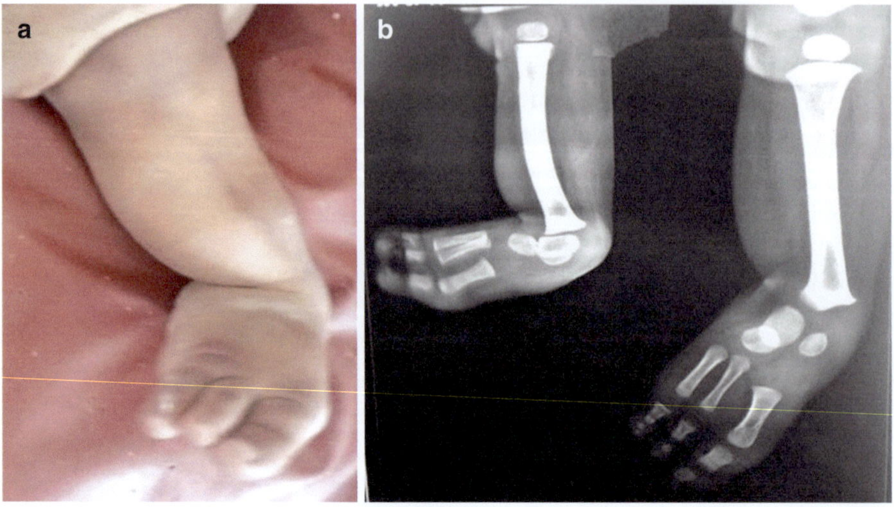

Fig. 19.45 (**a**) Clinical picture of the right leg reveals congenital calcaneovalgus deformity. Note the oligodactyly. (**b**) The roentgenogram reveals fibular hemimelia. The ankle and subtalar joints are malformed. The tarsal bones are hypoplastic

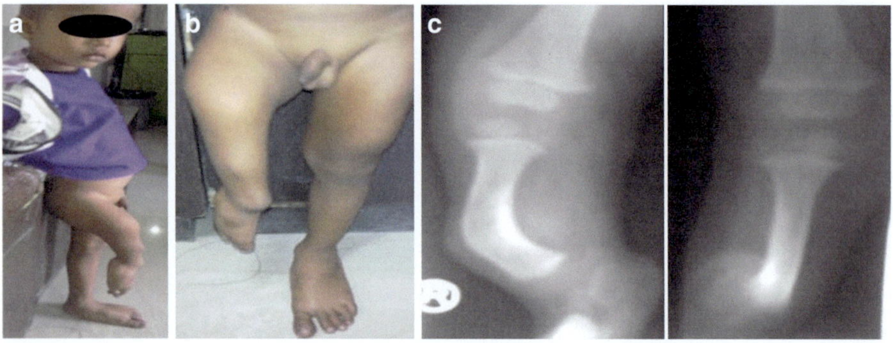

Fig. 19.46 (**a, b**) A 3-year-old boy with fibular hemimelia showing short and deformed leg with calcaneovalgus deformity. (**c**) Roentgenogram reveals absence of fibula with anterior bowing of the tibia. Note the hypertrophy of the tibia

The child with complete absence of the fibula clinically presents with an bowing of the tibia, an equinovalgus deformity of the foot, and a tarsal coalition (Figs. 19.47a, b and 19.48a–f). The tibiotalar joint is usually malformed, with fused talocalcaneal having a flat upper surface that articulates with the tibia in a valgus and equines position. The foot may be missing one or two lateral rays. There is a consensus that complete fibular hemimelia is best treated by performing an ankle disarticulation in early childhood and fitting a Syme-type prosthesis.

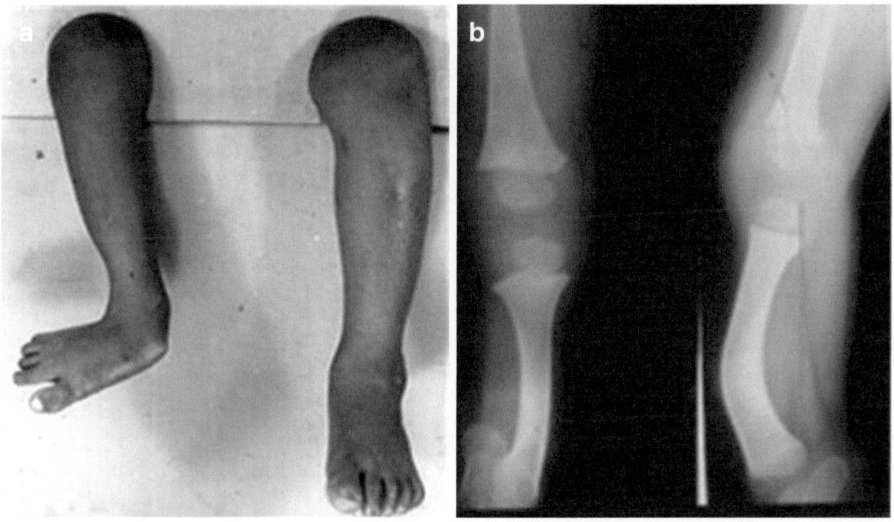

Fig. 19.47 (**a**) Clinical picture of a 25-year-old lady with fibular hemimelia on the right side with valgus foot. (**b**) Roentgenogram revealing the classical deformity already described

In order to obtain a stable axial position of the foot beneath the tibia, the procedure consisted of soft tissue releases together with Double Obligue Diaphyseal Osteotomy (DODO) of the tibia. The patients are able to walk well with their own feet with sensation and proprioception (Fig. 19.49a–t). This type of approach should be considered in preserving the foot rather than performing an amputation, which might be necessary in only selected cases (Fig. 19.50a–i).

Abnormal feet associated with fibular aplasia and deformities included calcaneovalgus, hind foot synostosis and deficient rays. The function of the foot depends on attaining a stable position beneath the tibia. Such children were operated on with early soft tissue release and tibiotalar arthrodesis. It is possible to preserve and lengthen the limb.It is vital that the foot is retained in a stable position beneath the tibia in order that satisfactory result is obtained. However, with a fear of an amputation some patients do not agree for any surgery and use a suitable orthotic appliance (Fig. 19.51a–d).

This patient was referred to Prof. G. Illazrov for management. Somehow his father reached me through one of my staff members at the Mullana Azad Medical College, New Delhi. I told him that the patient could be managed here, if he wants and he readily agreed. Destiny Prevails.

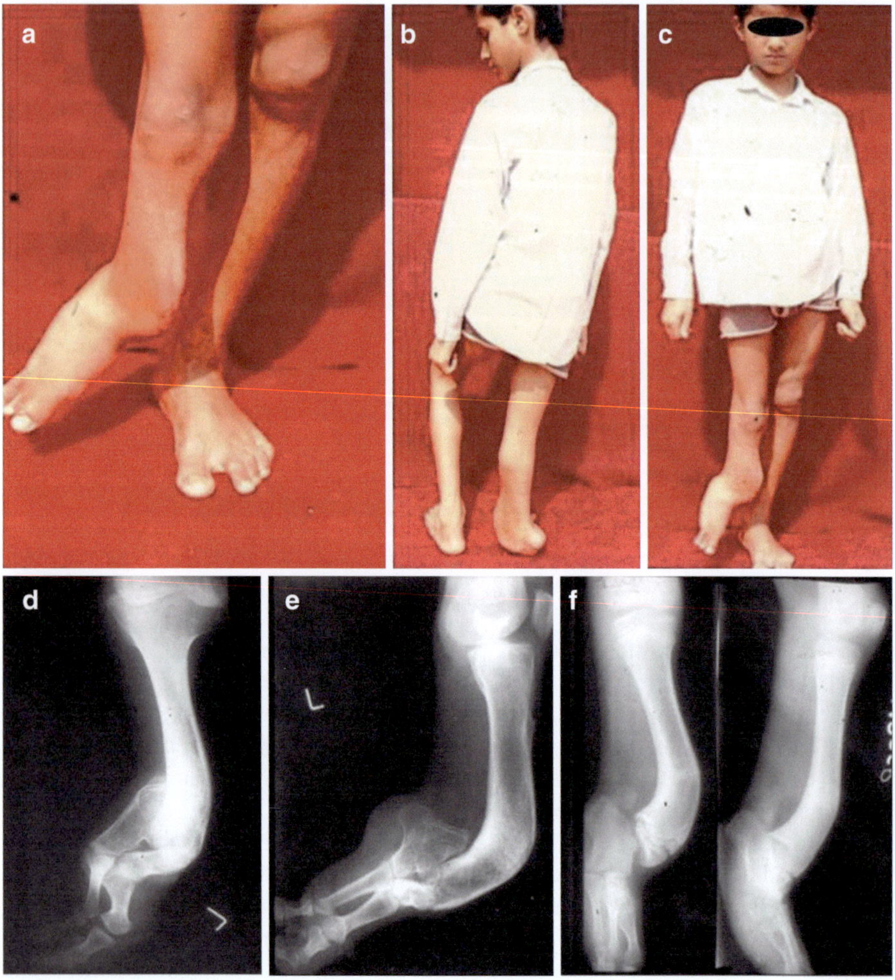

Fig. 19.48 (**a, b, c**) Clinical picture of a 15-year-old boy with fibular hemimelia revealing kyphosis at the distal leg, valgus of the right foot with oligodactyly. (**d**) Roentgenogram of the same patient showing the deformity and two medial rays. (**e, f**) Roentgenogram of the patient reveals grossly disorganized ankle. The talus is absent and the calcanium has posterior elongation (Due to the pull of the Tendo-Achilles). The second ray is fused with the calcaneus

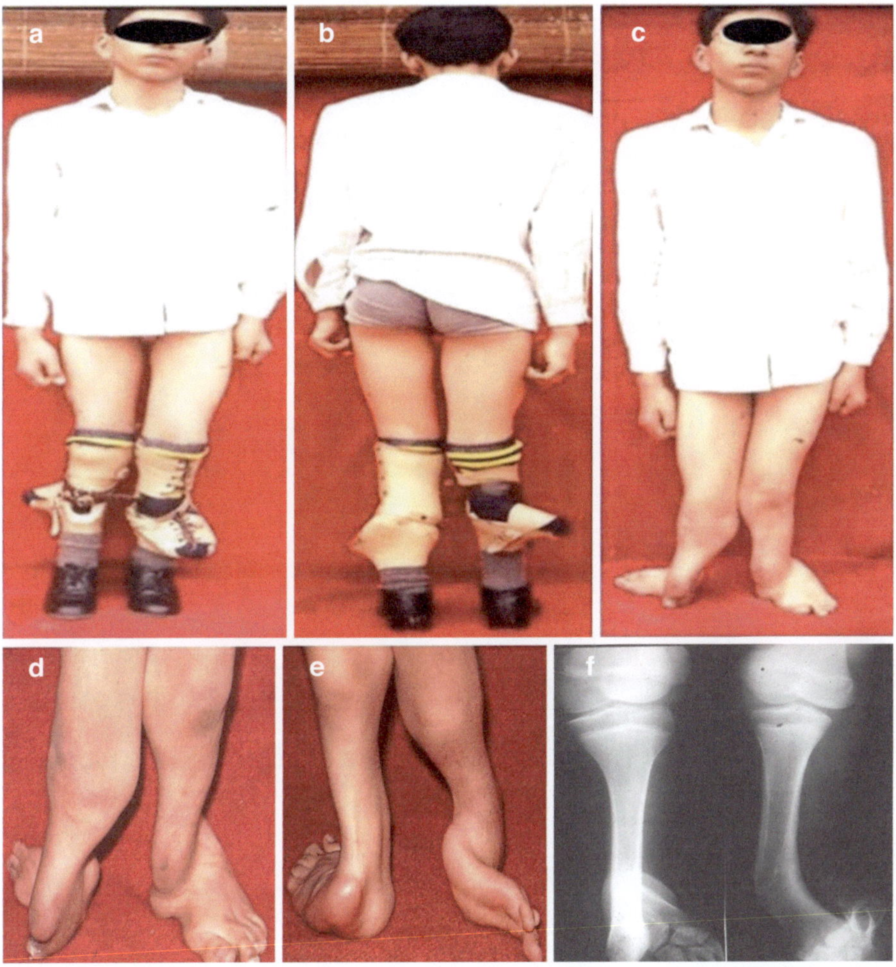

Fig. 19.49 (**a, b**) A 16-year-old boy with short and deformed legs, came walking on orthotic devices. (**c, d, e**) He had bilateral fibular hemimelia with severe deformities of the legs, ankles and the feet. Anatomically it was difficult to define the abnormalities—a very rare presentation. (**f**) Roentgenogram on the left side reveals severe kyphotic deformity of the distal tibia with disorganization of the ankle and foot. Note the elongated and deformed calcaneus. (**g**) Angiogram of the right leg reveals only posterior tibial artery. (**h**) Extensive soft tissue release and DODO. Lengthening and bilateral ankle arthrodesis. (**i**) Roentgenograms reveal straightening of the left leg after DODO, skeletal traction and application of POP Cast. (**j**) The final roentgenogram of the left leg with fusion of the ankle and tarsal joints. (**k**) The patient stands with left side corrected with plantigrade foot. (**n**) On the right side he still stands on orthotic appliance (To be corrected). (**l, m**) Roentgenogram of the right leg revealing the DODO and the Steinmann pins passed for traction on either side of the tibia (DODO). (**n**) Final picture of the leg after union of the DODO. (**o**) After the management is completed, the patient stands on his own legs and feet with sensations and proprioception. Note the congenital absence of three lateral rays on the right side (Oligodactyly). (**p, q, r**) The presentation after a follow-up of 25 years. The result did not deteriorate with time. Note the bi-fid hand on the left side. (**s, t**) The roentgenogram reveal hypertrophy of the tibia on either side and sound arthrodesis at the ankle and subtalar joints. The patient also had Oligodactyly of the left hand with the absence of two medial rays

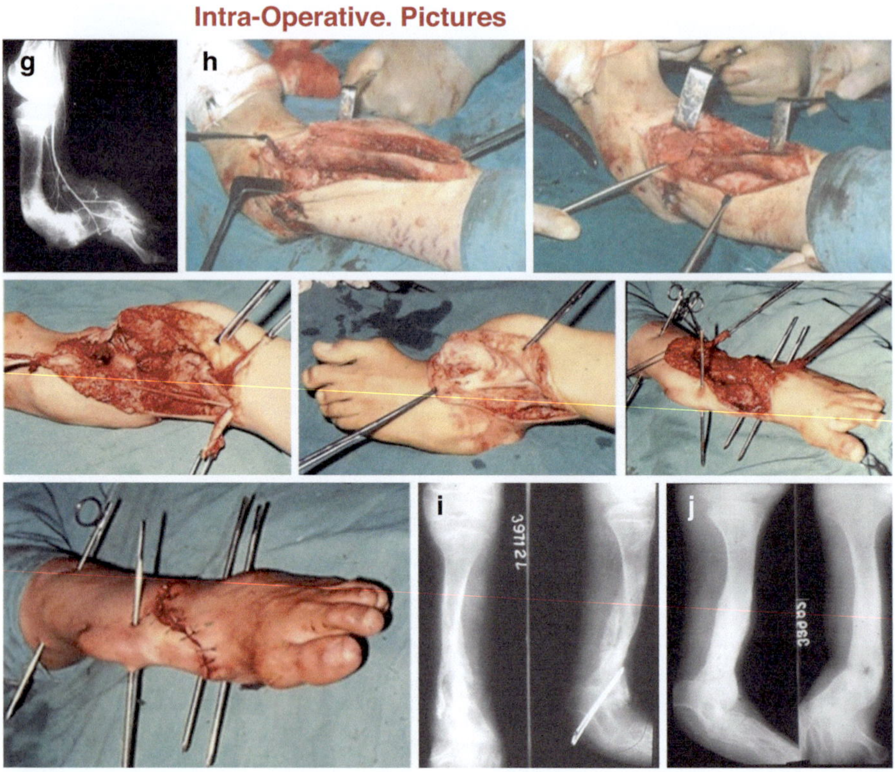

Fig. 19.49 (continued)

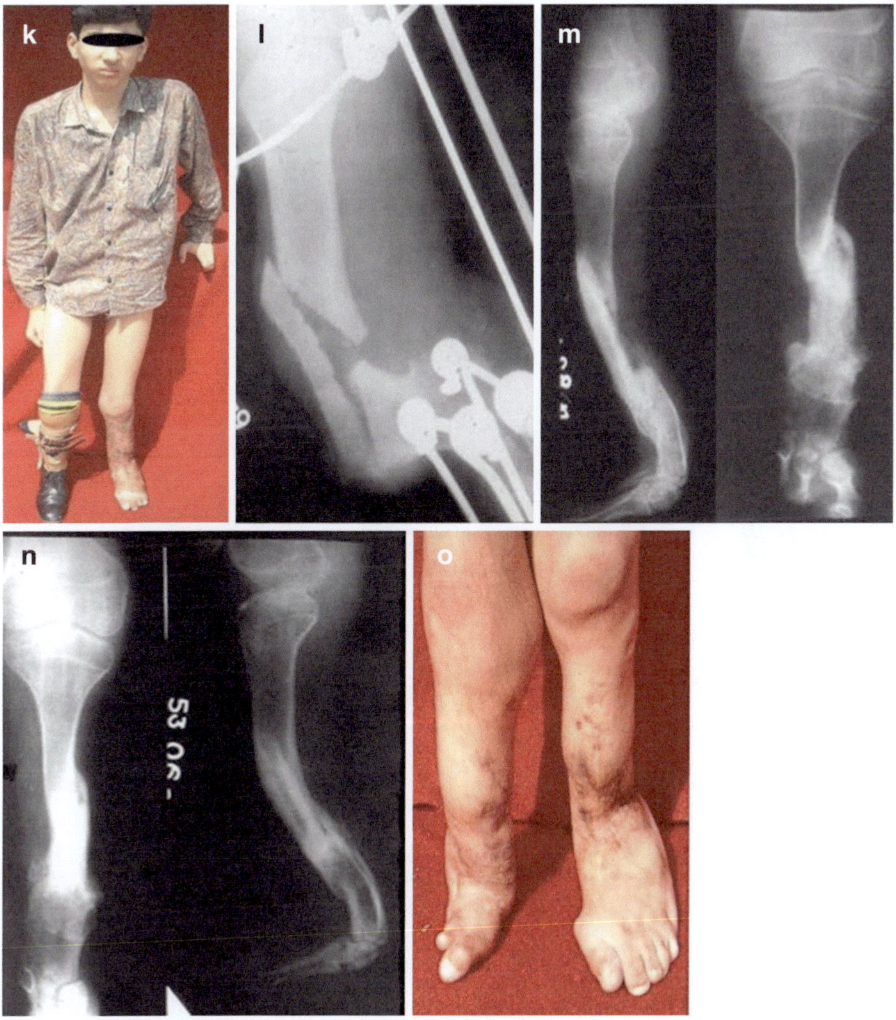

Fig. 19.49 (continued)

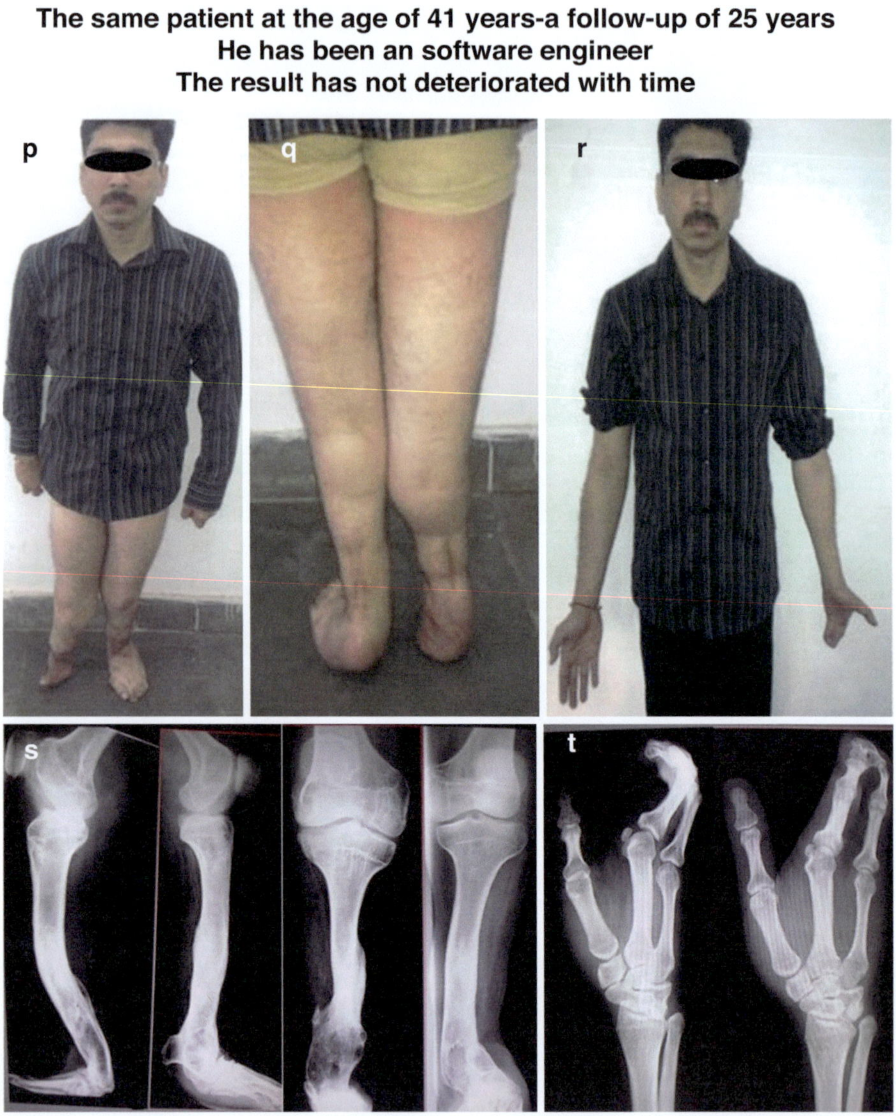

Fig. 19.49 (continued)

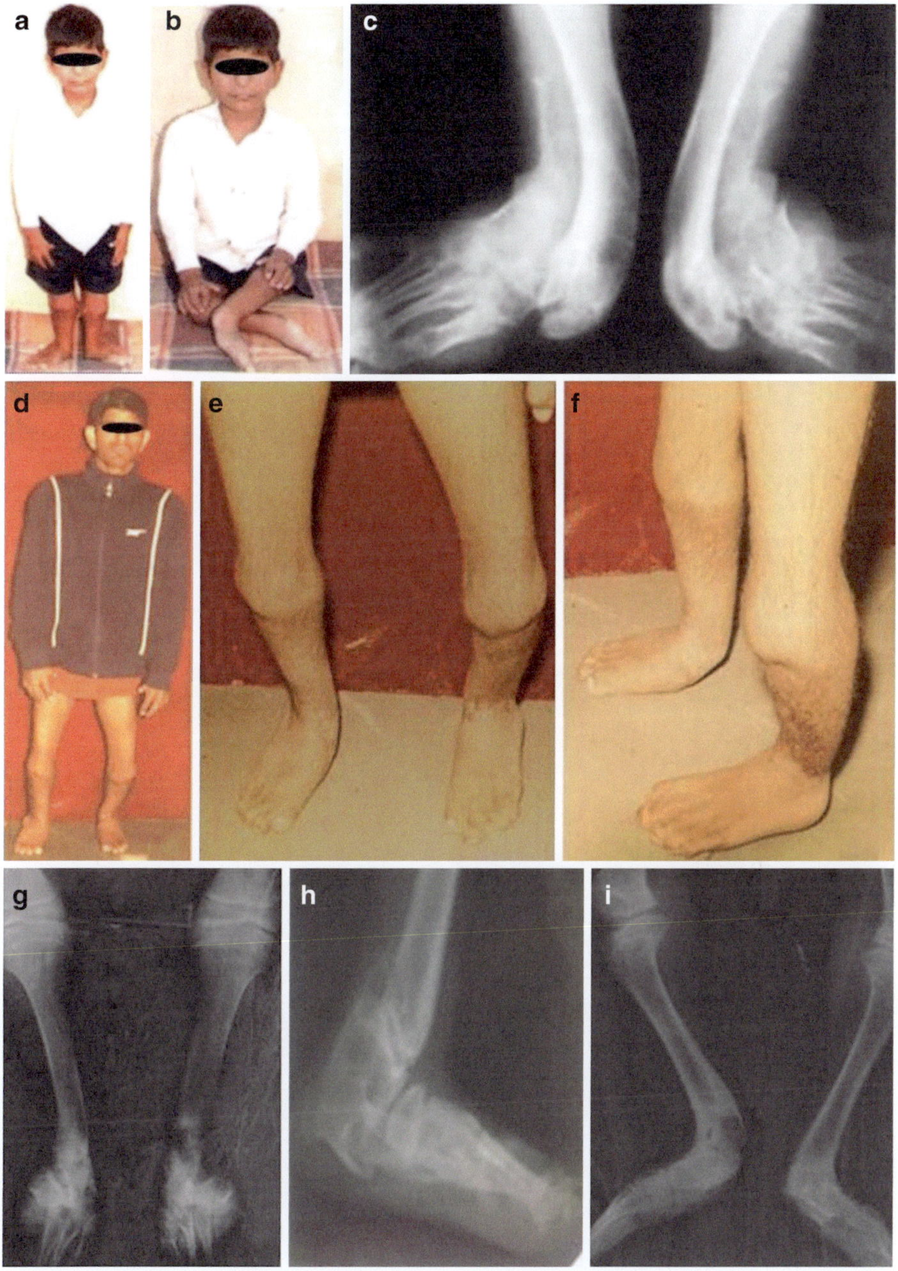

Fig. 19.50 (**a, b**) Clinical presentation of a severe case of bilateral fibular hemimelia revealing the feet in extreme valgus. The patient has been walking on the distal tibial ends. (**c**) Roentgenogram reveals the feet are in extreme valgus position. Note the bilateral analog as a band of fibrous tissue. (**d, e, f**) Post-operative picture of the same patient (AP & Lat. views) reveal fully plantigrade feet after extensive soft tissue release and a well-planned pan-talar arthrodesis on both sides. (**g, h, i**) Roentgenogram of the both the ankle and feet revealing healed pan-talar arthrodesis. The patient could walk well after a single and definitive surgical procedure

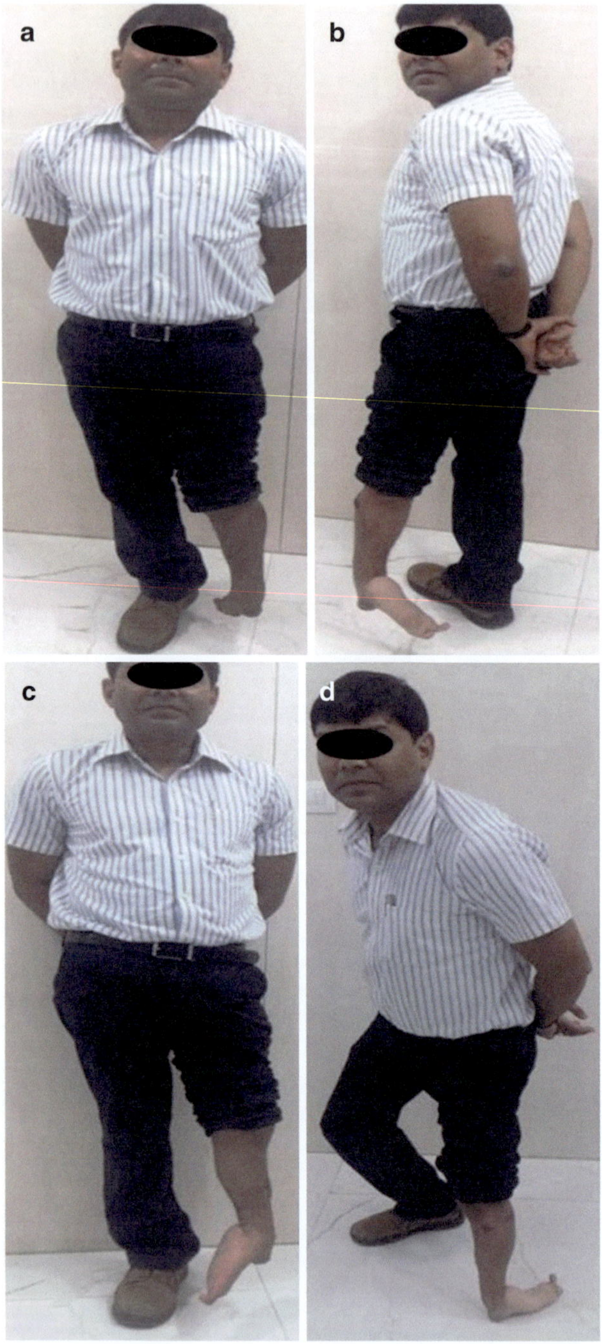

Fig. 19.51 (**a, b, c, d**) A 40-year-old patient with fibular hemimelia on the left side revealing the deformity of the ankle and foot, shown at different angles

19.12 Congenital Pseudoarthrosis

Congenital pseudarthrosis of long bones occurs most commonly in the tibia and fibula. It may also affect other tubular bones including the femur, radius and ulna; it may also occur in the clavicle. It is characterized by progressive anterior angulation of the distal portion of the leg and is associated with bone fracture and non-union. Neurofibromatosis or skin manifestations of this condition occur in up to 50 percent of patients with pseudarthrosis. The defect may be noted at birth but usually presents later in infancy or early childhood. Some cases develop in late childhood or adolescence. In most cases there is a pathological fracture, bone pain and deformity. The cortex is thin. An area of radiolucency is seen at the site pseudarthrosis. Characteristically, there is progressive tapering of the bone ends towards the site of pseudarthrosis.

Fibrous dysplasia and congenital ring syndrome also have been described in coincidence with congenital tibial pseudarthrosis. Neurofibromas are rarely found at the site of pseudarthrosis. MR imaging of congenital pseudarthrosis provides valuable information on the extent of the disease and can be accurately evaluated.

19.12.1 Congenital Pseudoarthrosis of the Radius

Congenital Pseudoarthrosis of the Radius is an extremely rare condition. It is usually associated with neurofibromatosis (von Recklinghausen's disease) or fibrous dysplasia. Various treatment options have been explored with varying degrees of success and the reports have demonstrated successful healing (Fig. 19.52a–h). Review of the literature discloses the extreme difficulties in treating this condition. It was also found that vascularized fibular graft has been used successfully.

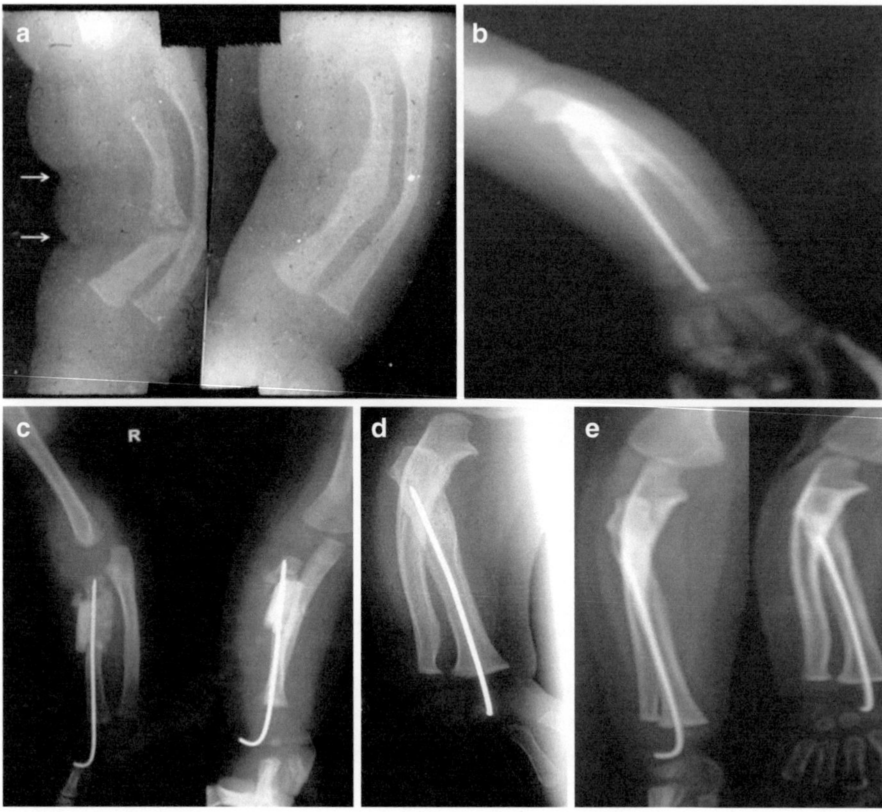

Fig. 19.52 (**a, b**) Roentgenogram shows congenital pseudoarthrosis of the right radial shaft. The ulna is also deformed. Note the congenital furrows in the soft tissues of the forearm which is diagnostic in such a case (Arrows). (**c**) Roentgenogram reveals excision of the pseudoarthrosis and reconstruction with maternal iliac crest graft fixed with a K-wire. (**d**) Roentgenogram showing complete union merger of the maternal graft with the host bone. (**e**) Roentgenogram showing complete merger of the graft. Note the remodelling. (**f, g**) Post-operative clinical pictures of the patient with straight right forearm which is fully functional. (**h**) Radiograph reveals complete union and remodelling of the radial shaft. The forearm has normal function

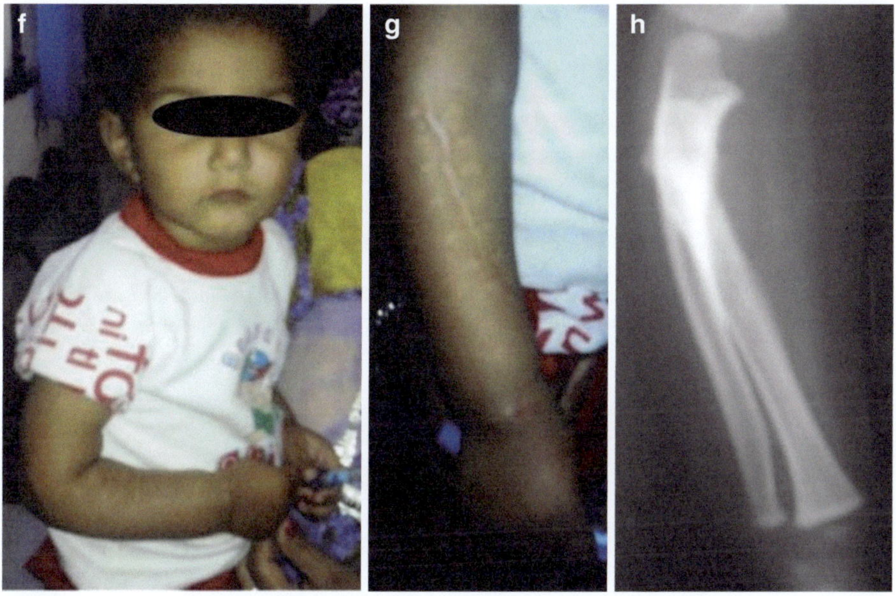

Fig. 19.52 (continued)

19.13 Congenital Pseudoarthrosis of Tibia

Congenital pseudoarthrosis of tibia is a rare disorder and poses one of the greatest challenges of orthopaedic practice. It refers to a spontaneous fracture which progresses to non-union. It is characterized by progressive anterior angulation of the distal portion of the leg and is associated with bone fracture and non-union (Fig. 19.53a, b). This is rarely present at birth but commonly develops during the first 18 months of life. The distal end of the tibia is the most frequently affected site associated with congenital anterolateral tibial bowing which progresses to pseudoarthrosis (Fig. 19.54a, b). There is no predominance of the right or left leg. Bilateral cases are very rare.

Several conditions may be associated with congenital pseudoarthrosis. Neurofibromatosis or skin manifestations of this condition (café au lait spots) occur in up to 55% of patients with pseudoarthrosis. Rarely fibrous dysplasia has also been reported.

Radiologically the level of bowing or pseudoarthrosis is at the junction of the level middle and distal thirds of the tibia. A fibular pseudoarthrosis is usually present. The appearance of the involved segment of the tibia may be constricted, cystic or sclerotic. Those tibiae presenting as cystic have a better prognosis than those which are dysplastic or fractured at birth.

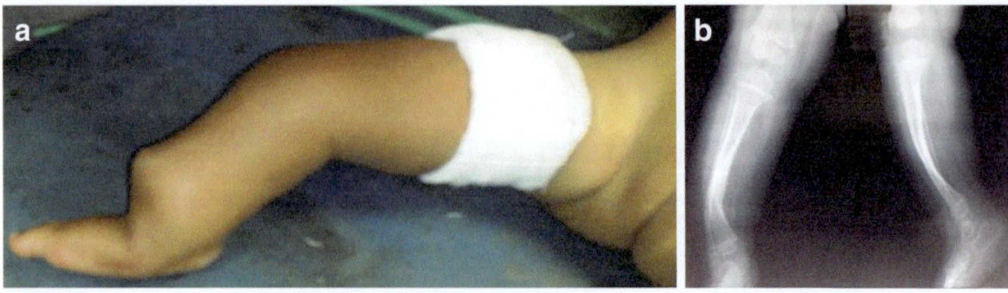

Fig. 19.53 (**a**) Clinical picture of a 2-year-old patient with congenital pseudoarthrosis of tibia and a Kyphosis deformity. (**b**) Roentgenogram revealing pre-pseudarthrosis tibia

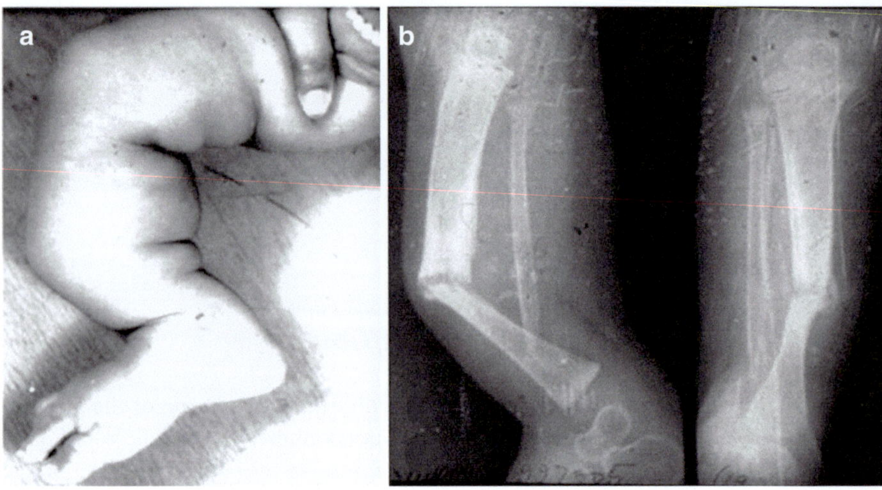

Fig. 19.54 (**a, b**) Clinical picture and the radiograph of a classical patient with congenital pseudoarthrosis of the tibia

Many methods have been described to manage this difficult condition. Intramedullary nail, plating, vascularised bone graft, external compression-distraction treatment and electrical stimulation have been recommended. However, there is no standard accepted procedure for its management.

The author has managed such cases by extensive excision of the involved tibia and reconstructed the gap by maternal non-vascularized fibular strut fixed in the proximally leftover tibia and distally into the body of the talus and calcaneus, with high rate of success (Fig. 19.55).

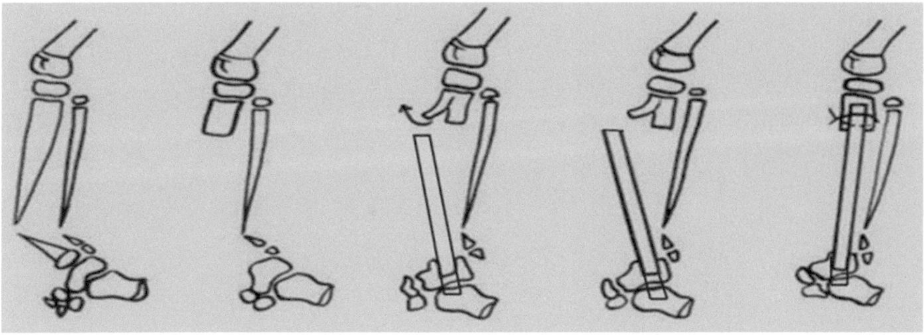

Fig. 19.55 Diagrammatic steps of the surgical technique described in the text. The diseased tibia has been excised and the maternal fibula has been used to bridge the gap, as described

19.13.1 The Author's Technique of Maternal Fibular Grafting

A suitable incision is made over the leg to expose the tibia and the pseudoarthrosis site (Fig. 19.56a). The pseudoarthrotic site is excised enbloc along with normal looking tibia (Fig. 19.56b). In case the distal end is small, it is excised totally. However, if it is long, the distal end of the tibia may be left behind. A long strut of non-vascularised maternal fibular graft of suitable length is fixed distally after making a slot in the body of the talus and calcaneous. Proximally a lid is made at the distal end of leftover tibia (Fig. 19.56c). The proximal end of the fibular strut is now placed into the tibial medullary canal and the lid is placed back in situ and tied with chromic catgut circumferential wire (Fig. 19.56d). If considered necessary to stabilize the graft, a long Kirschner wire may be passed from the heel to the calcaneus, talus, through the graft and the proximal end of the tibia (Fig. 19.56e). The wound is closed in layers. An above knee POP cast is applied to provide further support to the construct (Figs. 19.57a–e, 19.58a–f, 19.59a–f, 19.60a–i, 19.61a–d, 19.62a–c, and 19.63a–c).

A rare case of bilateral congenital pseudoarthrosis of the tibiae managed with maternal fibular graft: a follow-up of 53 years.

This patient has bilateral congenital pseudoarthrosis of both tibiae and was operated in three different centres before she came to our outpatient department (Fig. 19.64a, b). The pseudoarthrosis on both sides persisted (Fig. 19.64c, d).

The patient was operated after excision of the pseudoathrotic site on both sides and the gap was bridged by maternal fibular grafts on either side. The left side had healed well. Although the right side had also healed but with anterior angulation (Fig. 19.64c, f). It was decided to perform a corrective osteotomy to treat the angulation. The corrective osteotomy had healed well and a straight leg was achieved (Fig. 19.64g–p).

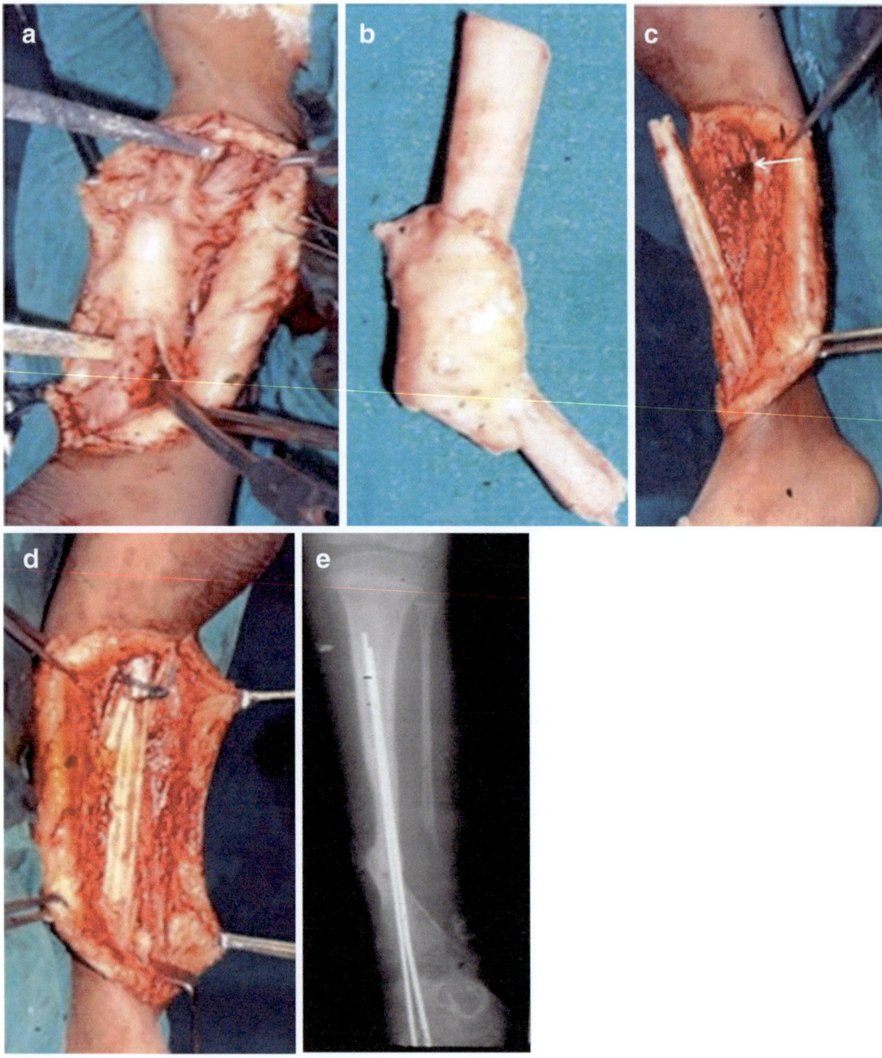

Fig. 19.56 (**a**) Intra-operative photographs of the surgical procedure—(**a**) The Exposure. (**b**) The excised specimen. (**c**) Maternal fibular strut being passed through the talus and calcaneus (arrow). (**d**) The proximal end of the fibula is stabilized in the tibial end. (**e**) Roentgenogram reveals a twin Kirschner wires passed through the fibular strut and the proximal tibial end. However, the wires were not removed for next 2 years as a support to the graft

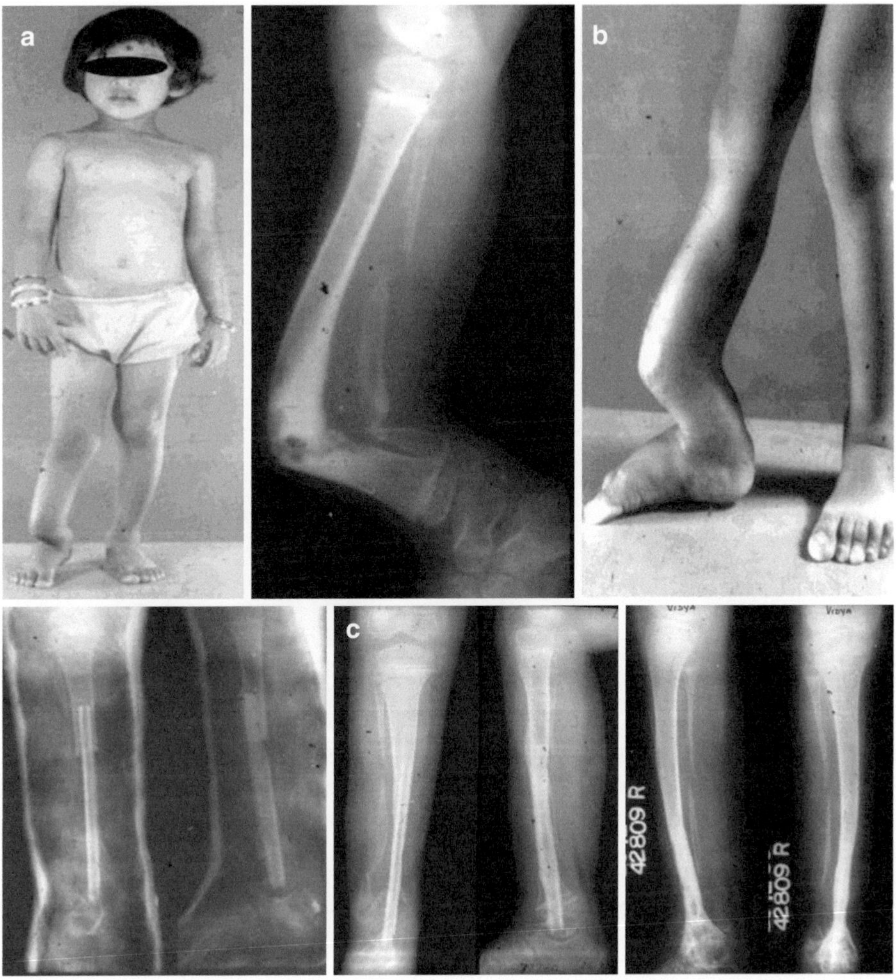

Fig. 19.57 (**a, b**) Clinical picture and the roentgenogram of a young girl with congenital pseudoarthrosis of tibia. (**c**) Post-operative roentgenograms taken at the various stages of management. (**d, e**) The roentgenogram reveals complete healing. However, the patient stands in a protective boot

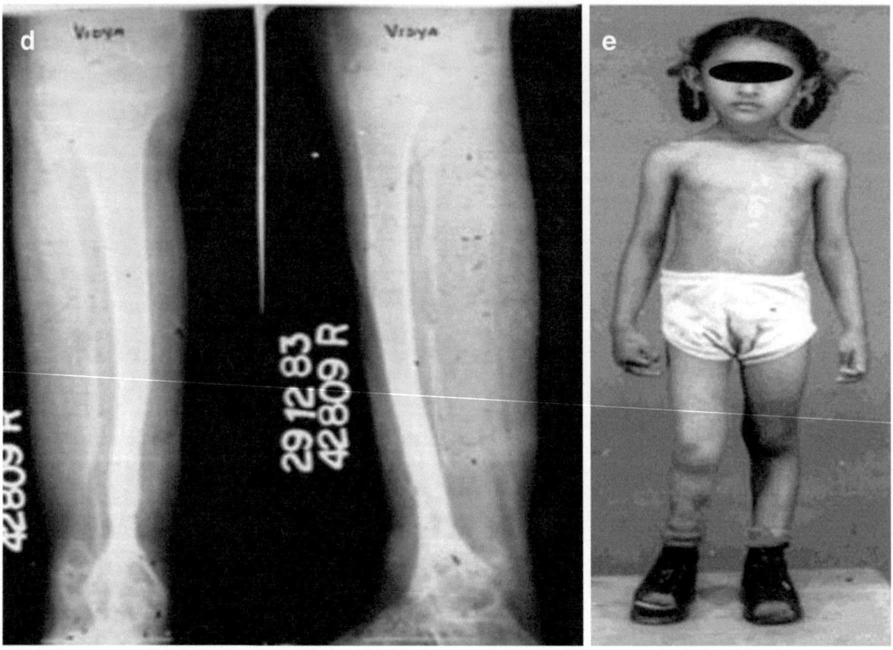

Fig. 19.57 (continued)

A 20-year-old boy was observed with dual congenital pseudoathrotic lesions of the right tibia. Roentgenogram revealed a double pseudoathrotic lesion, a very rare anomaly. Patient did not agree for treatment (Fig. 19.65a, b).

19.13.2 Congenital Postero-Medial Bowing of the Tibia

The term "Congenital kyphoscoliotic tibia" has been used for congenital angulation of the tibia with anterolateral or postero-medial curvature (Fig. 19.66a–e). The postero-medial bowing is an entity which is more uncommon than anterolateral. Postero-medial bowing of the tibia procedures a striking radiographic and clinical appearance, but usually the bowing spontaneously resolves. The dorsum of the foot usually touches the lower third of the leg. There may be associated calcaneovalgus and other congenital anomalies of the ipsilateral extremity.

A conservative approach to the management is advocated. It is a benign condition which responds adequately to corrective plaster casting and bracing (Fig. 19.67a–c). Residual leg-length discrepancy may occur. Only a few cases may require surgical correction.

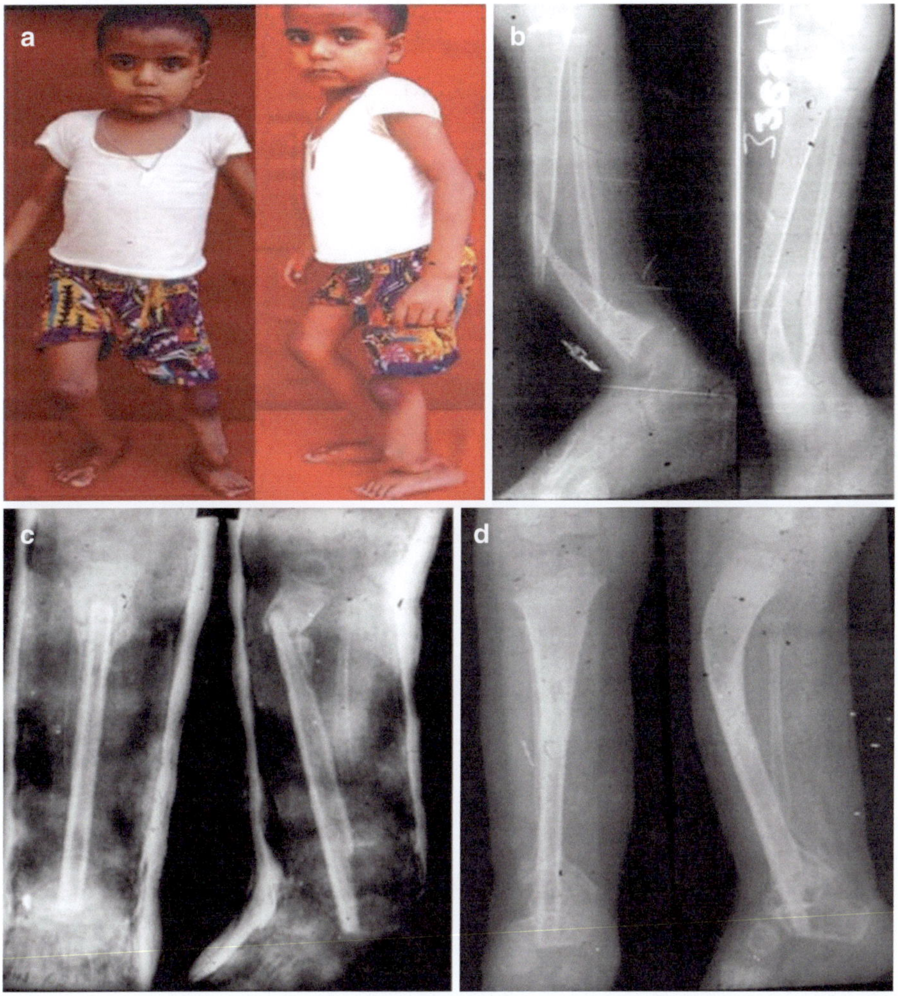

Fig. 19.58 (**a**) A child with classical congenital pseudoarthrosis of the left tibia. (**b**) The roentgenogram showing the lesion. Note the tapering ends of the tibia and fibula at the site of pseudoarthrosis. (**c, d**) Roentgenograms taken at various stages of the treatment. The final picture shows complete healing with union at both the ends. Note the remodelling of the host-graft junction at the proximal end. (**e**) Roentgenogram reveals complete healing of the maternal fibular graft at both its ends. (**f**) An angiogram taken after the healing reveals near normal vasculature

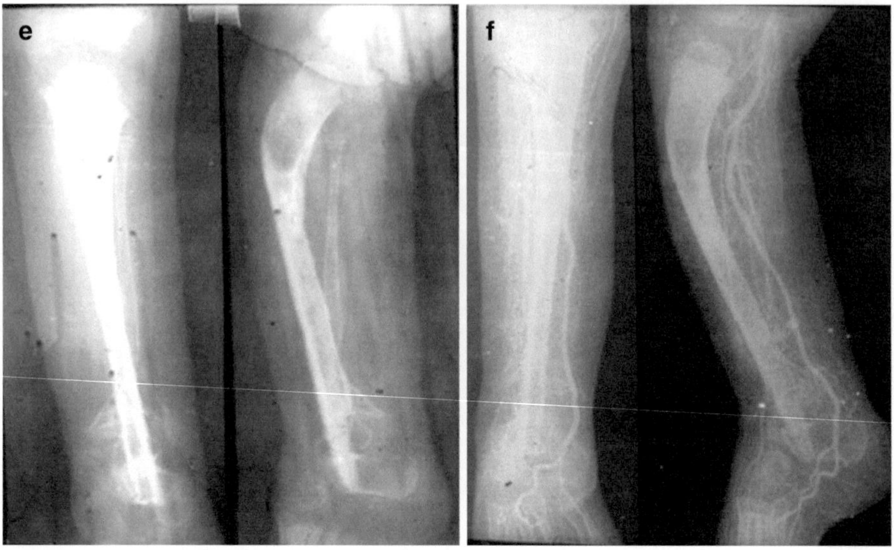

Fig. 19.58 (continued)

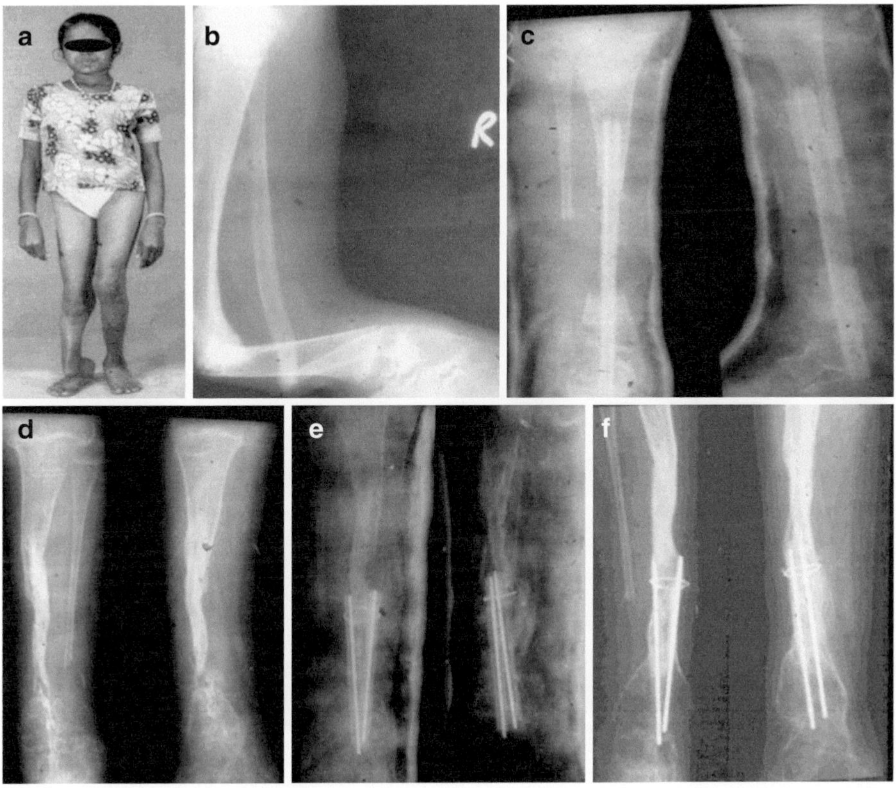

Fig. 19.59 (**a, b**) Clinical picture and roentgenogram of a 14-year-old girl revealing congenital pseudoarthrosis of the tibia. (**c**) Post-operative roentgenogram. (**d**) The patient had a fracture at the sight of pseudoarthrosis. (**e**) Two struts of contra lateral fibula with inlay Kirschner wires stabilized with circumferential wires and POP cast. (**f**) The roentgenogram of the construct shows complete healing

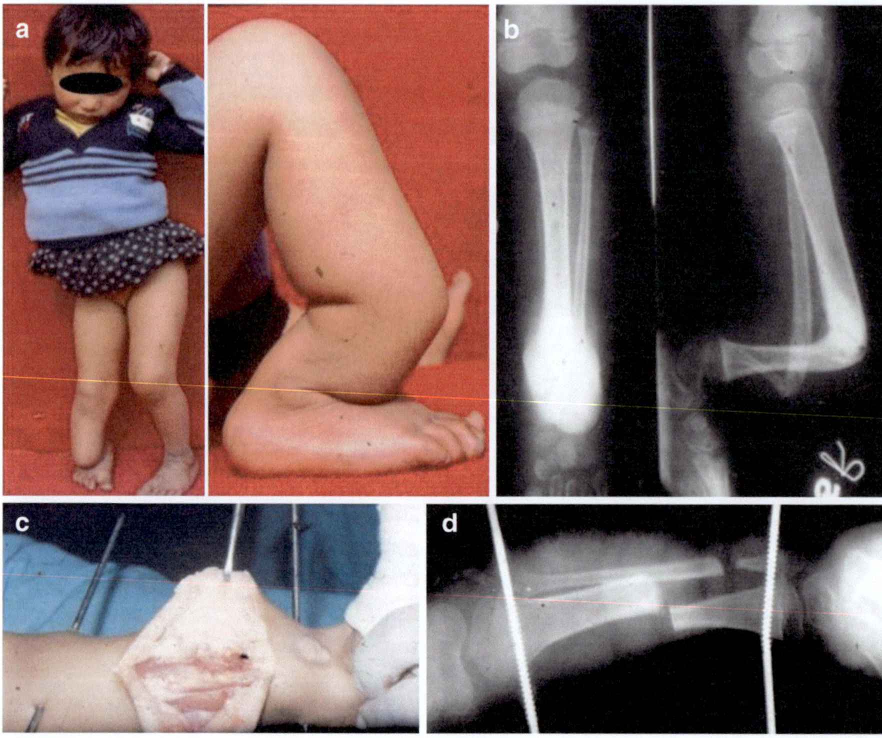

Fig. 19.60 (**a**) Clinical picture of a 3-year-old girl with severe deformity of pseudoarthrosis of the right tibia. (**b**) Roentgenogram shows classical pseudoarthrosis of tibia. (**c**) Intra-operative picture shown after excision of the lesion. (**d**) Roentgenogram reveals the excision after which the leg had become very short. (**e, f, g**) Tibial traction on either side of the tibia subsequently lead to the lengthening of the leg. Intra-operative picture of the second surgery reveals bridging by maternal fibula supported with extra-long Kirschner wires of suitable diameter. (**h, i**) Roentgenograms reveal healing. However, the wires were not removed for next 2 years as a support to the graft

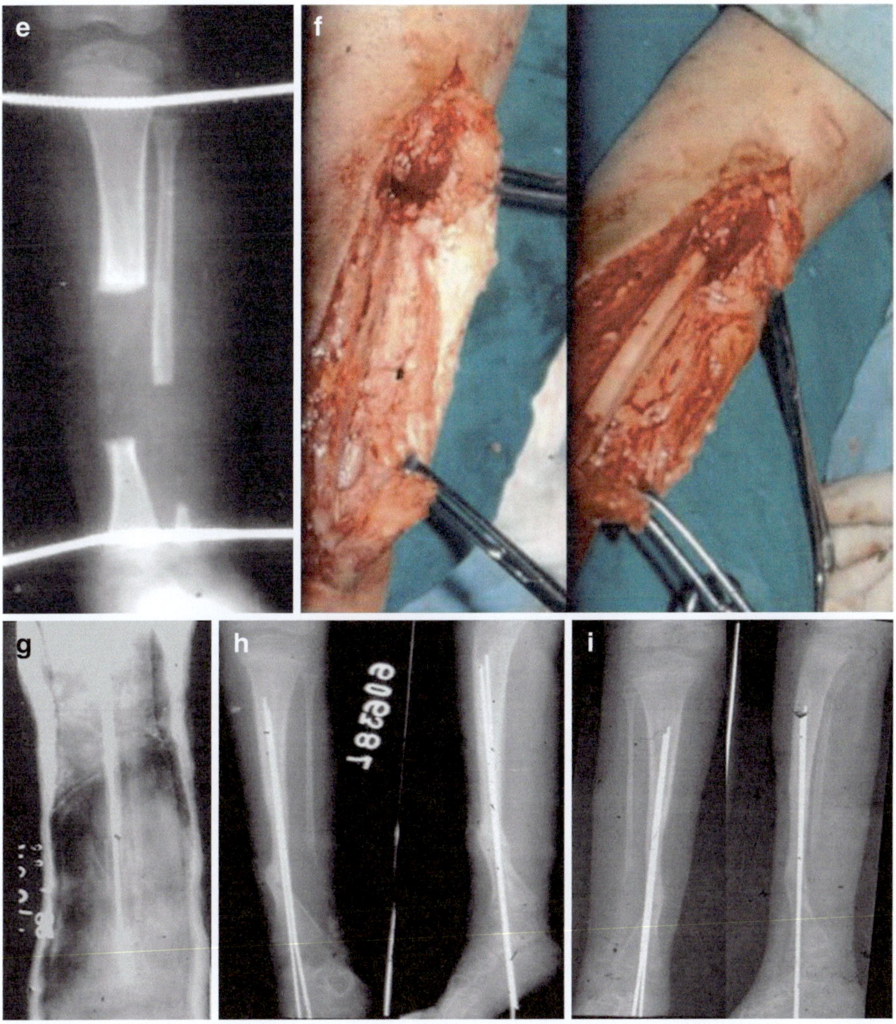

Fig. 19.60 (continued)

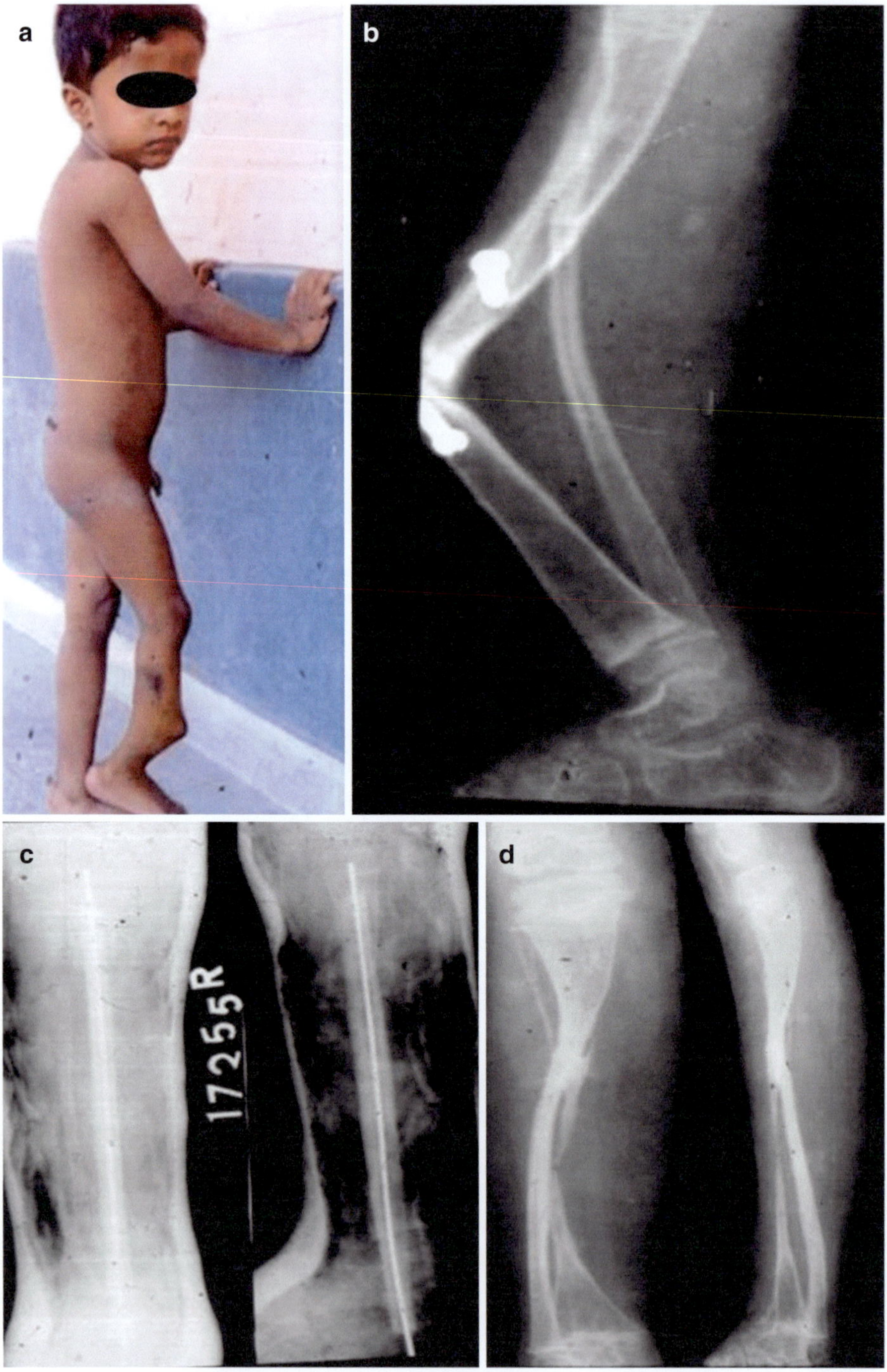

Fig. 19.61 (**a**) A 6-year-old boy with congenital pseudoarthrosis of tibia right side. (**b**) Roentgenogram shows failed surgery which was done outside. (**c**) Roentgenograms revealing excision of the lesion and gap bridge by mother's fibula stabilized by a long Kirschner wire. (**d**) The final picture revealed tibia fibular union. The extremity needed protective brace

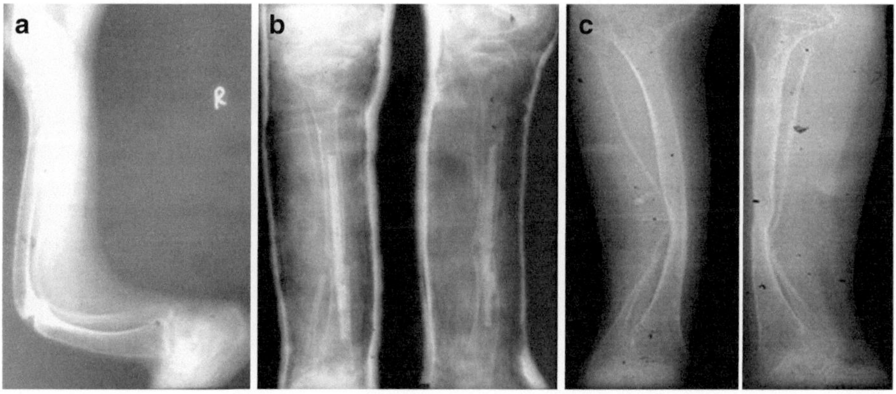

Fig. 19.62 (**a**) Roentgenogram revealing right sided congenital pseudoarthrosis of tibia. (**b**) Managed by maternal fibula. (**c**) The final roentgenogram showing healing, though in mild valgus

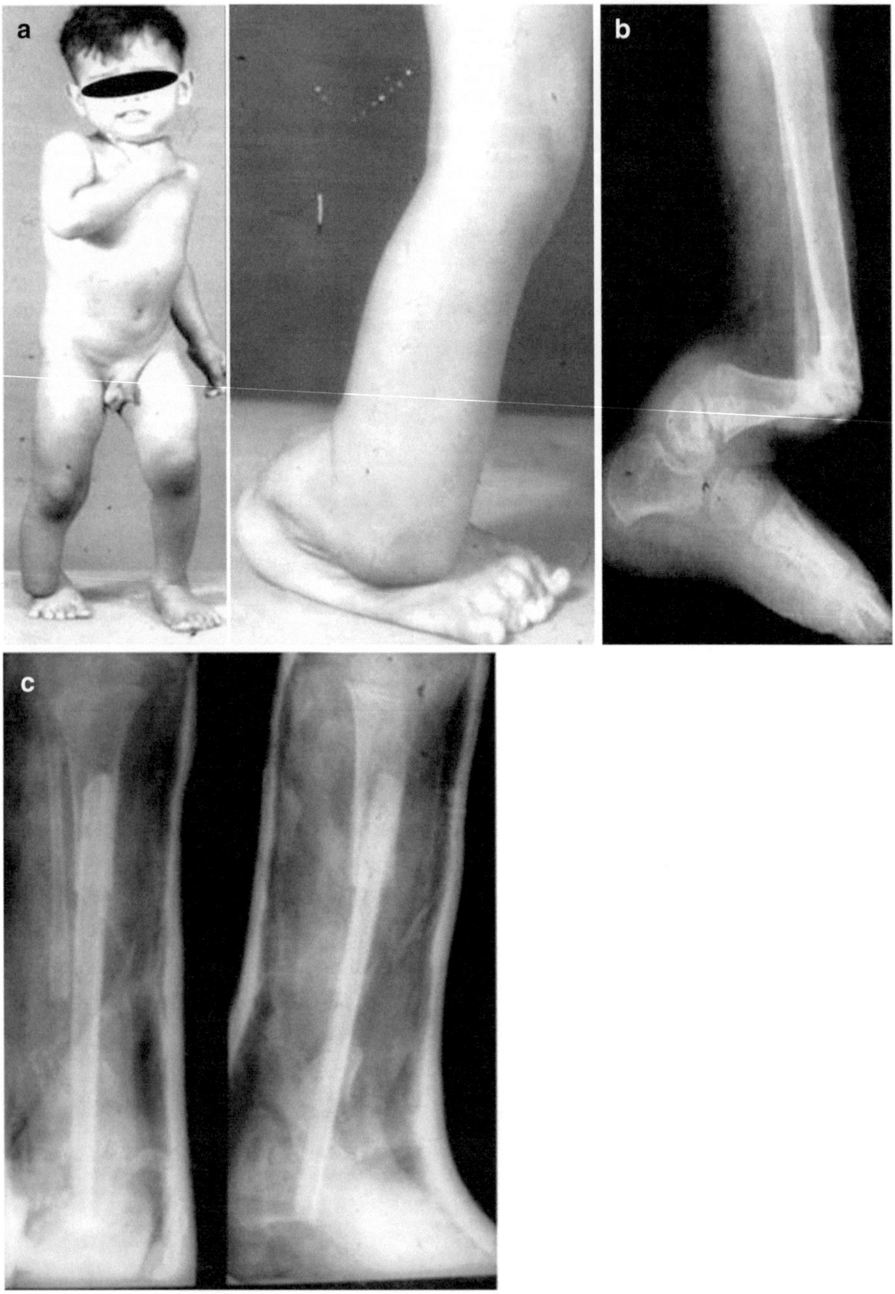

Fig. 19.63 (**a**, **b**) Congenital pseudoarthrosis of the right tibia with severe angulation. (**c**) Excision of the lesion bridged by the mother's fibula, stabilized in the POP cast

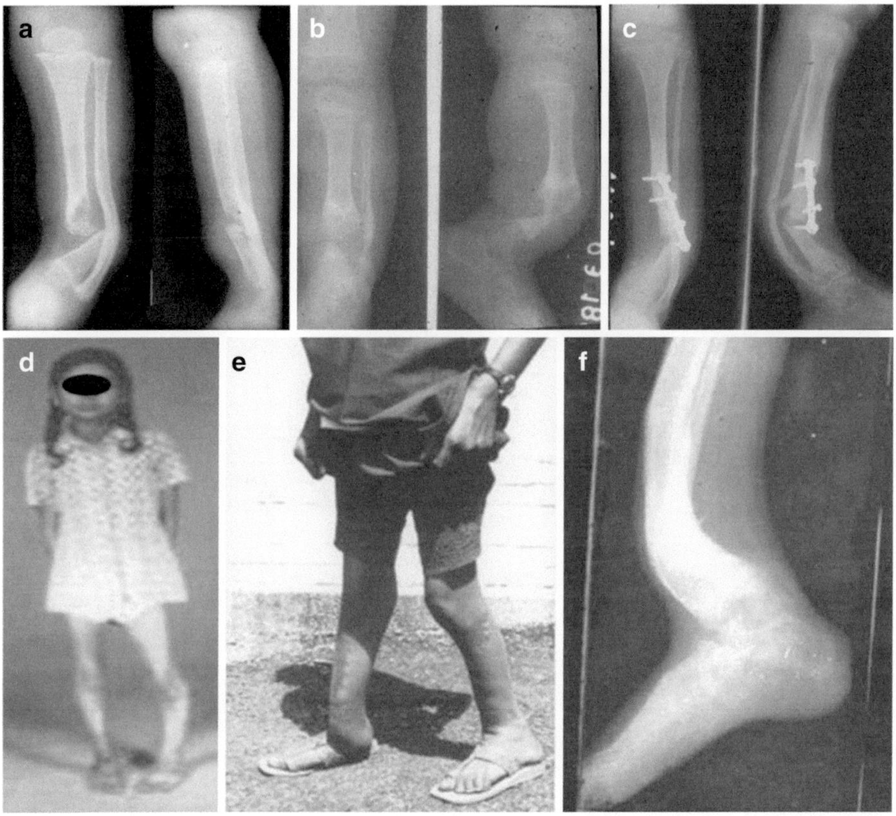

Fig. 19.64 (**a, b**) Roentgenogram of a patient with bilateral congenital pseudoarthrosis of the tibiae—a very rare presentation. (**c**) The patient had many operations in other centres but failed. The roentgenogram reveals failed plating. (**d**) The patient stands on ununited pseudoarthrosis with recurvatum and the left knee. (The patient had come to us at this stage). (**e**) The patient revealed union and the distal third of tibia but had severe anterior angulation. (**f**) The roentgenogram show complete healing but with angulation. (**g**) A corrective osteotomy was performed which had healed well. Note the signs of the healed osteotomy (arrow). (**h**) The patient stands with both the legs straight and treated. (**i**) Roentgenograms taken after sound healing of the corrective osteotomy on the right side. (**j**) A healed fibular graft can be seen on the left side. (**k, l**) The clinical picture of the patient at the age of 15 years. (**m, n**) Roentgenograms of both the legs showing complete healing. The signs of a corrective osteotomy can still be observed on the right side (Arrow). She is married and has kids. (**o, p**) Patient and radiographs at the age of 53 years

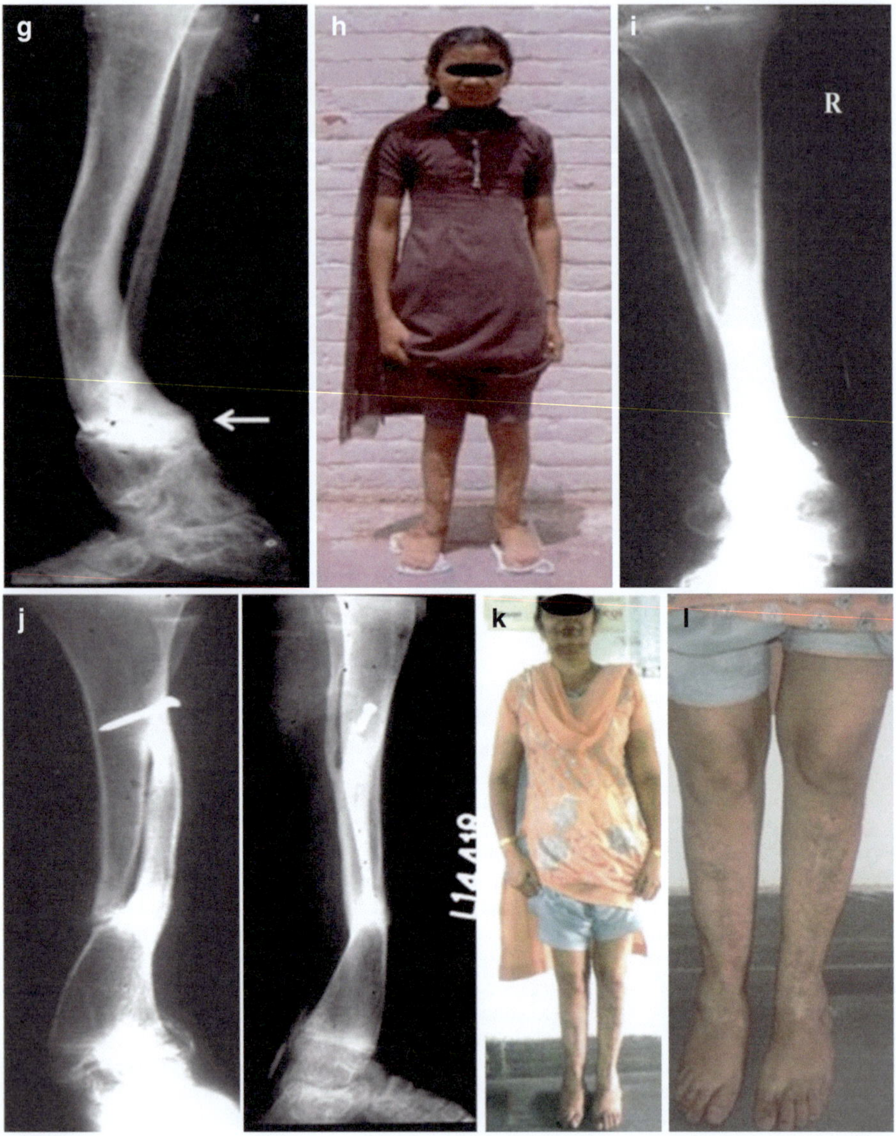

Fig. 19.64 (continued)

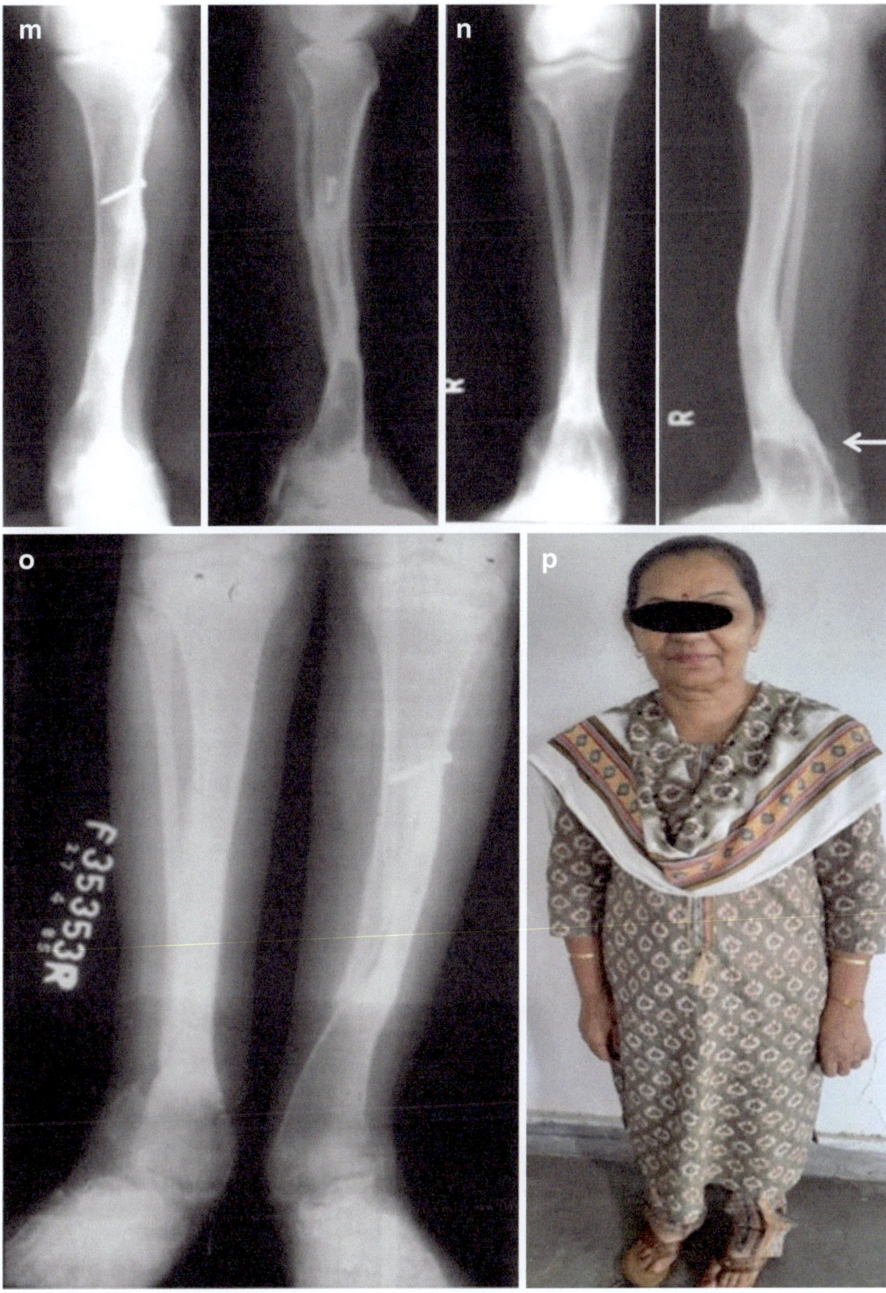

Fig. 19.64 (continued)

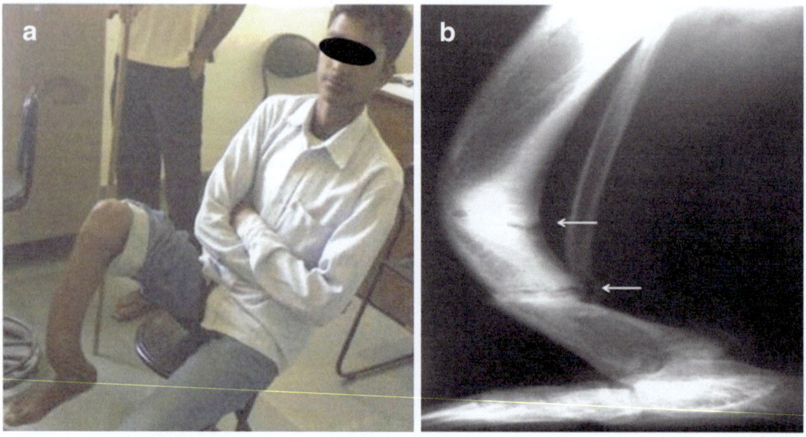

Fig. 19.65 (**a**) Clinical picture of congenital pseudoarthrosis of the right tibia in a 20-year-old boy (arrows). (**b**) Roentgenogram revealed a double pseuthoarotic lesion, a very rare anomaly. (Patient did not agree for treatment)

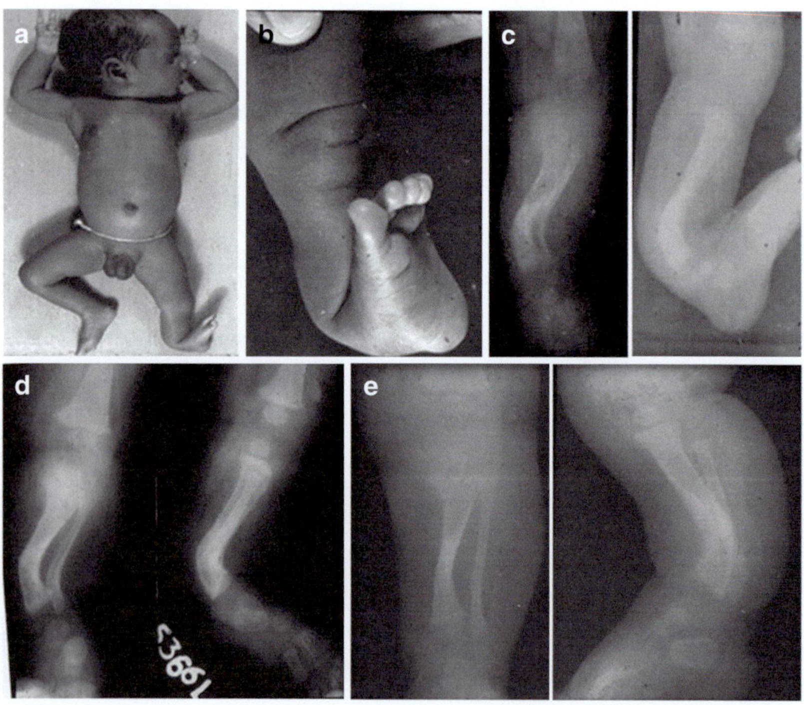

Fig. 19.66 (**a, b**) A 6-months-old boy showing classical postero medial bowing of the left leg. The deformity reveals the dorsum of the foot touching the anterior aspect of the leg. (**c**) The roentgenogram shows typical postero medial bowing of the tibia. (**d, e**) Radiograph reveals postero-medial bowing which can well be corrected by serial corrective plaster casts

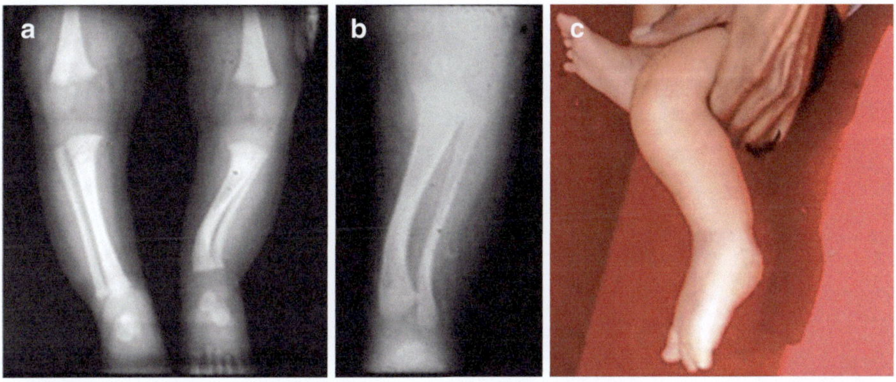

Fig. 19.67 (**a, b**) The roentgenogram reveals postero-medial bowing of the left leg. Note the correction of the deformity by a series of corrective POP casts. (**c**) Clinical picture of the patient after treatment showing correction of the deformity

19.14 Congenital Talipes Equinovarus (Pes Equinovarus Congenitus)

Congenital talipes equinovarus (Pes equinovarus congenitus, clubfoot) is a complex deformity with many components: hindfoot equinus and varus, forefoot adduction and talo-navicular subluxation (Figs. 19.68a, b and 19.69a, b). The incidence may be 1/1000 live births. Although many theories have been put forward but the cause of CTEV remains unclear. The physical examination should confirm that the foot has a fixed deformity in which the hindfoot is in equinus, heel in varus and the forefoot is adducted. Cavus may also be evident. A clubfoot is smaller than the contralateral normal foot. The calf is small and the involved leg may be shorter. The feet usually respond to periodic correction POP cast.

The resistant congenital clubfoot cannot be brought to neutral and has creases medially, transversally across the arch (Figs. 19.70a–d). The cause of the deformity is idiopathic and it may be associated with other congenital anomalies like—arthrogryposis and myelomeningocele. A routine X-ray, CT scan and ultrasound may be useful.

The goal of treatment of CTEV is a functional, pain-free, plantigrade foot with mobility. A near normal range of motion is desirable. Serial plaster casting at frequent intervals has been the main stay of treatment since long. The earlier it is begun, the greater is the likelihood of success. The technique attempts to correct the deformity around the talus. The forefoot is gradually abducted and the hindfoot is everted. The equinus should be corrected only after the foot is straight. Care must be exercised to avoid a rocker-bottom-foot, which has been considered as one of the worst complications of the treatment.

Ponsety described the final points of the manipulative management of clubfoot and mentioned that 85–90% of clubfeet could be treated successfully with proper

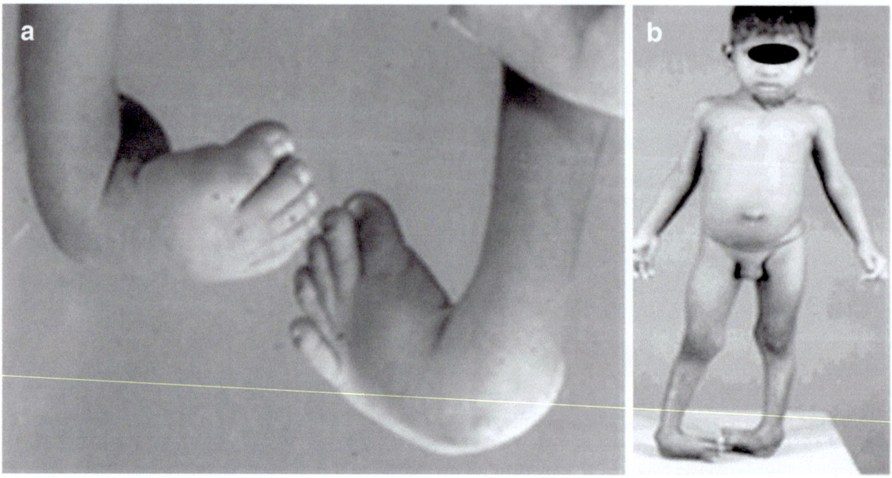

Fig. 19.68 (**a**) A severe case of bilateral congenital talipes equinovarus (CTEV). (**b**) A neglected case of bilateral CTEV

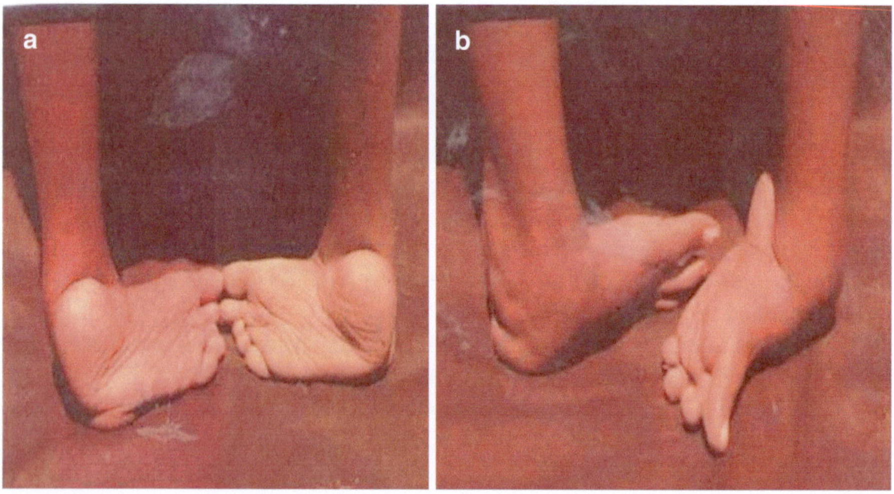

Fig. 19.69 (**a, b**) Severe bilateral neglected CTEV in a 12-year-old girl. Note the callosity on either side

manipulation, casts, splints and corrective shoes. He also recommended that a relapsed clubfoot could also be treated by casting, Dennis Browne splints. Complications of close treatment include increased cavus deformity, rocker bottom deformity and flattening of the proximal surface of the talus with lateral rotation of the ankle and increased stiffness. Care must be exercised to prevent skin complications.

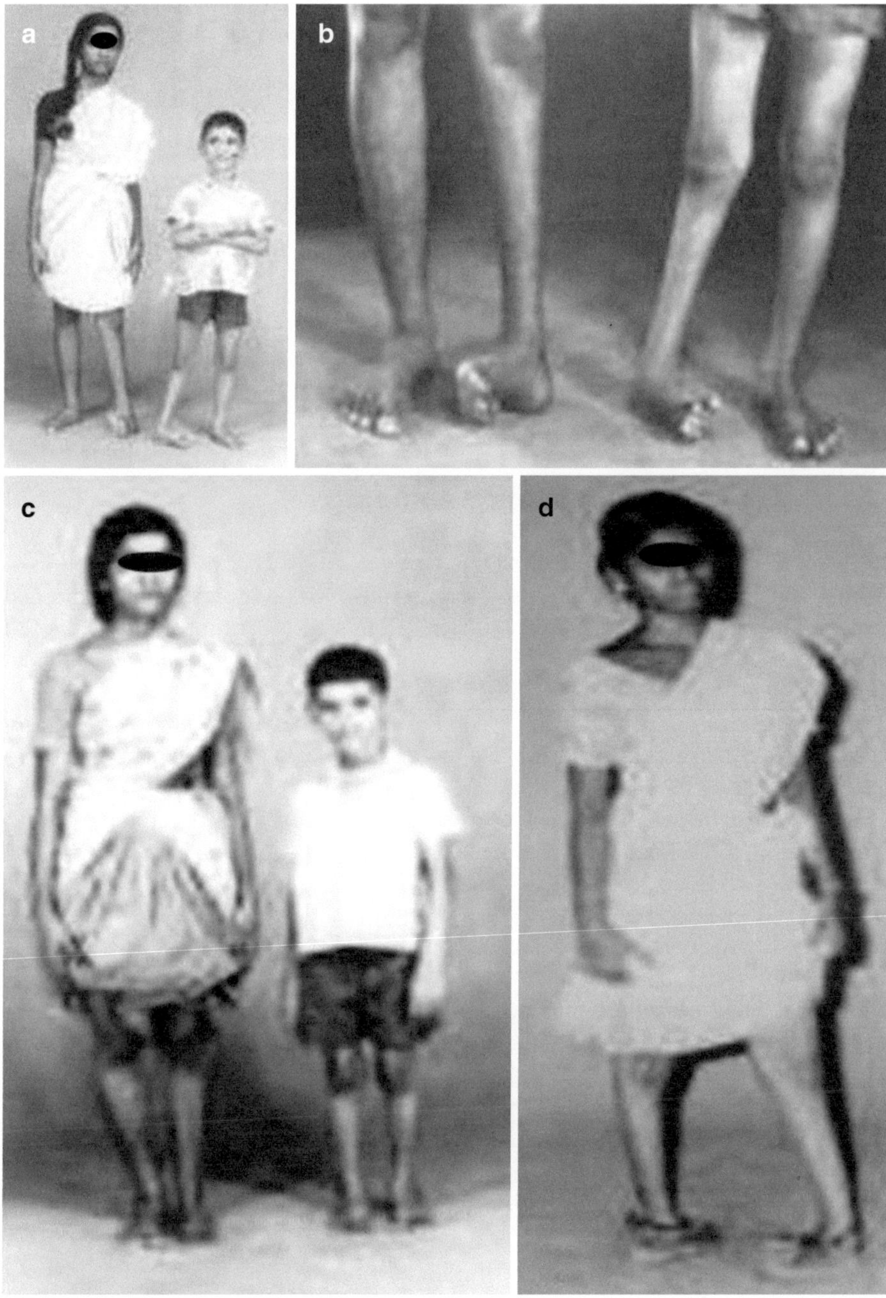

Fig. 19.70 (**a**) A sister and brother show congenital talipes equinovarus (CTEV) on the left and right side, respectively. (**b**) A close-up view of the deformed feet. (**c**) Complete correction of the deformity after surgery. The involved feet are smaller in size. (**d**) A 20-year-old girl showing bilateral neglected CTEV. She had designed shoes for her deformed feet

19.14.1 Surgical Treatment

Surgery is indicated for a failure is achieve satisfactory result by a close methods. There is no unanimity about when surgery should be performed (Figs. 19.71a–c and 19.72a–c). If indicated, surgery should be performed in earlier stages. The commonly advocated steps in the surgical procedure are:

1. Z-Plasty of the tendo-achilles.
2. Z-Plasty of tibialis posterior.
3. Posterior capsulotomy of the talocalcaneal joint.
4. Release of the Henryknot and Open reduction of talonavicular dislocation.
5. A Plantar release.
6. Any other soft tissue structure needing correction.

19.14.2 Complication of Surgical Approach

Skin problems/necrosis and infection have been the most common complication of the surgical treatment. Deep infection may result in osteomylitis of the bones and damage to the joints. This boy was referred to us had severe infection of the left ankle with complete rotation of the foot with heel in front and toes behind (Figs. 19.73a–e and 19.74a–e).

The feet with congenital talipes equinovarus (CTEV) and other deformities showing different presentations in different age groups which needed surgical correction (Figs. 19.75a–c, 19.76a–d, and 19.77a–d).

Severe deformities of the feet needing suitable surgical correction (Conti.) (Figs. 19.78a, b, 19.79a, b, and 19.80a–c**).**

19.14.3 Congenital Tendo-Achilles Contracture: Equinus Foot

Congenital Tendo-Achilles contracture is a rare bilateral anomaly. The child walks on the forefoot and heel does not touch the ground. The deformity is due to the contracture of the Tendo-Achilles (Fig. 19.81a, b). The calf is wasted and cylindrical. The child is unable to squat. In neglected cases the forefoot is fanned out and the heel remains small (Fig. 19.82a, b). If treated early by Tendo-Achilles lengthening and posterior capsulotomy of the ankle and subtalar joints, the deformity is fully correctable. However, in neglected cases the treatment depends on the age of the patient.

19.14.4 Congenital Calcaneus Deformity: Calcaneocavovalgus

Congenital calcaneus deformity is a rare congenital anomaly and is observed when the child starts walking. The patient starts walking on heal which is broad. The

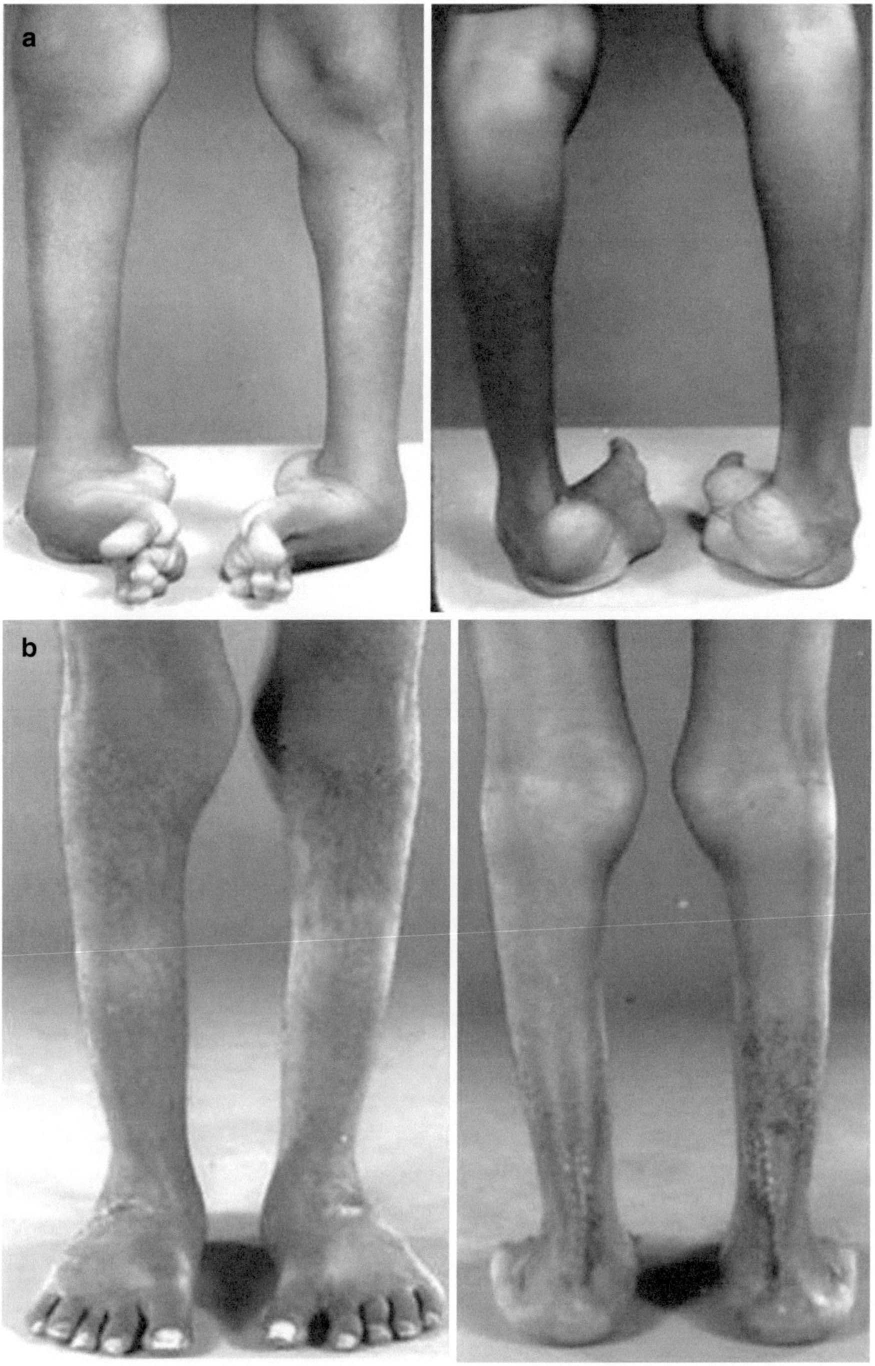

Fig. 19.71 (**a**) A 16-year-old girl with bilateral severe—neglected CTEV—front and back views. (**b**) The deformity has been corrected completely after extensive postro-medial release, dorso lateral wedge resection and triple arthrodesis

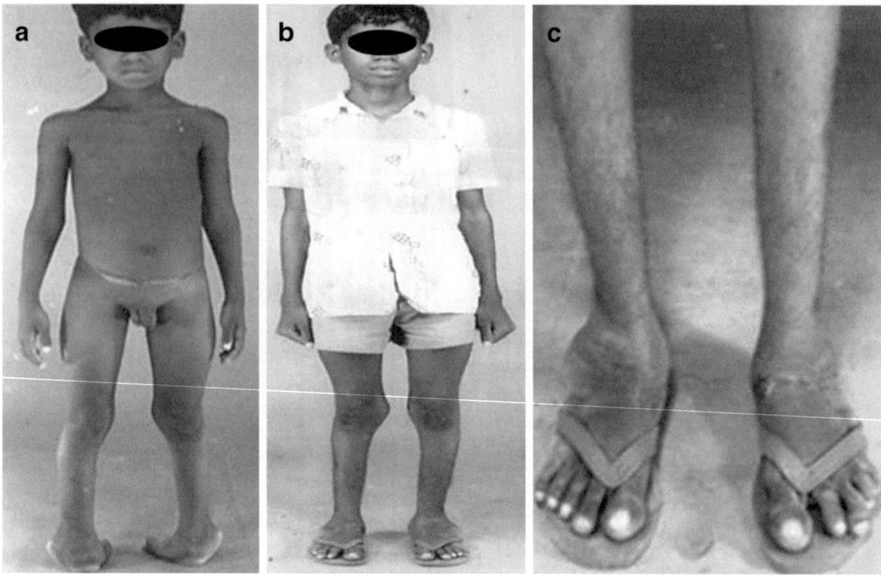

Fig. 19.72 (**a**) A 6-year-old boy with severe bilateral deformed feet. The right foot is in posteriorly supinated position with toes pointing towards the wall. The patient has been walking on the head of the talus on the right side. (**b, c**) A 5-year follow-up of the same patient, managed by extensive planter soft tissue release and a suitable dorsal bony wedge resection

forefoot usually does not touch the ground. The deformity occurs because of Tendo-Achilles paralysis and is very difficult condition to treat (Fig. 19.83a–c). In early cases corrective POP cast is recommended. However, most of the cases need surgical intervention which depends on the age of the patient. Soft tissue release with tendon transfer is indicated. However, in later stages bony operations are indicated. Patients older than 12 years of age usually have fixed bony deformity at the ankle. The roentgenogram reveals the calcaneus in vertical position. Soft tissue release and triple arthrodesis/pantalar arthrodesis may be required. The triple arthrodesis in such cases is difficult to perform, because the calcaneus must be brought to a horizontal position. As an acquired condition calcaneus deformity is seen in patients suffering from post-polio residual paralysis.

19.14.5 Congenital Calcaneovalgus Deformity

Congenital Calcaneovalgus (pes planus, planovalgus, flatfoot) is a common presentation in almost all the infants. It is reported in 30–50% patients. The deformity is usually supple and foot can be passively brought into planter flexion and supination.

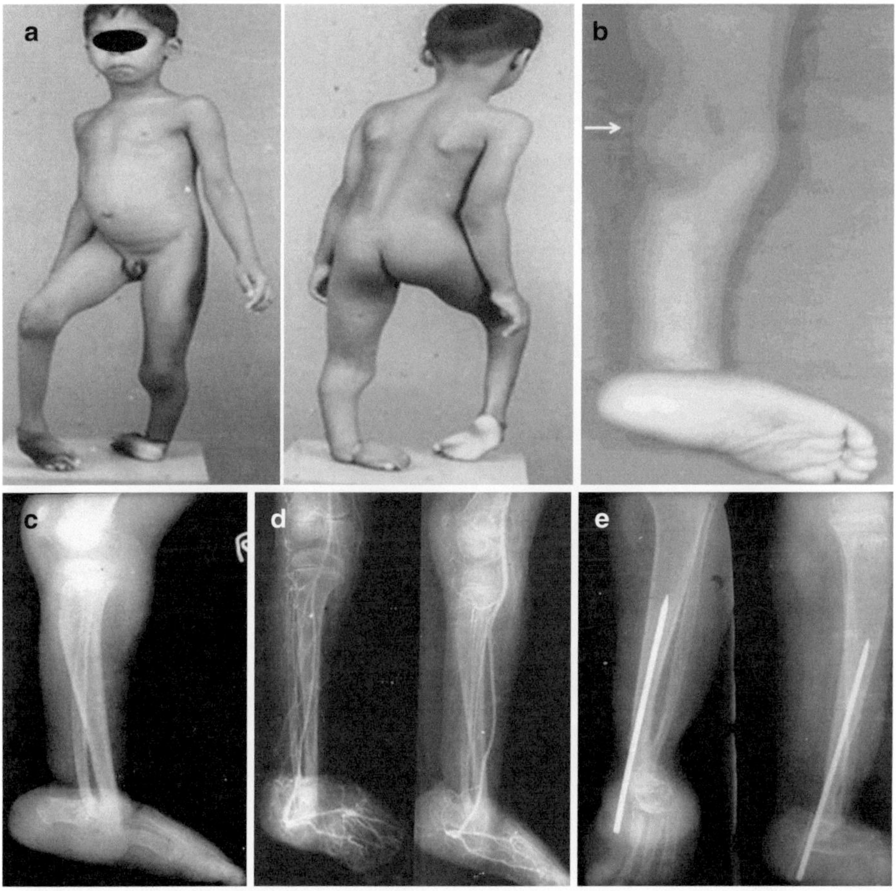

Fig. 19.73 (**a**) A 7-year-old boy with bilateral CTEV. He was referred to us after surgery on the left side which got infected and had resulted in a severe deformity with heel in front and toes behind. (**b**) The position of the foot shows heal front in toes behind. Note the anterior placement of the heel in line with patella (arrow). (**c**) The roentgenogram of the left foot showing 180° rotation at the ankle. Note the position of patella and the toes. (**d**) Roentgenogram of the leg and foot reveals the position of the arteries which is important in further surgical management. (**e**) Roentgenogram shows re-rotation of the foot at the ankle and fixed with a Steinman pin, which finally resulted in healing

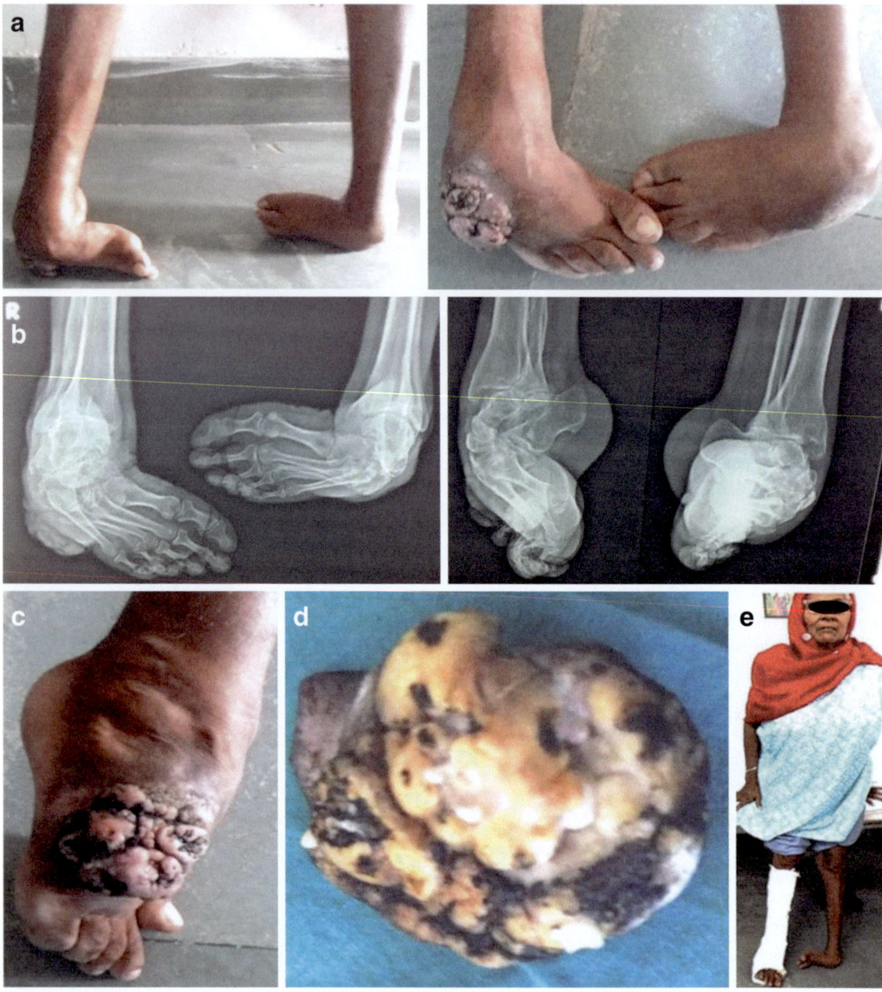

Fig. 19.74 (**a**) A 60-year-old lady presented with bilateral severe CTEV with large callosities. The one on right side was infected and fungated. (**b**) Roentgenograms reveal severe developmental changes in the tarsal bones. (**c**) A close-up view of the fungated callosity on the right side. (**d**) The excised specimen of the callosity. The biopsy did not reveal any malignant change. (**e**) Postoperatively the patient stands on a plantigrade right foot in POP. The left side is yet to be operated

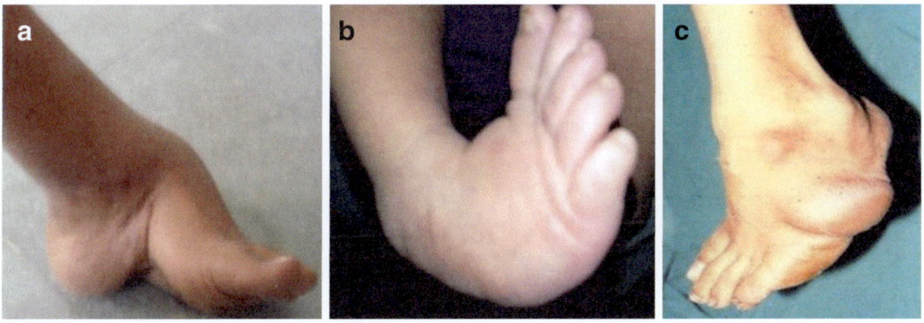

Fig. 19.75 (**a, b**) Note the crease medially at the mid foot. (**c**) A large callosity in a neglected case

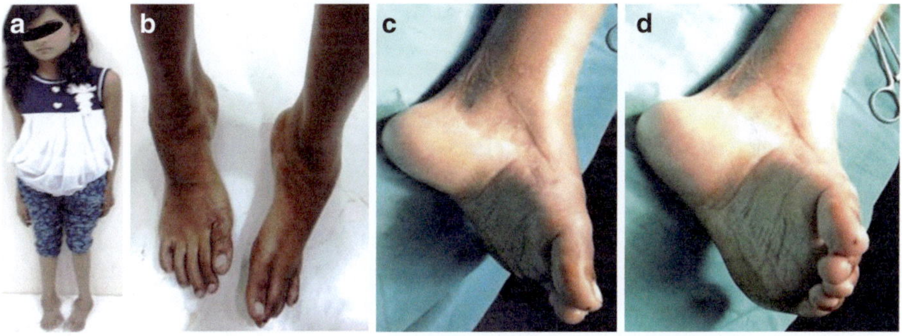

Fig. 19.76 (**a, b, c, d**) Presentation of equinocavovarus deformity

The heel valgus may be increased. If considered necessary a series of corrective casts can be used in early cases to achieve a normal position. Late cases may require surgical management (Figs. 19.84a, b and 19.85a, b).

Calcaneovalgus feet have to be differentiated from congenital vertical talus, in which there is a fix deformity with a rounded sole. The heel is in equinus position. When painful a suitable medial shoe support has been recommended to relieve the pain. In severe and late cases with pain, surgical intervention may be necessary according to age of the patient.

19.14.6 Congenital Vertical Talus

Congenital vertical talus is rare disorder of the foot and presents in the form of rigid flatfoot (a rocker-bottom foot) which requires early identification and aggressive

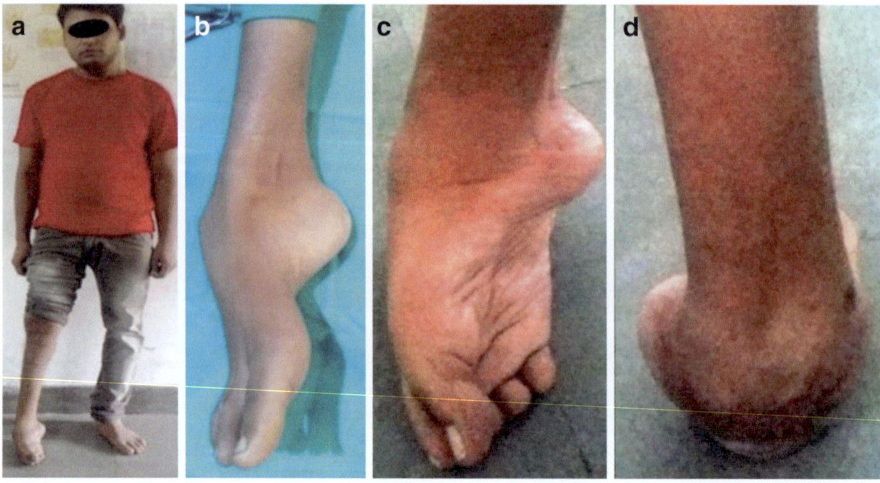

Fig. 19.77 (**a, b**) Presentation of severe equinocavovarus deformity. (**c, d**) Severe equinus deformity of the right foot. The left foot is severely supinated with toes pointing towards the wall (The forefoot is not even visible)

treatment (Fig. 19.86a, b). It is because of planter dislocation of the head of the talus that the foot is rigid and painful. The clinical appearance of a foot in such cases is diagnostic (Fig. 19.87a, b). The heel is in fixed equinus with very tight tendo-achilles. The head of the talus is prominent and palpable medially in the sole. The hind foot is in valgus with the forefoot abducted and in dorsiflexion at the mid tarsal joint. The navicular remains dislocated on the dorsum of the talus.

Radiologically the diagnosis can be confirmed by observing that there is a talo navicular dislocation with the forefoot displaced dorsally on the talus. The calcaneus is in fixed equinus. Only a small portion on the posterior aspect of the talus is in articulation with the tibia. In extreme cases, the talus lies parallel to the tibia.

In early stages, the anomaly may be corrected by application of serial corrective plaster casts. Rigid cases require surgical correction. In neglected cases the deformity becomes severe and the patient walks on the heel (Fig. 19.88a). Roentgenogram in such cases reveals completely disorganized tarsal bones (Fig. 19.88b). Another case of neglected congenital vertical talus has been observed in a 50-year-old lady (Fig. 19.89a). The roentgenogram in this case also revealed disorganization of the tarsal bones to the extent that they cannot be defined (Fig. 19.89b).

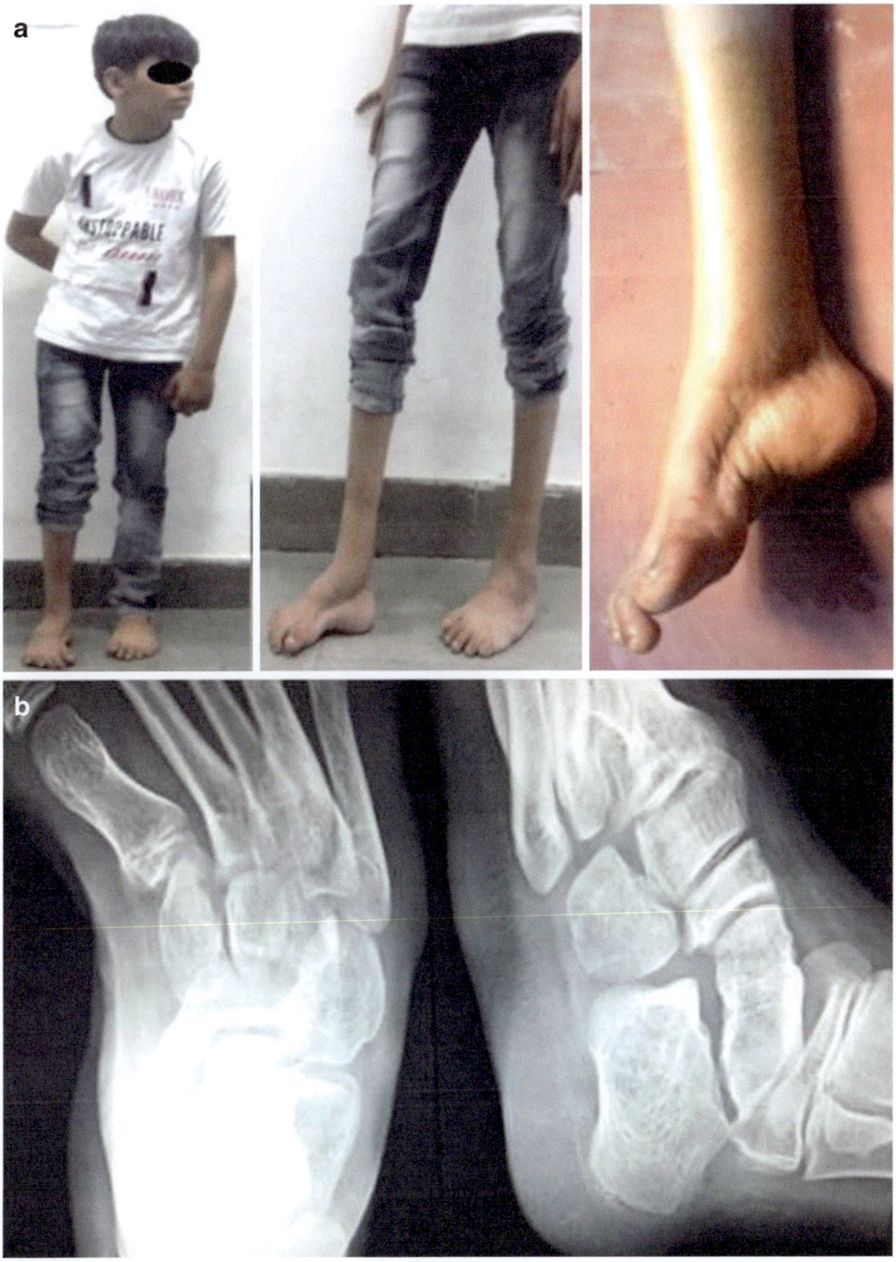

Fig. 19.78 (**a**) A 13-year-old boy with bilateral CTEV was operated in other centre on both sides at the age of 3 years. The left side was reasonably corrected. The right side revealed forefoot adduction, cavus and inversion of the hind foot. Note the deep incision mark of the previous operation. (**b**) The roentgenogram revealed adduction of the forefoot and developmental anomalies of the tarsal bones

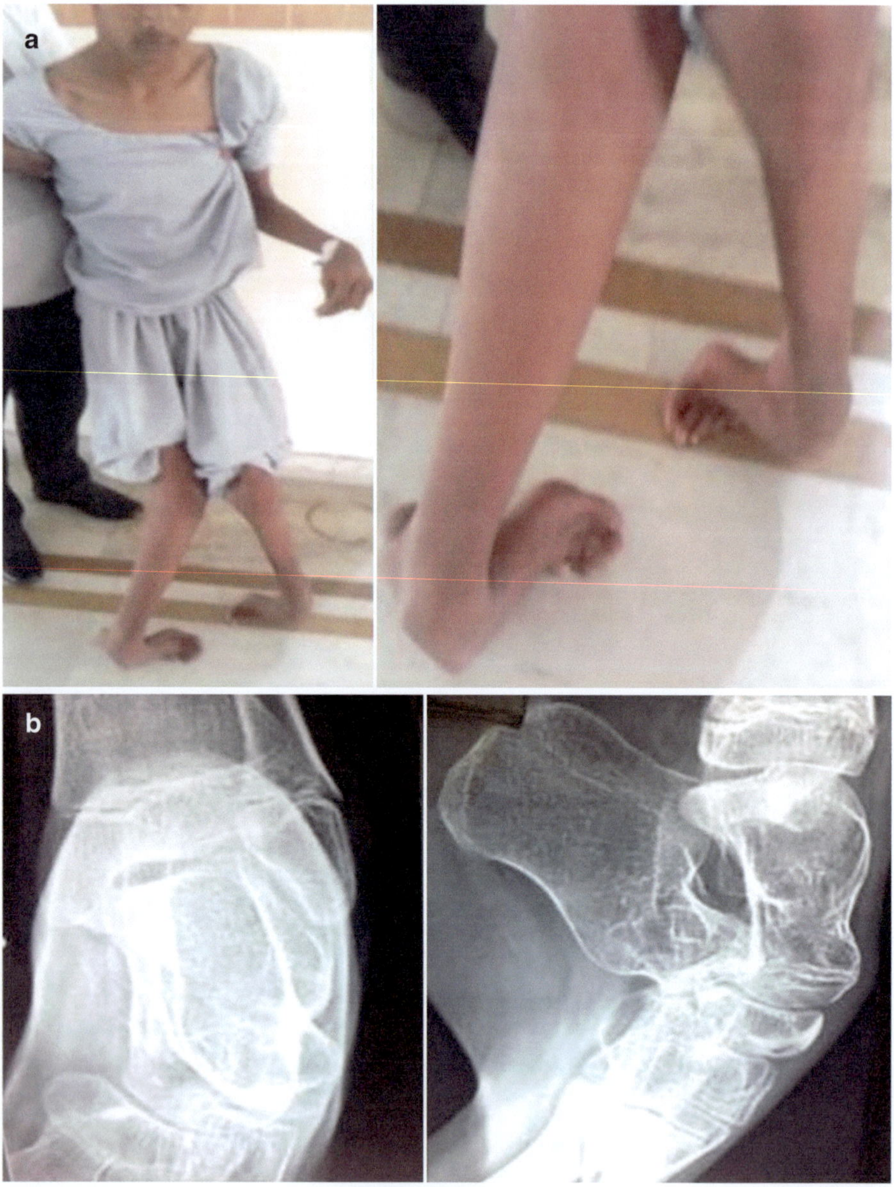

Fig. 19.79 (**a**) A case of bilateral, neglected and severely deformed feet. (**b**) Roentgenogram reveals severe equinovarus deformity and abnormal growth of the tarsal bones, more so of the talus

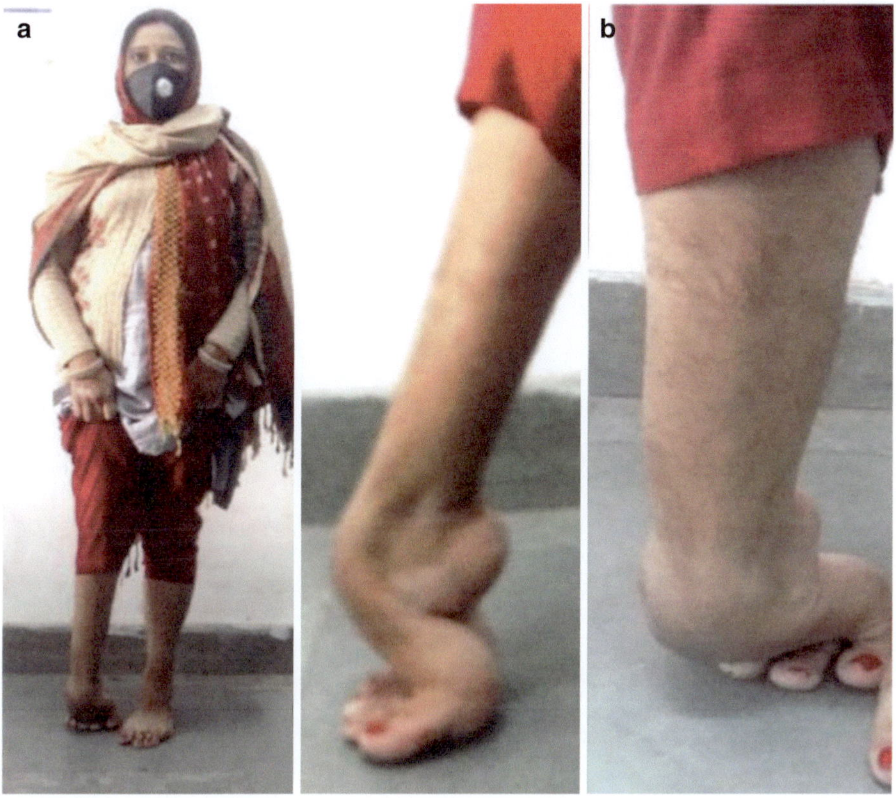

Fig. 19.80 (**a, b**) Clinical picture of a lady with severe equinocavovarus deformity of the right foot. The foot shows planter transvers crease and dorsal prominence. The patient has been walking on the heads of metatarsals with toes dorsiflexed. (**c**) Roentgenogram of the same patient showing gross deformity of the tarsal bones. The foot can be surgically corrected in plantigrade position with well-planned extensive planter soft tissue released suitable dorsal wedge resection and a pantalar arthrodesis

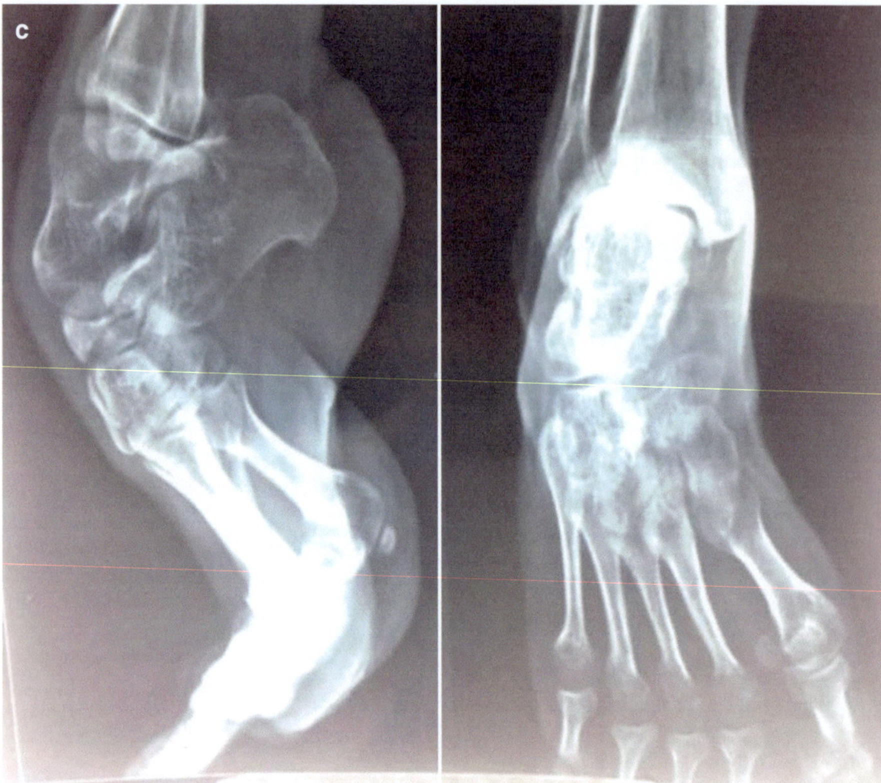

Fig. 19.80 (continued)

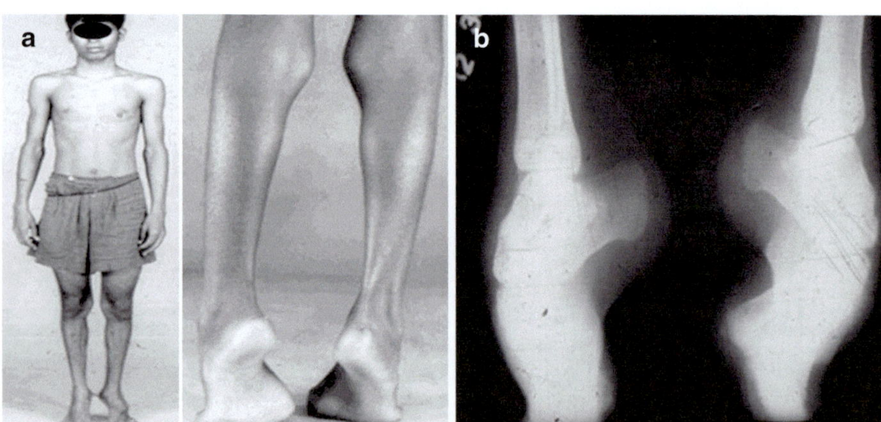

Fig. 19.81 (**a, b**) A 14-year-old boy with bilateral severe congenital equinus deformity. The deformity was corrected by Tendo-Achilles lengthening along with extensive soft tissue release, dorsal wedge resection and triple arthrodeses

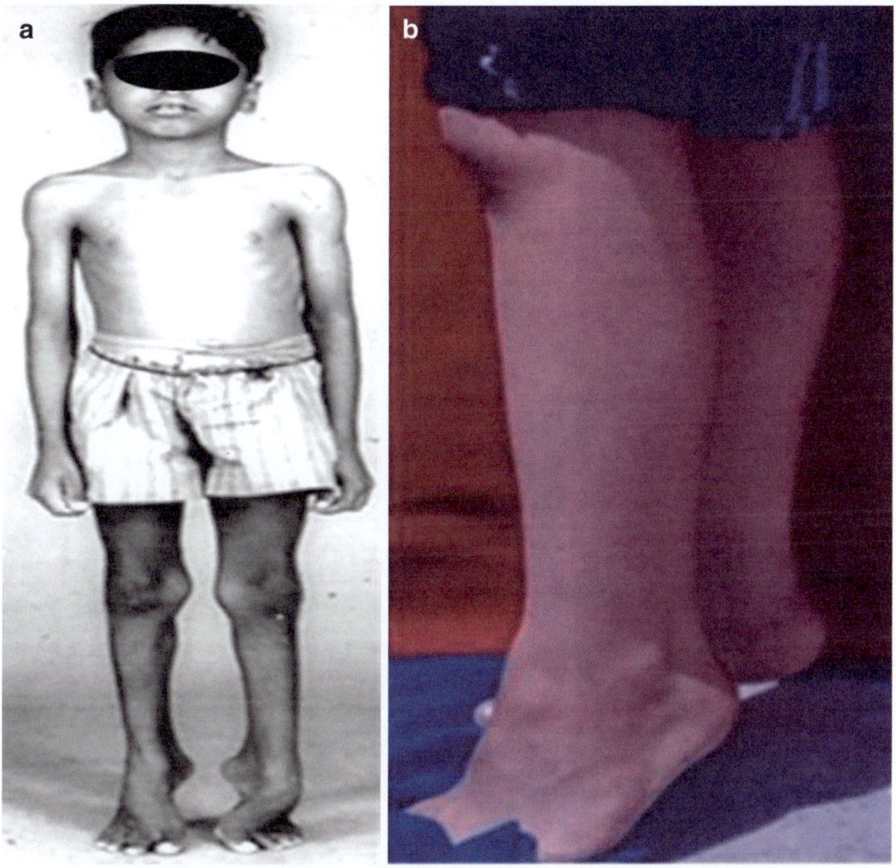

Fig. 19.82 (**a**) A 10-year-old boy with bilateral severe congenital equinus deformity. The deformity was corrected by Tendo-Achilles lengthening along with wide soft tissue release, dorsal wedge resection. (**b**) Another patient had the same congenital deformity and was corrected, as described above

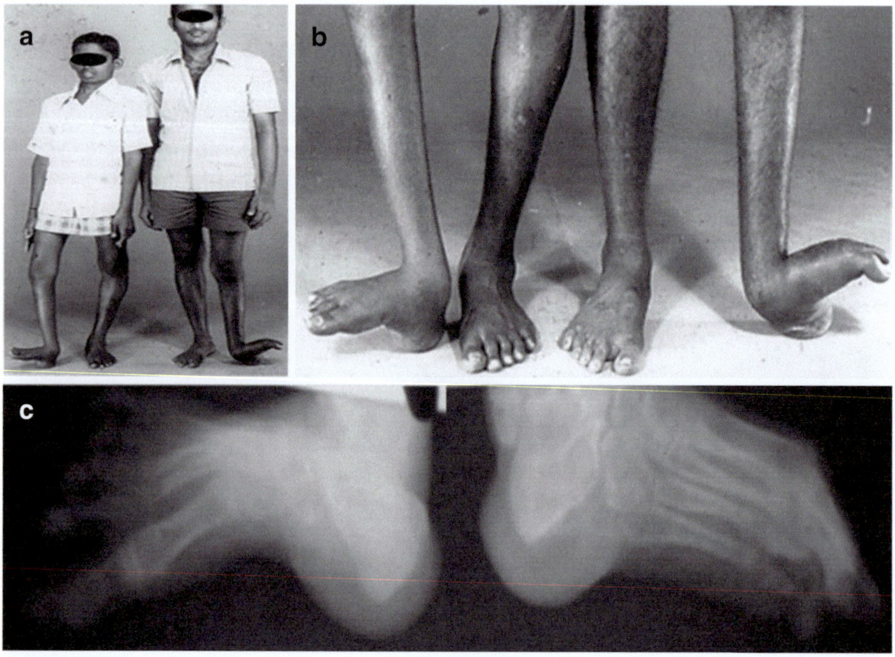

Fig. 19.83 (**a, b**) Two brothers born with severe calcaneo-cavo-valgus deformity of the feet—the elder one the left side and the younger on the right side. A close-up view of the severe deformities. Note the broad heel and a dorsi-flexed forefoot. (**c**) The roentgenograms reveal classical calcaneo-cavo-valgus deformity of the feet. The heels reveal vertical position of the calcanai

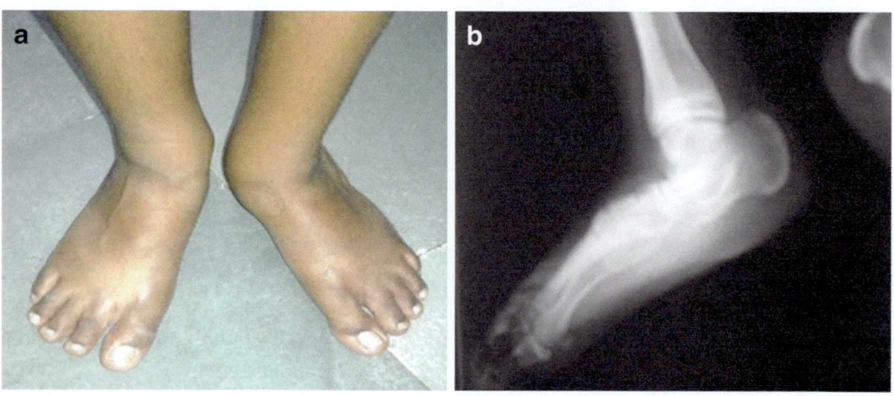

Fig. 19.84 (**a**) Clinical presentation reveals calcaneovalgus deformity, more so on the left side. (**b**) Roentgenogram shows complete loss of the arch of the foot

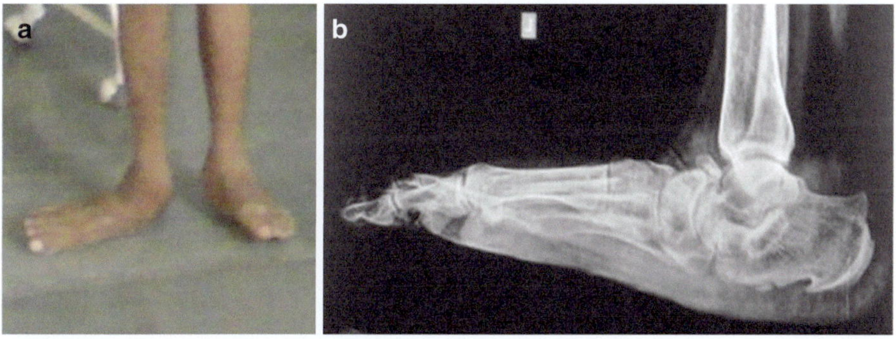

Fig. 19.85 (**a**) The patient reveals severe calcaneovalgus deformity of the right foot. (**b**) Roentgenogram of the right foot shows complete sagging of the mid foot and may need surgical intervention to relieve pain

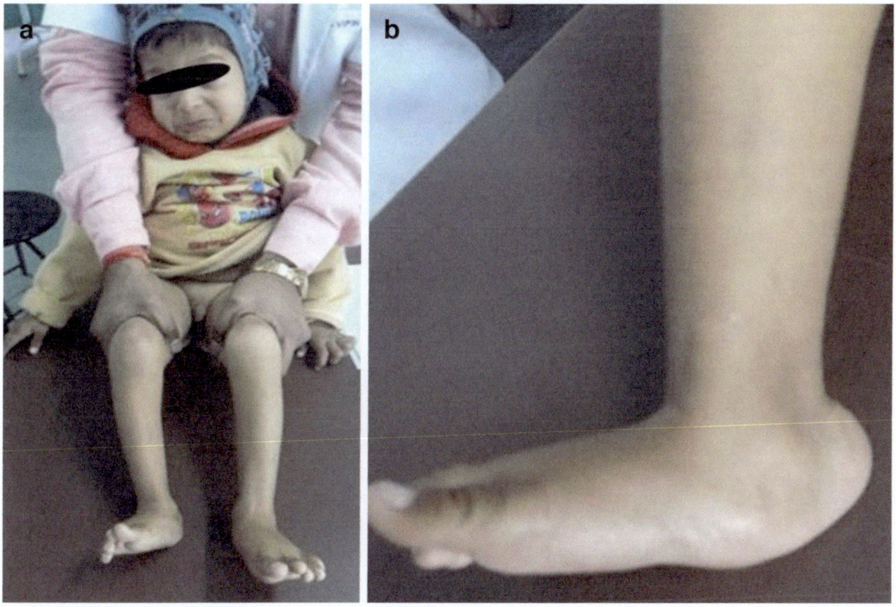

Fig. 19.86 (**a**) A child with cerebral palsy showing congenital vertical talus on the right side. (**b**) The head of the talus could be felt under the sole

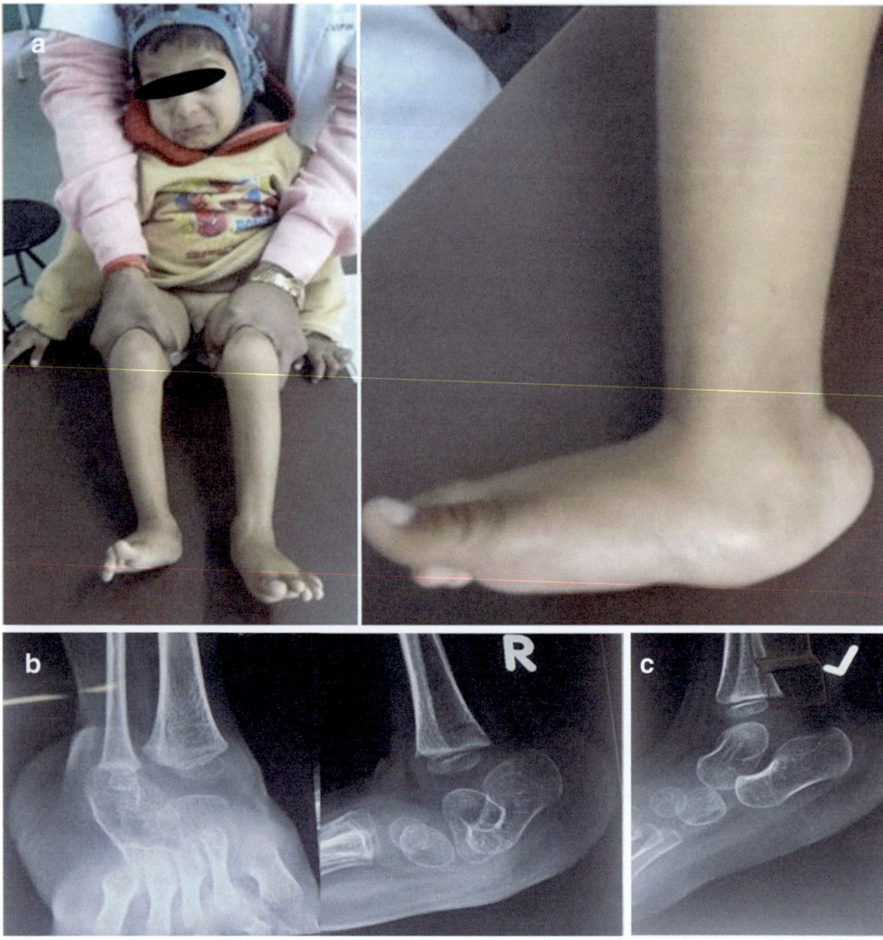

Fig. 19.87 (**a**) Roentgenogram of the same patient revealed the talar head pointing down and the calcaneous in equinus. There is dorsal dislocation of the navicular on the talus. (**b**) The roentgenogram of the left (normal) foot of the patient revealed the normal anatomical position of the talus and calcaneum

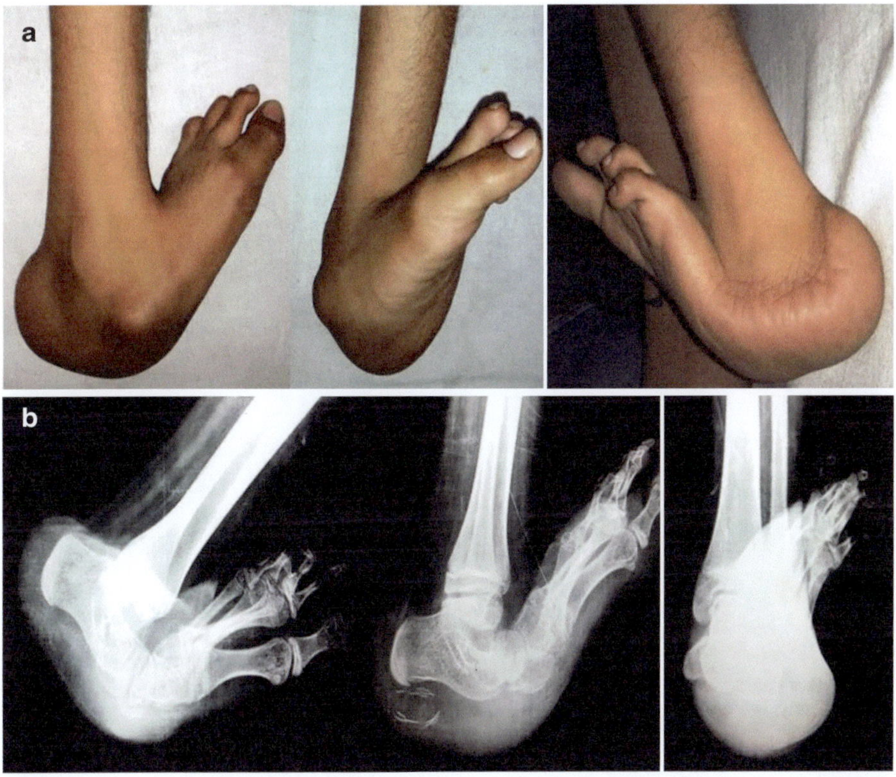

Fig. 19.88 (**a**) Clinical picture of the left foot of a 29-year-old boy showing neglected and severely deformed left foot with calcaneovalgus deformity, dorsi-flexion and eversion of the forefoot with severe rocker-bottom deformity. A clinical diagnosis of neglected dislocation of the talus can be made. (**b**) Roentgenogram of the same foot showing neglected congenital vertical talus and dislocation at the talo-navicular joint along with dorsi-flexion deformity of the forefoot. Note the other abnormalities of the tarsal bones and joints. The patient came for consultation but did not agree for any attempt to surgical correction, fearing an amputation

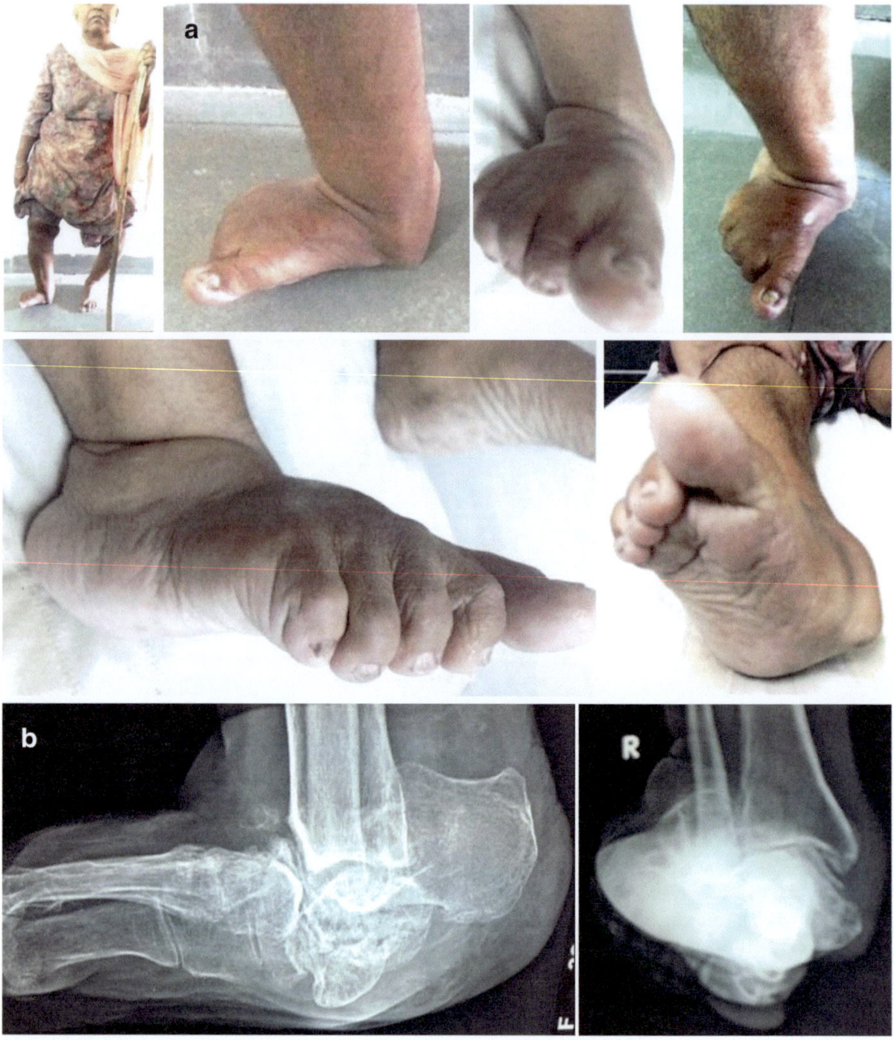

Fig. 19.89 (**a**) A 50-year-old lady came with severe calcaneovalgus and rocker bottom deformity on the right side. She has been having progressive deformity since childhood and occasional pain around the deformity. The presentation appeared as a neglected case of congenital vertical talus. (**b**) The roentgenogram revealed complete disorganization of the ankle and sub-talar joints with deformed tarsal bones. This could simulate a neuropathic presentation. The patient refused further investigations and surgical management

19.15 Congenital Coalition of Tarsal Bones

Congenital coalition of tarsal bones is not a rare condition. The most common coalitions involve the calcaneonavicular and talocalcaneal joints (Fig. 19.90). Tarsal coalition can be fibrous, cartilarginous or bony connection of two or more tarsals. The aching pain is usually vague and insidious in onset. Onset may coincide with increased activity or trauma. The pain is relived at rest. The patients with peroneal spastic flatfoot often have pain over the site of the coalition and range of the movements is limited. There is loss of subtalar movement which becomes more definite at the time of the ossification of the coalition which takes place during maturity.

Lateral radiograph of the foot in a patient with talo-calcaneal coalition shows a shortened talar neck, talar beaking and elongation of the sustentaculum tali. A CT scan is more informative and accurate. It provides precise delineation of the anatomy when surgery is indicated.

In case conservative treatment fails, surgery is contemplated. For calcaneo navicular bars, resection and interposition of the extensor degitorum bravis has been the treatment of choice. Talo-calcaneal coalitions are difficult to diagnose and treat. A CT scan is essential in these cases. Calcaneal osteotomy to bring the heel out of valgus has been reported. However, triple arthrodesis is indicated for patients who have unresectable coalitions and pain that does not respond to conservative treatment. It is also the treatment of choice in the presence of degenerative changes (Fig. 19.91).

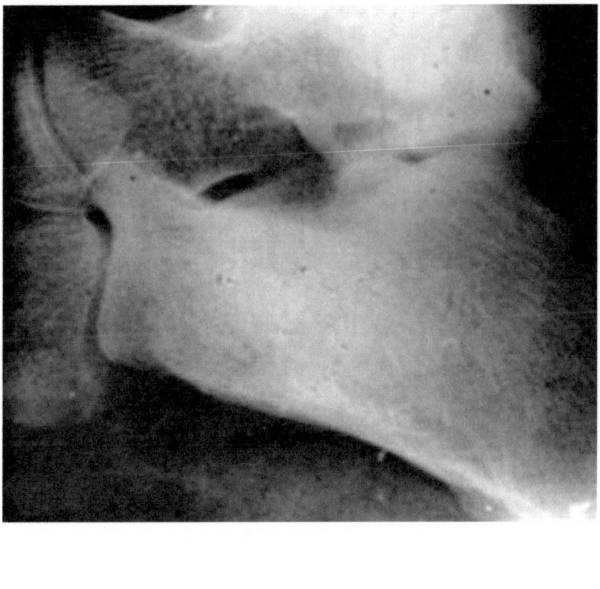

Fig. 19.90 Roentgenogram reveals a talo-calcaneal bar which also appears to be involving the navicular

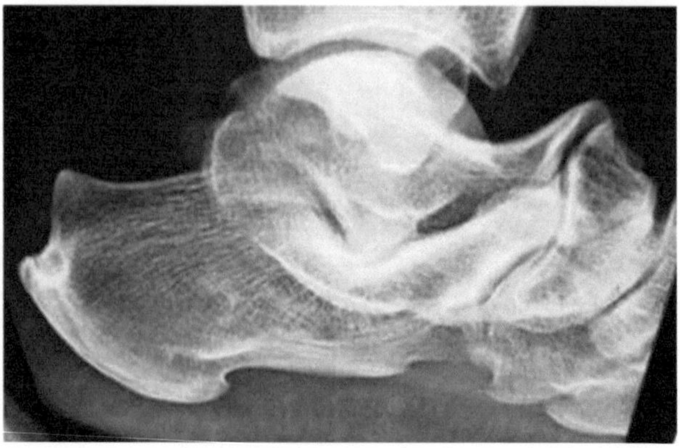

Fig. 19.91 Roentgenogram reveals talo-calcaneo-navicular-coalition. Note the advanced osteoarticular changes in a late case

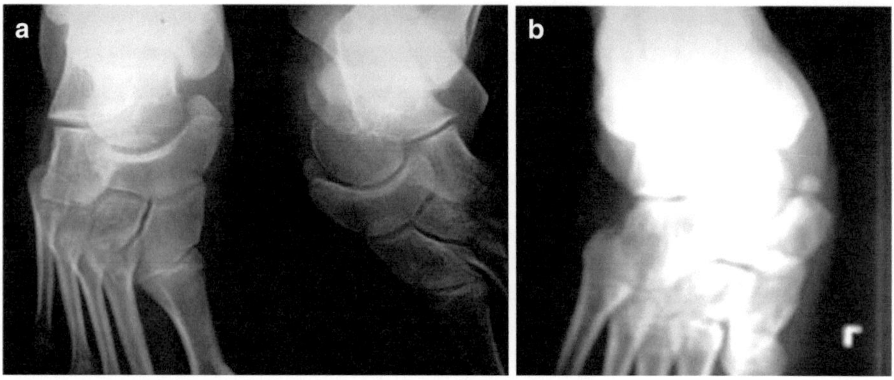

Fig. 19.92 (**a, b**) The roentgenograms show accessory navicular type-I

19.15.1 Accessory Navicular

Accessory navicular (Os Tibiale Externum) when present is localized medially to the navicular bone and is usually symptomatic in the form of pain. This is the most common accessory bone in the foot. There is palpable prominence at the inner boarder of the foot, along the insertion of tibialis posterior tendon. In a symptomatic patient the other findings include tenderness over the prominence and pain may be elicited by foot inversion against resistance.

Radiologically accessory naviculars have been classified by their radiographic appearances. Type I is small ossicle within the substance of the tibialis posterior tendon (Fig. 19.92a, b). It is posterior and medial to the body of the navicular. Type-II is triangular, larger than type I and joint to navicular by a cartilaginous tissue (Fig. 19.93a, b). Injury in the form of repeated stresses through this cartilaginous bridge may account for pain. Type III is a cornuate navicular resulting from fusion of the accessory navicular with the navicular (Fig. 19.94a, b).

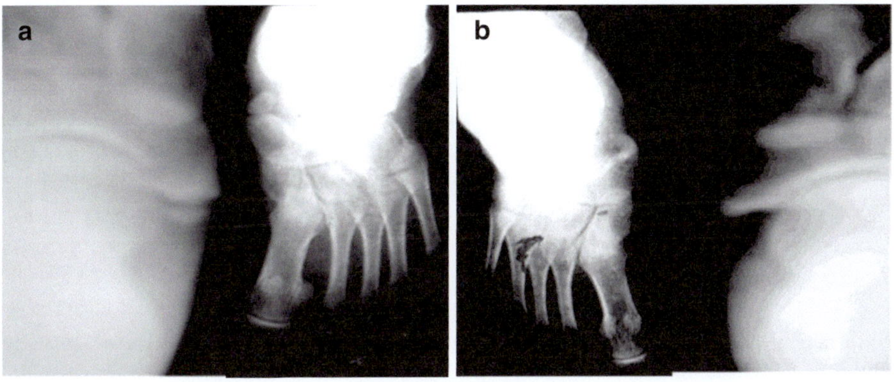

Fig. 19.93 (**a**) The roentgenogram shows accessory navicular type II. (**b**) The roentgenogram reveals accessory navicular bone type II. Note the fusion of the accessory navicular with the navicular type III

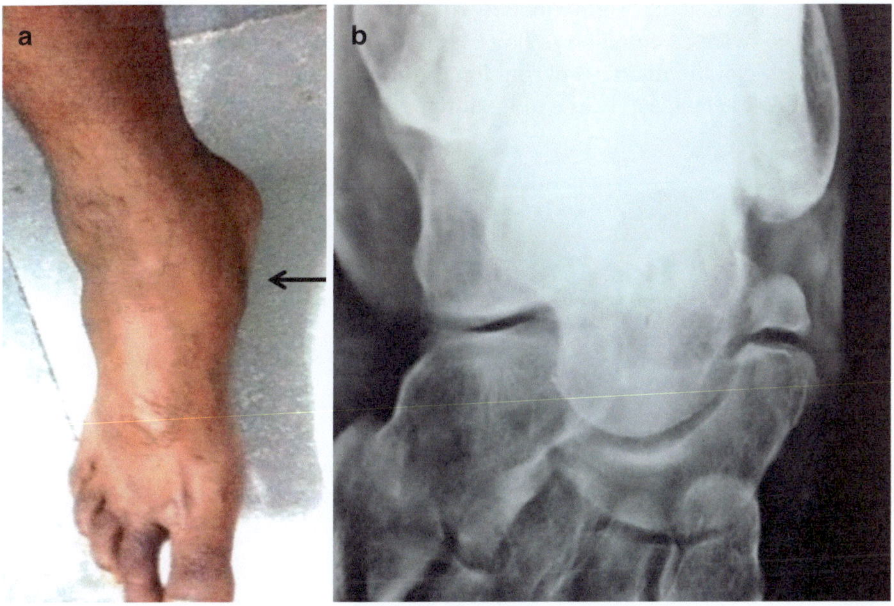

Fig. 19.94 (**a**) Patient shows prominence over the navicular (arrow). (**b**) Roentgenogram reveals a well-developed accessory navicular bone

The management of accessory navicular is mostly conservative. If the patient does not respond to conservative treatment, simple excision of the prominent ossicle seems to be the surgical procedure of choice.

19.15.2 Congenital Short Metatarsal (Brachymetatarsia)

Congenital short metatarsal (Brachymetatarsia) is a common observation which is often asymptomatic (Figs. 19.95a, b and 19.96a, b). The anomaly may be associated with other congenital defects (Fig. 19.97). The treatment depends on the presentation of the individual case. Ordinarily it needs no treatment.

19.15.3 Syndactyly of the Foot

Syndactyly means webbed finger/toe and the deformity is the commonest in hand and foot (Fig. 19.98a, b). Syndactyly may be complete or incomplete. The complete syndactyly is webbing of the fingers from the route of the fingers to the finger tips. Incomplete syndactyly has webbing extending distally, usually in the middle of the proximal phalanges and not up to the end of the digit (Fig. 19.99). The syndactyly is simple if skin alone is joined and complex if the bones of the adjacent finger are fused (Fig. 19.100a–c). Approximately one half of the cases are bilateral and symmetric. It is twice as common in boys as in girls. The incidence of involvement is highest between the middle and ring finger followed by ring and little fingers. The more simple the syndactyly, the more normal the joints.

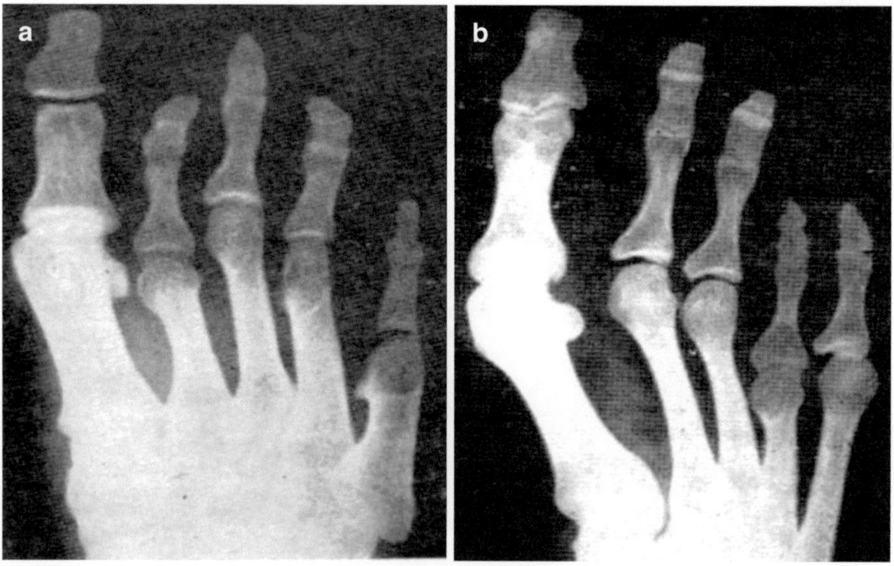

Fig. 19.95 (**a**) Roentgenogram shows brachymetatarsia of the seconds and the fifth ray. (**b**) Roentgenogram shows congenital shortening of the fourth and fifth metatarsals

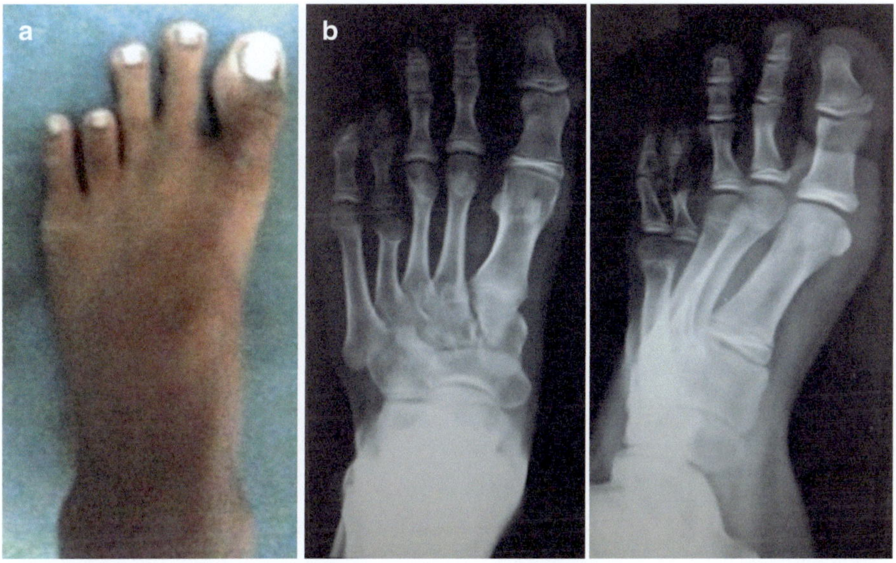

Fig. 19.96 (**a**) The clinical picture of the left foot shows short fourth and fifth toes. (**b**) The roentgenogram reveals short fourth and fifth rays

Fig. 19.97 Roentgenogram of both feet showing polydactyly with shortening of the second metatarsals on either side

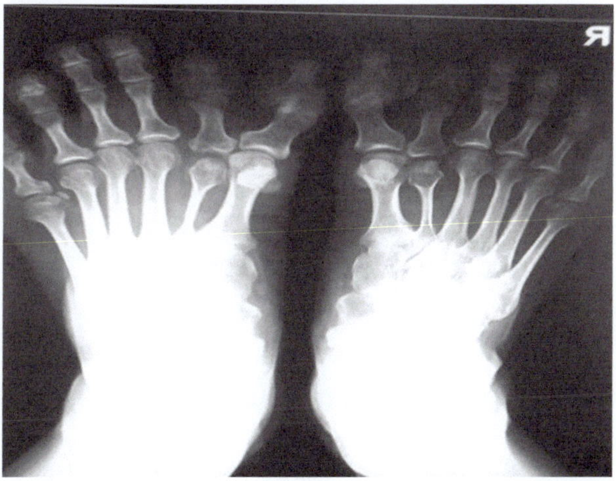

Acrosyndactyly is an adjacent fusion at their distal ends. There is usually proximal fenestration between the digits distal to the normal web space. Brachysyndactyly is referred to shortened fingers that are joint adjacent sites.

The treatment of the syndactyly requires good expertise. Improvement in function and a greatly enhanced cosmetic appearance can be achieved. A z-plasty can adequately deepen the cleft between the involved fingers/toes.

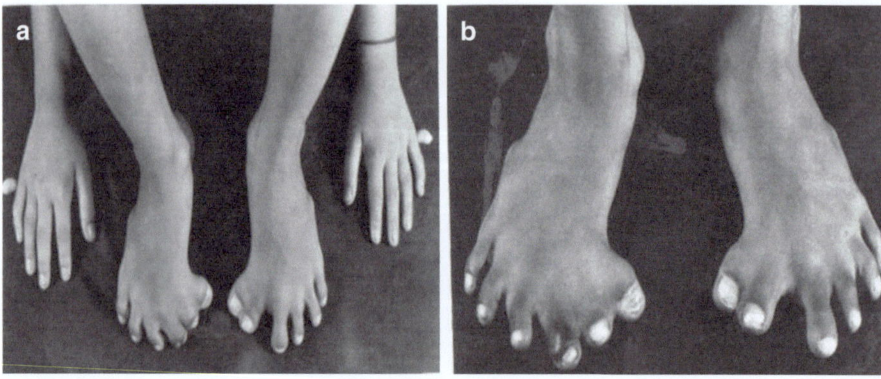

Fig. 19.98 (**a, b**) The feet reveal syndactyly-polydactyly between the first and the additional medial toes. The hands reveal a tag of skin at the base of little fingers on both sides

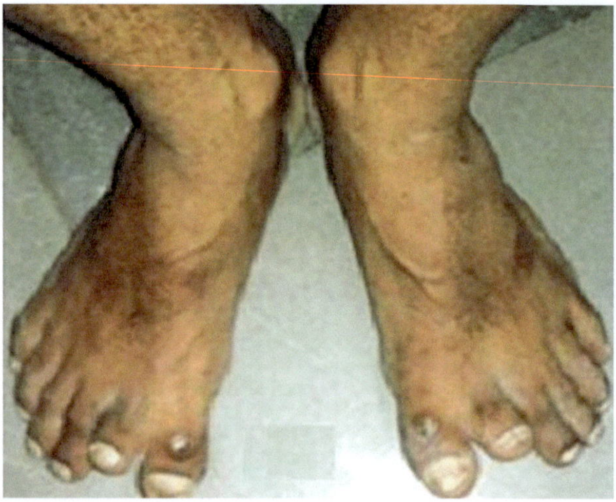

Fig. 19.99 Clinical picture shows bilateral incomplete polydactyly and bi-fid big toes with incomplete syndactyly

19.15.4 Polydactyly (Duplication) of the Foot

Polydactyly of the foot is a common congenital anomaly characterized by more than five digits and can be associated with other anomalies especially with polydactyly of the hand (Fig. 19.101). Polydactyly has an increased familial incidence. It generally involves the lateral side of the foot and rarely on the tibial side (Figs. 19.102a, b and 19.103a–c).

To contain the foot in the shoe, the extra digits can be excised. However, a decision must be made about which ray or toe is to be removed (Fig. 19.104).

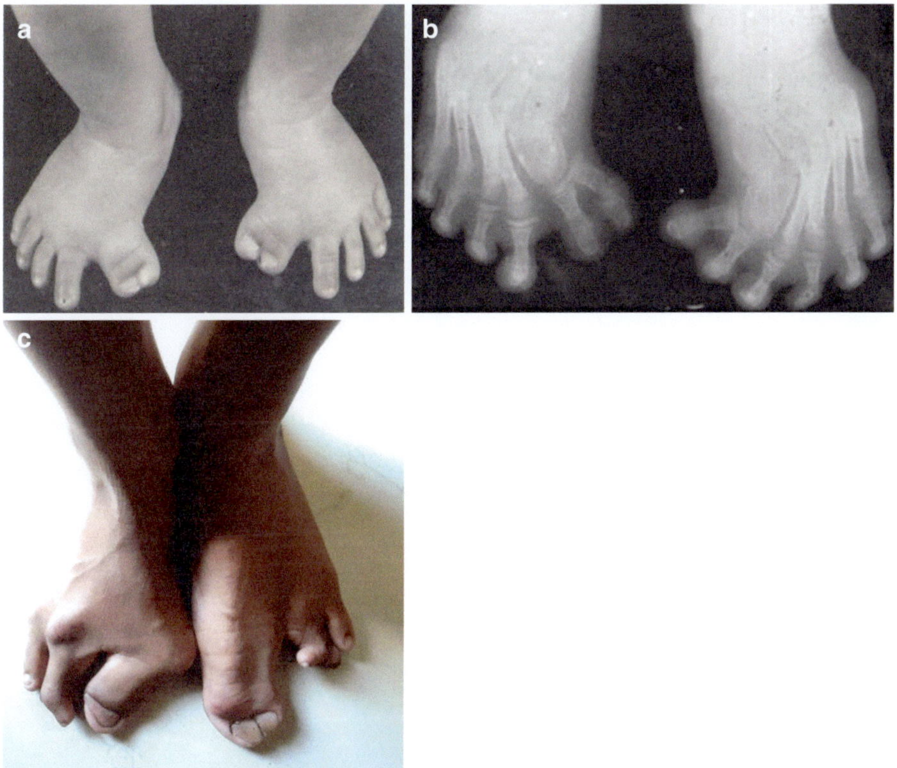

Fig. 19.100 (**a**) Both feet reveal syndactyly of the big toes with the accessory toes on the medial side. (**b**) Roentgenogram shows bifurcation of the big toes with hypertrophied first metatarsal. Note the fusion of the second and the third metatarsals on both sides. (**c**) Unusual clinical presentation of the feet-Right foot showing oligodactyly with hallux-valgus. Left showing bi-fid foot with absence of the middle ray and syndactyly of the great and the second toe

19.15.5 Oligodactyly

Oligodactyly is characterized by less number of toes. The lateral digits are relatively more involved (Fig. 19.105a, b). The big toe is rarely involved. This disorder can impair shoe-fitting but generally does not cause pain. Familial occurrence has been observed. In a case of fibular hemimelia, the lateral rays and in tibial hemimelia the medial rays are usually absent. It can also associate with other anomalies, especially with oligodactyly of the hand or it can present as an isolated malformation.

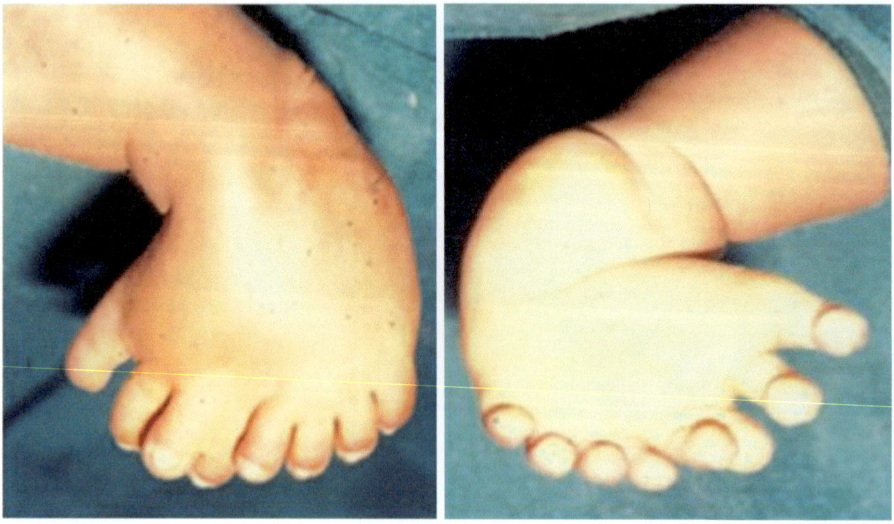

Fig. 19.101 Clinical presentation of the left foot showing polydactyly with eight digits. It is associated with congenital talipes equinovarus. The big toe is not well developed. Note the syndactyly at the site of big toe and the additional three medial toes

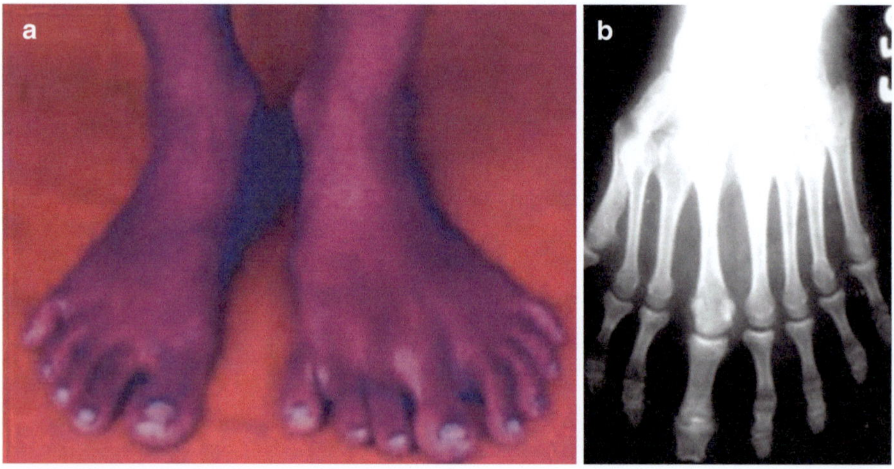

Fig. 19.102 (**a, b**) Another patient with polydactyly of the left foot with three extra medial digits which could well be excised. This will enable the patient to put on shoes

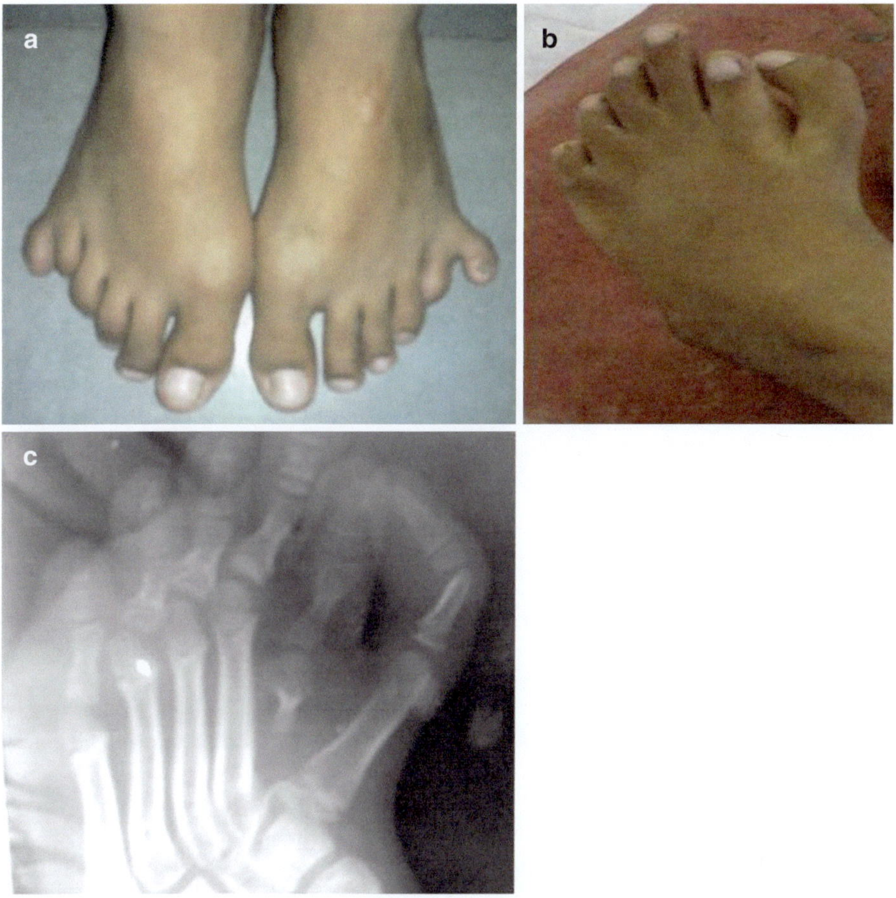

Fig. 19.103 (**a**) The feet show bilateral additional digits close to the little toes, needed excision. (**b**) The left foot reveals extra digit close to the hypoplastic big toe which can be excised that will enable to put on shoes. (**c**) The roentgenogram of the same patient reveals hypoplastic first ray. Excision of the additional medial ray can be excised which will enable the patient to put on shoe

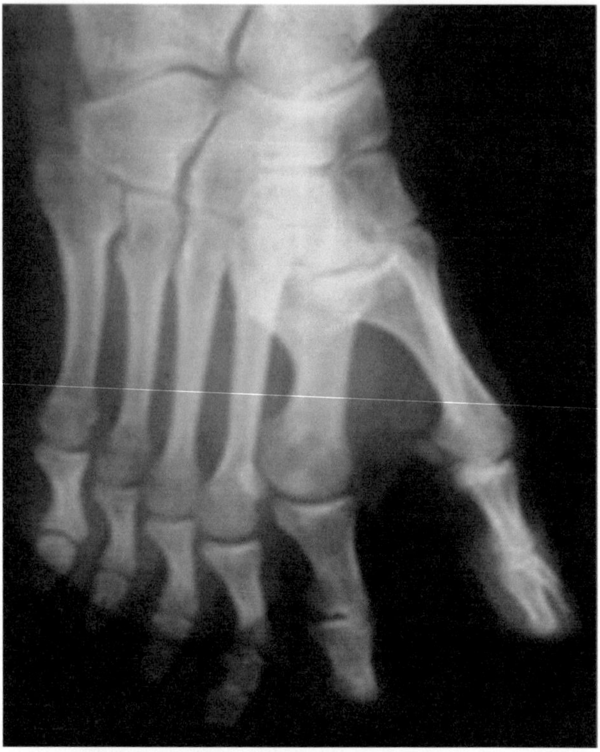

Fig. 19.104 Roentgenogram reveals an extra medial ray where the base of the metatarsal appears fused with the base of the first metatarsal. Excision of the ray could be done. With this the patient could put on the shoe. Note the bifurcation of the distal phalanx

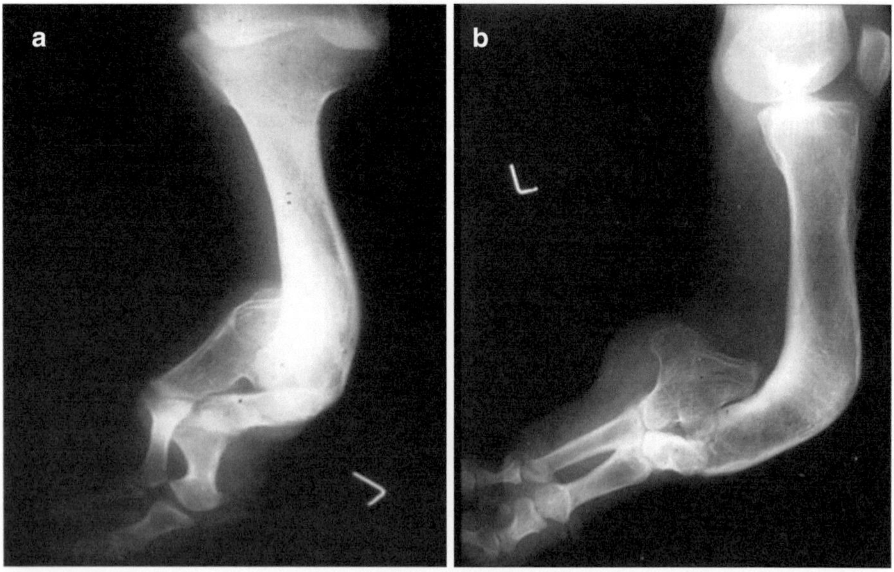

Fig. 19.105 (**a**) Roentgenogram of a case with fibular hemimelia showing absence of three lateral rays. Note the anomaly at the ankle joint and the foot. (**b**) The talus appears absent and the calcaneum is out of shape, may be the two bones are fused

19.15.6 Macrodactylia

Macrodactylia or gigantism (isolated overgrowth of the toe or toes) is a rare anomaly of a toe or toes (Figs. 19.106a–c, 19.107a, b, 19.108a–c, 19.109a–c, 19.110, and 19.111a–c). Macrodactyly is one of the rarest anomalies of the lower limb. It may be associated with other systemic defects. Nearly 10% of the cases are known to be associated with syndactyly. The deformity involves all the structure of the digit such as the phalanges, the tendon, the nerves, the blood vessels, the fat, the finger nails and the skin.

Macrodactyly may the primary of secondary. In primary or true macrodactyly all tissues are enlarged and the digit appeared elongated and broadened which may subsequently grow (Fig. 19.112a, b). The rate of growth in the progressive form is much greater than that of the other digits. Differences have been observed in growth behaviour, extent of bony involvement and presence of neural involvement. The progressive type is more common than the static type. The observations suggest that this rare, usually isolated, congenital anomaly may be two distinct entities. Such a presentation of primary macrodactyly should be distinguished from secondary type which may result from disorders like neurofibroma, haemangioma, lymphangioma, arteriovenous fistula, fibrous dysplasia and lipoma.

The involved digit is cosmetically unsatisfactory as well as stiff. It is because of inability to put on the shoes and cosmetic appearance that it needs surgical treatment in the form of excision of the digit or the ray (Figs. 19.113a–c and 19.114a–e).

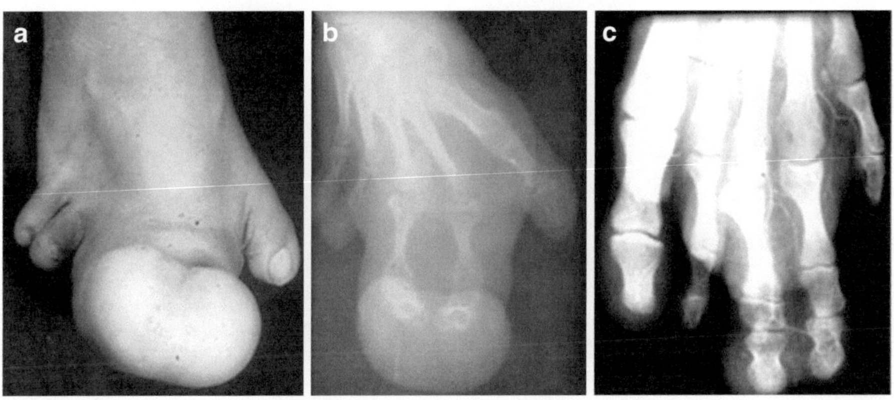

Fig. 19.106 (**a, b**) Massive congenital hypertrophy of the second and third digits of the right foot. (**c**) Osseous hypertrophy of the third and fourth toes of the left foot in a case of macrodactyly

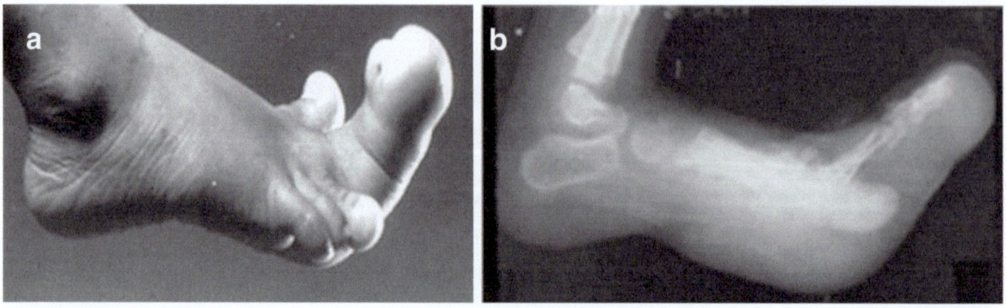

Fig. 19.107 (**a**) Elongated hypertrophy of the second toe of the right foot. Note the dorsiflexion of the toe. (**b**) Radiograph reveals soft tissue and bony enlargement of the second toe

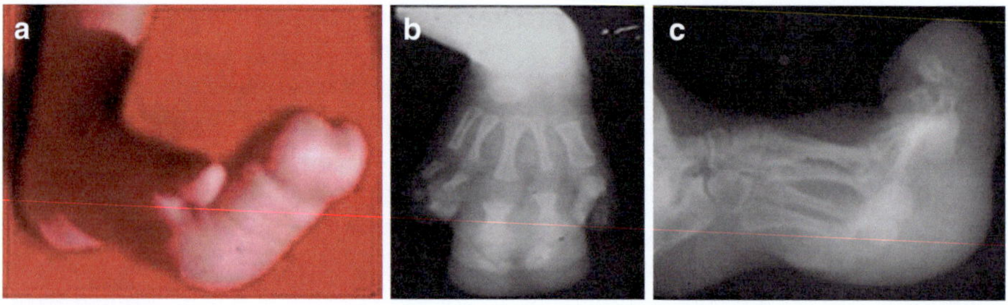

Fig. 19.108 (**a**) Massive enlargement of the second and third toes of the right foot. (**b**) Roentgenogram reveals hypertrophy of the phalanges and the soft tissues. (**c**) Roentgenogram reveals enlargement of bones and soft tissue. Note the dorsiflexion deformity of the toes

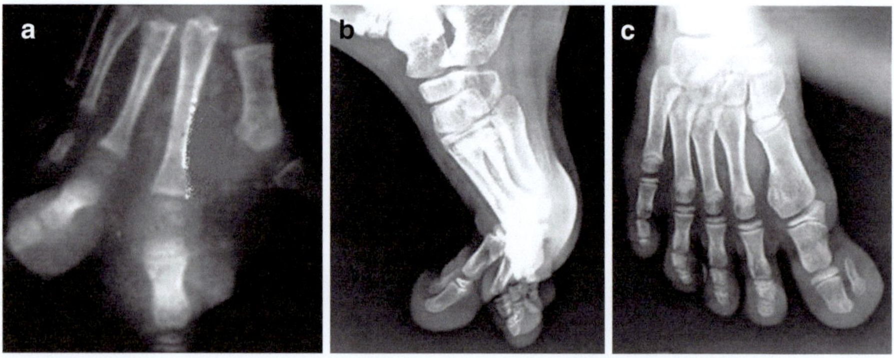

Fig. 19.109 (**a**) Macrodactylia of the second and third ray on the right side. (**b, c**) Roentgenogram reveals macrodactyly of the great toe. Note the linear bone by the side of the distal phalanx of the great toe which is in dorsi flexion

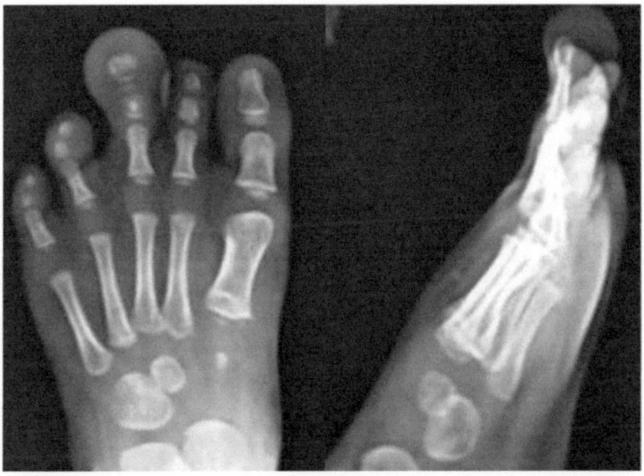

Fig. 19.110 Roentgenogram reveals macrodactyly of the third toe

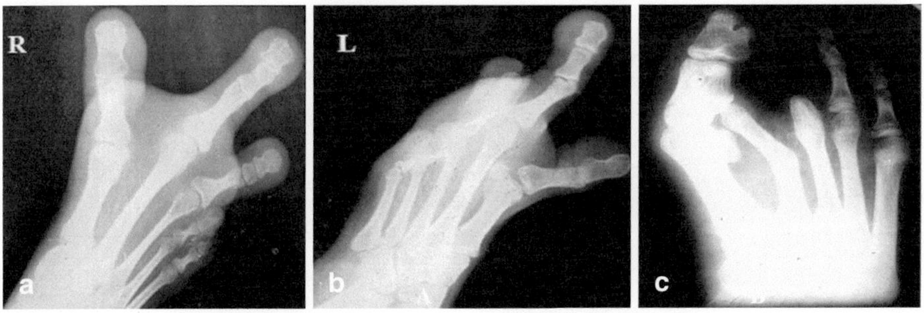

Fig. 19.111 (**a, b**) Roentgenograms reveal macrodactyly involving two medial rays on both sides. Corrective surgery can be carried out which will enable the patient to put on shoes. (**c**) Roentgenogram reveals congenital anomaly of the medial three rays. The second metatarsal is angulated to articulate with the hypertrophied proximal phalanx of the big toe. It appears that the phalanges of the second toe have merged with those of the big toe. The third toe is absent. Note the bifurcation at the tip of distal phalanx of the big toe. A rare anomaly

19.15.7 Congenital Hallux Valgus

A diagnosis of hallux valgus is made when the axis of the hallux is deviated laterally and develops slowly along with the drop of the transverse arch (Fig. 19.115a, b). The angle between the first and second metatarsals increases. The tendons, on the planter side of capsule dislocate laterally or dorsally along with the sesamoid bones (Fig. 19.116a, b). The capsule of the first metatarsophalangeal joint shrinks laterally and expands medially. Severe cases may present with pain at the first metatarsophalangeal joint and difficulty in putting on shoes (Fig. 19.117a–e).

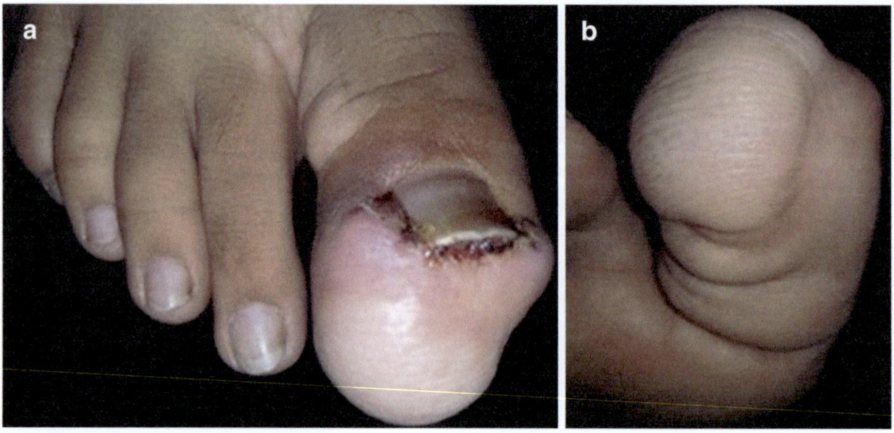

Fig. 19.112 (**a, b**) Dorsal and planter views of a mega big toe on the right side with infected nail

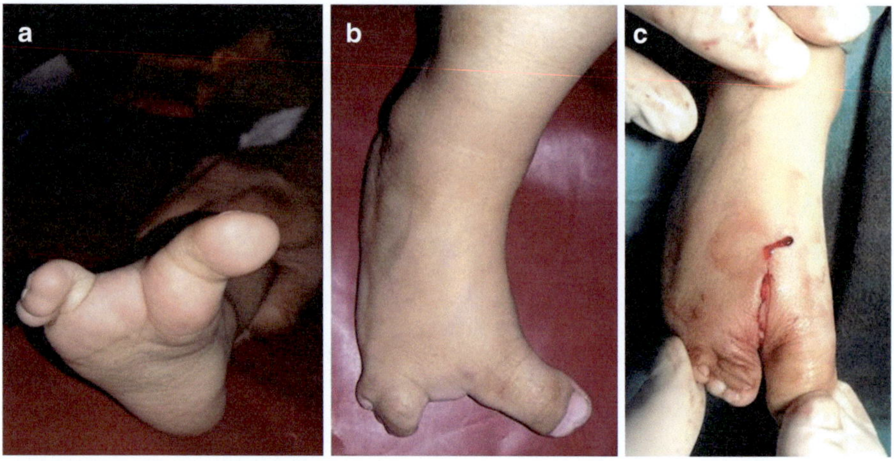

Fig. 19.113 (**a, b**) Clinical presentation of macrodactyly of the right big toe with loss of second toe. (**c**) Intra-operative picture after excision of the second meta-tarsal. The web could be closed. Post-operatively the patient could put on shoes

In interphalangeal hallux valgus the proximal and distal joint surfaces of the base phalanx of the hallux are not parallel, they have an angle opening medially. The end phalanx is thus positioned in valgus. Such cases may need surgical intervention.

19.15.8 Congenital Hallux Varus

Primary hallux varus is rare (Fig. 19.118). Congenital hallux varus consists of medial deviation of the great toe at the first metatarsophalangeal joint. In some cases

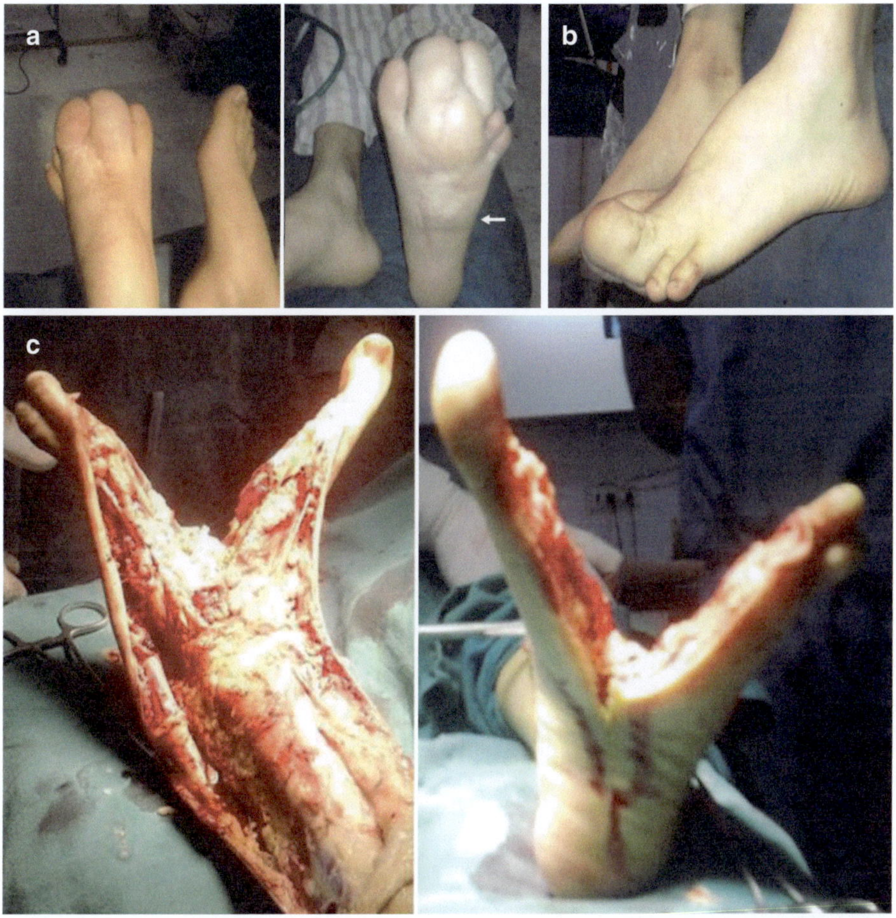

Fig. 19.114 (**a**) Dorsal and planter views of massive congenital hypertrophy and syndactyly of the second and third toes of the left foot. She could not put on shoes more so because of proximal extension over the sole (arrow). (**b**) Lateral presentation of the same foot. (**c**) Intra-operative dorsal and planter view after the excision of the abnormal second and the third rays. (**d**) Dorsal and planter views after closing the wound. Note the proximal extension of the abnormal soft tissues in the excised specimen. (**e**) Post-operative roentgenogram of the same patient in POP cast. The first and the fourth meta-tarsals are approximated with a circulage wire. The patient could put on the shoes and was satisfied

the first metatarsal is short. It usually interferes in putting on the shoes. Severe cases may need surgical intervention (Figs. 19.119a, b and 19.120a–c). Adequate soft tissue operation must be carried out early in life. The correction should be maintained till the big toe is stable. This condition is often associated with other congenital anomalies such as syndactyly, polydactyly, cleft hand, Apert hand and ulnar club-hand. The hallux varus deformity may be treated by capsulorrhaphy stabilized with Kirschner wire.

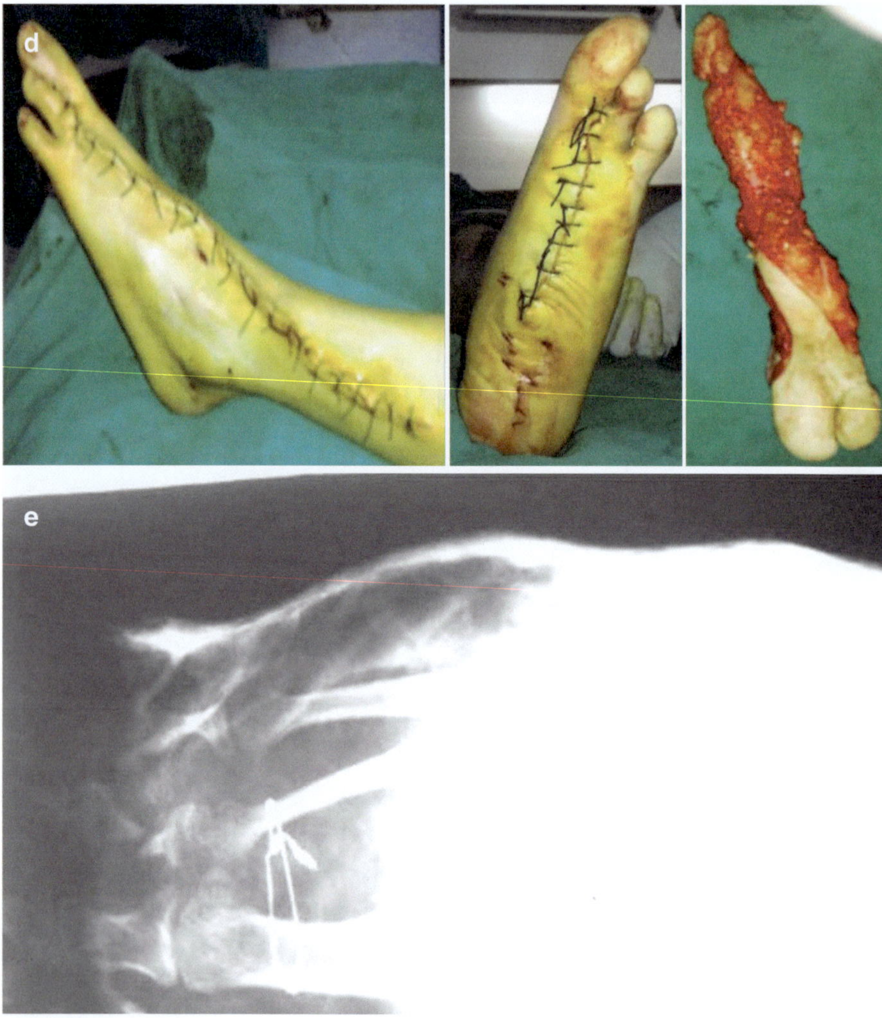

Fig. 19.114 (continued)

19.15.9 Rigid Hammer Toe

In rigid hammer toes there is flexion contracture of the PIP and/or DIP of the small toes and are not fully correctable either actively or passively (Fig. 19.121). Deformities with long duration will reveal painful callosities. This are usually associated hallux valgus. There may be associated hallux valgus.

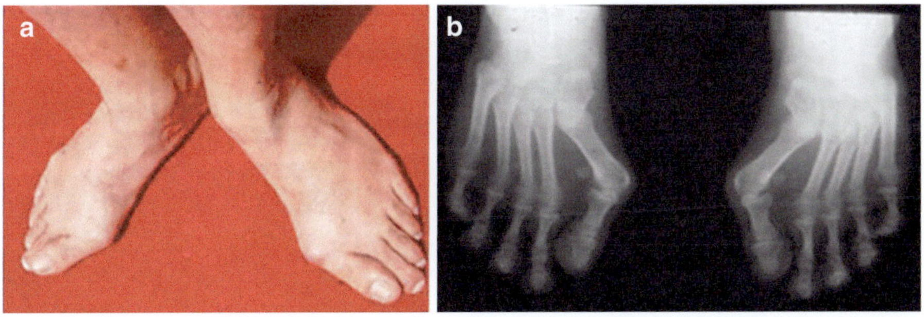

Fig. 19.115 (**a**) Clinical picture shows evidence of moderate hallux valgus. Note the bilateral pes planus. (**b**) Roentgenogram of another patient reveals severe hallux valgus on both sides with subluxation of the metatarsophalangeal joint

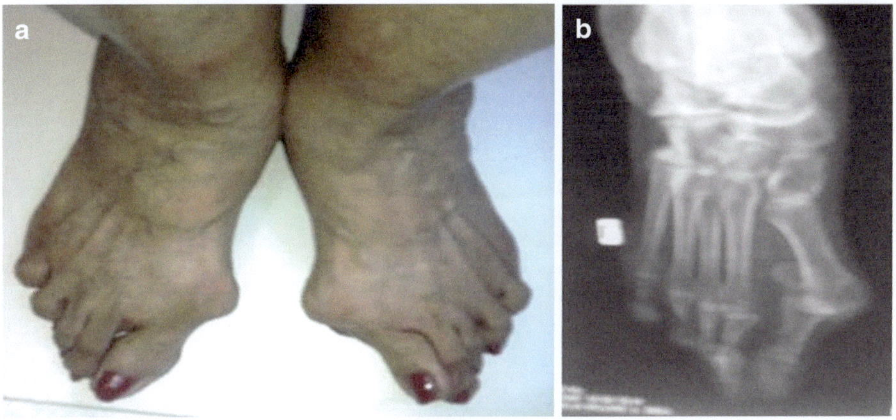

Fig. 19.116 (**a**) Severe bilateral hallux valgus deformity which is the cause of painful forefeet. (**b**) Roentgenogram reveals signs of osteoarthrosis of the first metatarsophalangeal joint. Note the lateral dislocation at the involved joint

19.15.10 Overlapping Second Toe

The second toe is usually positioned over the big toe (Fig. 19.122a, b). It may be fifth toe which is positioned over the fourth toe. The main complaint in these cases is that the shoe rubs the toe. Under the circumstances a callosity often develops over the riding toe.

19.15.11 Bi-Fid Foot: Split Foot

Split foot, Ectrodactyly and Central hypoplasia are the terms used to describe a cleft foot. The cleft foot is usually bilateral and of those affected, about 60% are boys

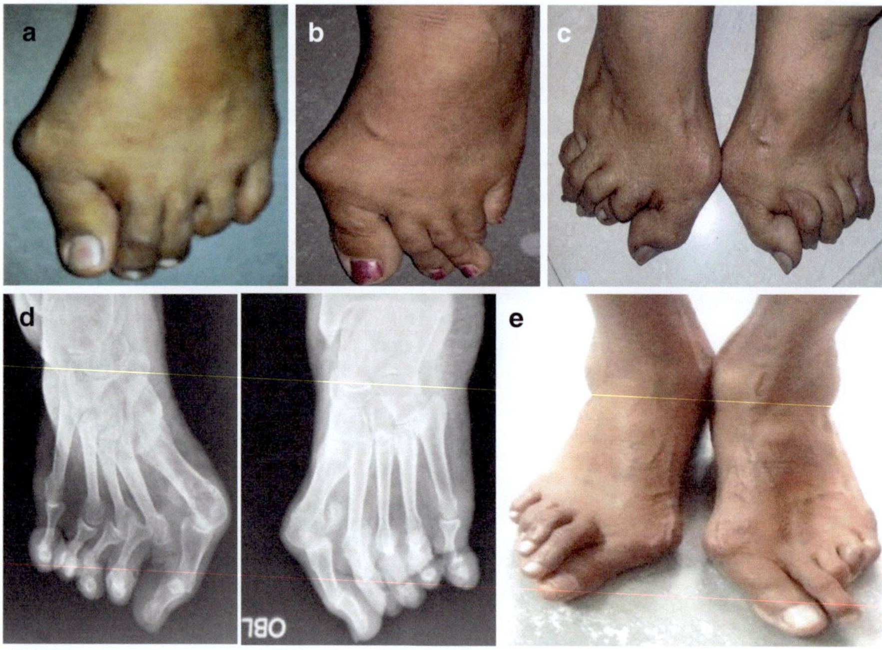

Fig. 19.117 (**a, b**) Hallux valgus on the left side of these two patients has been the cause of pain at the first metatarsophalangeal joints. (**c**) A 40-year-old patient revealing bilateral hallux valgus and hammer toes. (**d**) Roentgenogram of the same patient showing dislocation and subluxation of the metatarsophalangeal and interphalangeal joints. (**e**) Clinical picture reveals bilateral severe hallux valgus

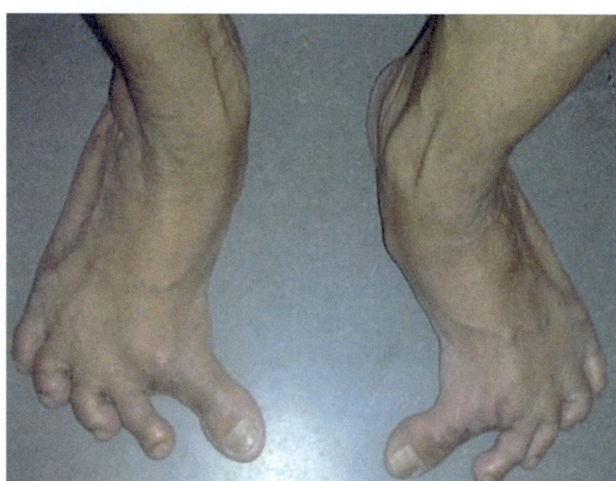

Fig. 19.118 A 45-year-old lady presented with severe bilateral hallux varus. The only drawback was that she could not put on shoes

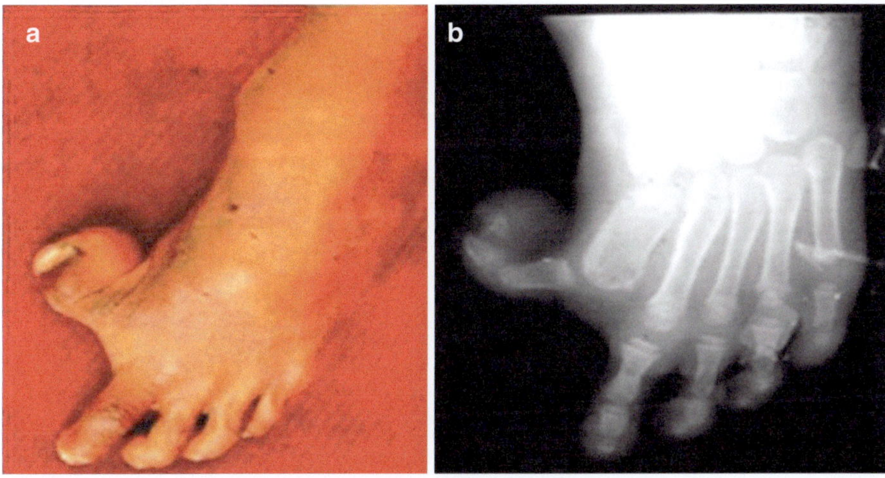

Fig. 19.119 (**a**) Left foot shows severe hallux varus deformity. (**b**) Roentgenogram shows severe congenital hallux varus with tri-phalangeal big toe. Note the ovoid first metatarsal

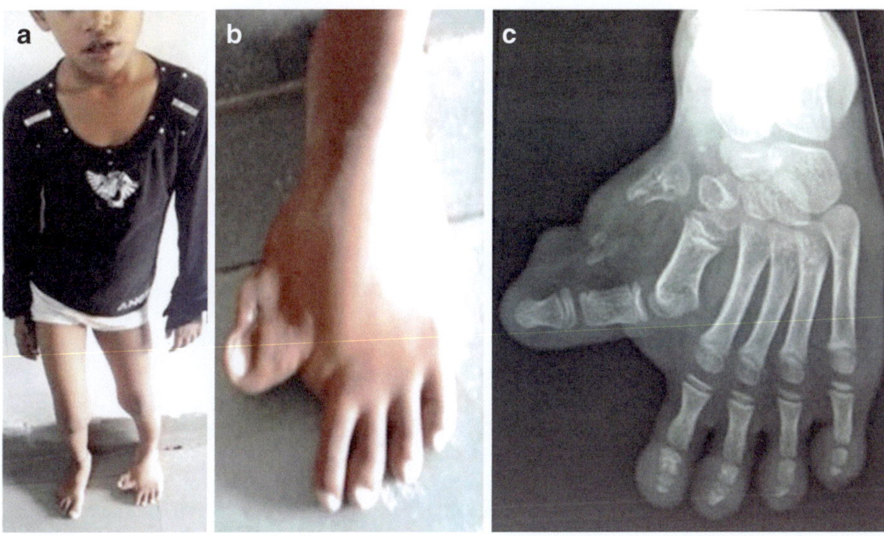

Fig. 19.120 (**a, b**) A 10-year-old boy presented with severe adduction deformity of the left big toe. (**c**) Roentgenogram reveals severe adduction of the big toe at the metatarsophalangeal joint with irregular segmentation of the big toe. Note the rudimentary bones in the medial skin fold and short first metatarsal

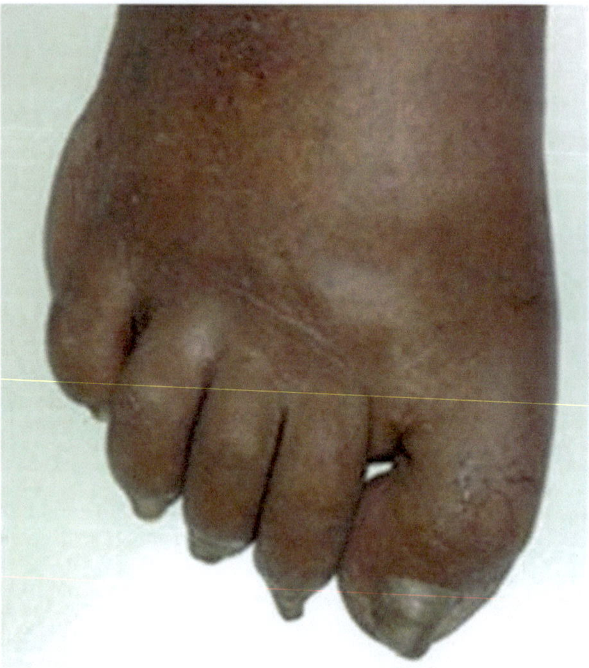

Fig. 19.121 Right foot shows hammer toes

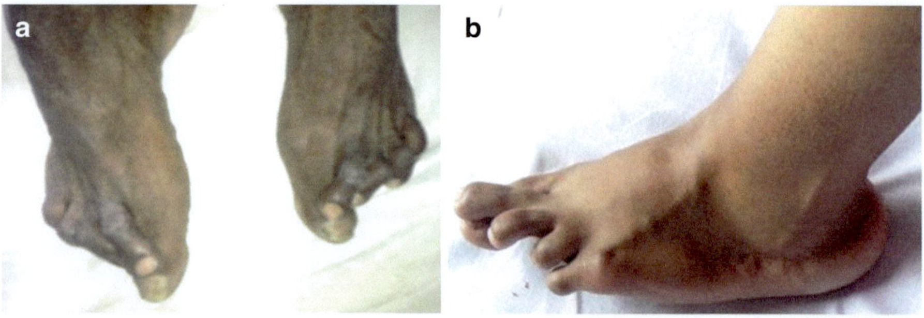

Fig. 19.122 (**a**) Clinical picture of bilateral overlapping second toe. Callosity over the PIP joint is visible. (**b**) Clinical picture of a boy with overlapping second toe of the left foot. The fourth and fifth toes are short

(Fig. 19.123a–i). The deficiencies may be classified into two patterns: typical and atypical. The typical is categorized as cleft foot whereas the atypical defect is known as symbrachydactyly. In symbrachydactyly of the foot often there are so many crossed bones and abnormal soft tissues which result in severe deformity of the foot (Fig. 19.124a–d).

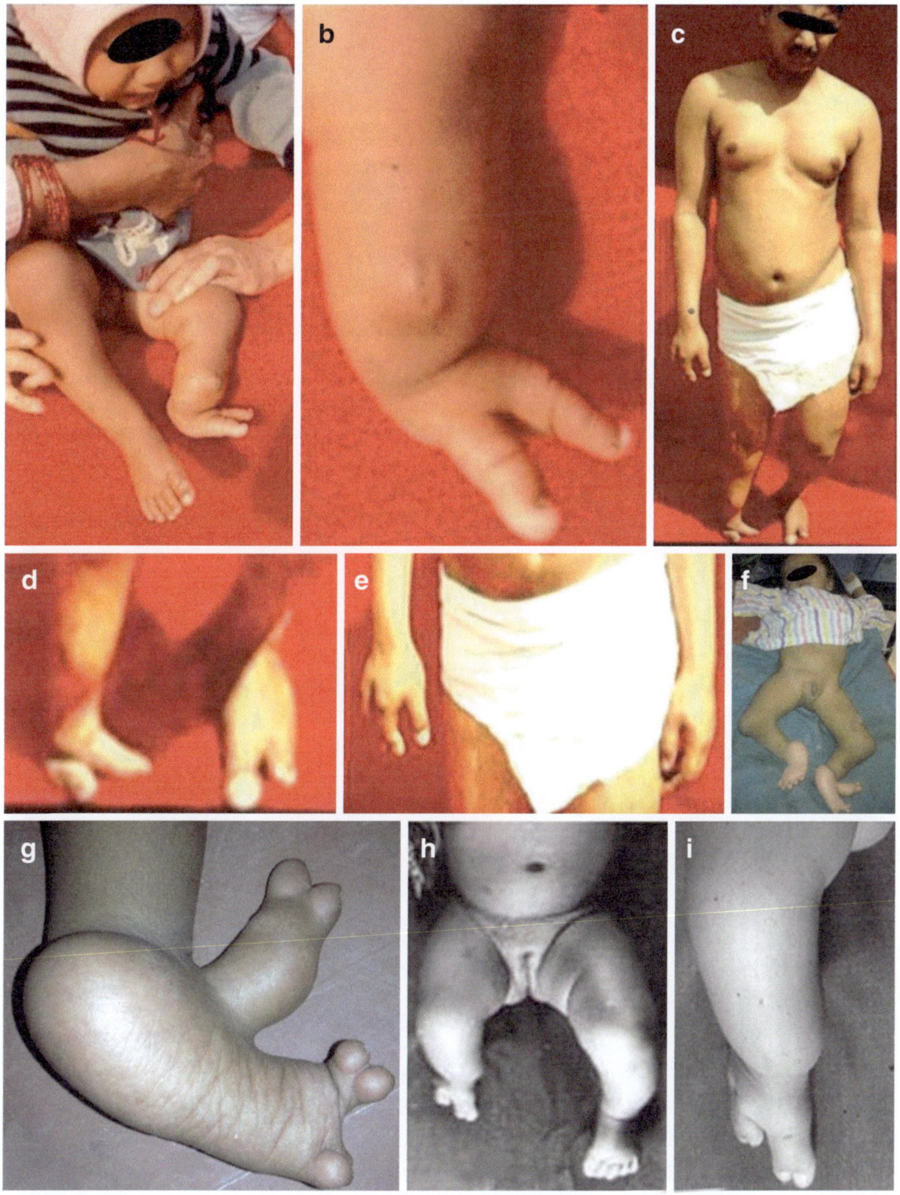

Fig. 19.123 (**a, b**) A 1-year-old child shows bi-fid left foot. The presentation is like fingers—a very rare anomaly. (**c, d, e**) A 32-year-old male shows bi-fid feet. The both feet and the right hand reveal oligodactyly. (**f, g**) A 2-year-old boy with bilateral tibial hemimelia type-I. Note the bi-fid foot on the left side. (**h, i**) A 2-year-old girl revealed bi-fid foot on the right side and syndactyly of the left toes

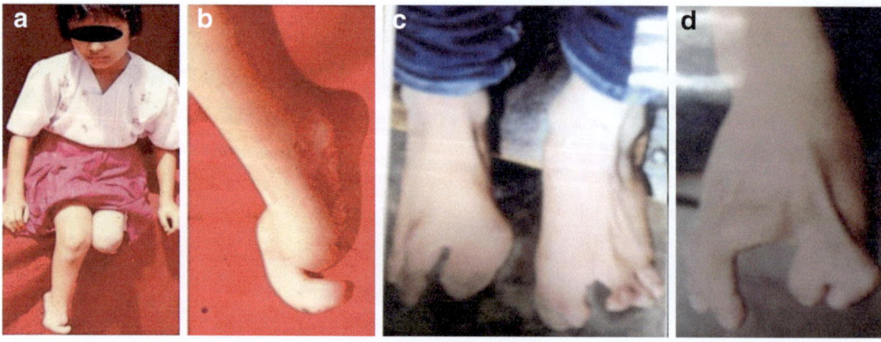

Fig. 19.124 (**a, b**) A 9-year-old girl had a hook-like presentation of the right forefoot (A very rare anomaly). She had congenital below knee amputation on the left side. (**c**) Clinical presentation of bi-fid feet. It is difficult to define the anomalies of the feet in this case. (**d**) Bi-fid foot with a deeper cleft (Rare presentations)

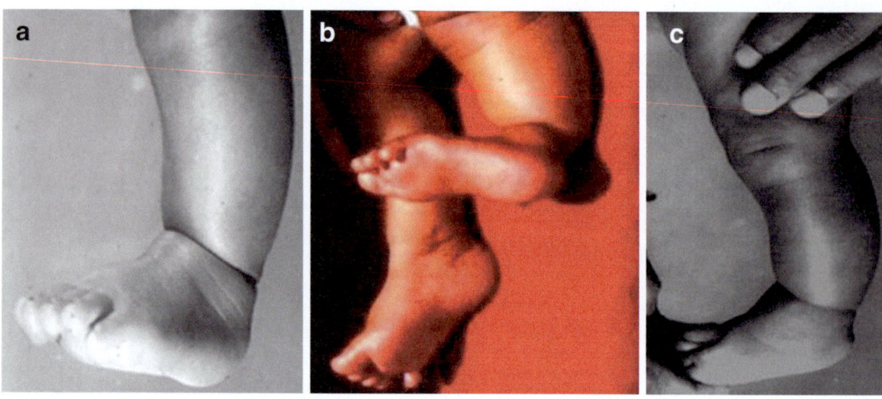

Fig. 19.125 (**a, b**) A 4-months-old child had circular constriction band at the left distal leg. (**c**) Another patient had similar constriction band just above the ankle

19.16 Congenital Amniotic Constriction Band Syndrome

The congenital constriction band syndrome is also known as amniotic band syndrome and pseudoainhum (Fig. 19.125a–c). It is most common cause of congenital malformation of the limb. There is no gender difference in incidence. Transverse terminal amputations are the most common component of the syndrome. Polydactyly is also seen in some patients with constriction bands. Clubfoot has also been observed in 25% patients (Fig. 19.126a, b). Cranial and facial malformations may also coexist. Clinically and radiologically the constriction bands can be identified with distal oedema (Fig. 19.127a–c). These bands may be released by excision and multiple Z-Plasties.

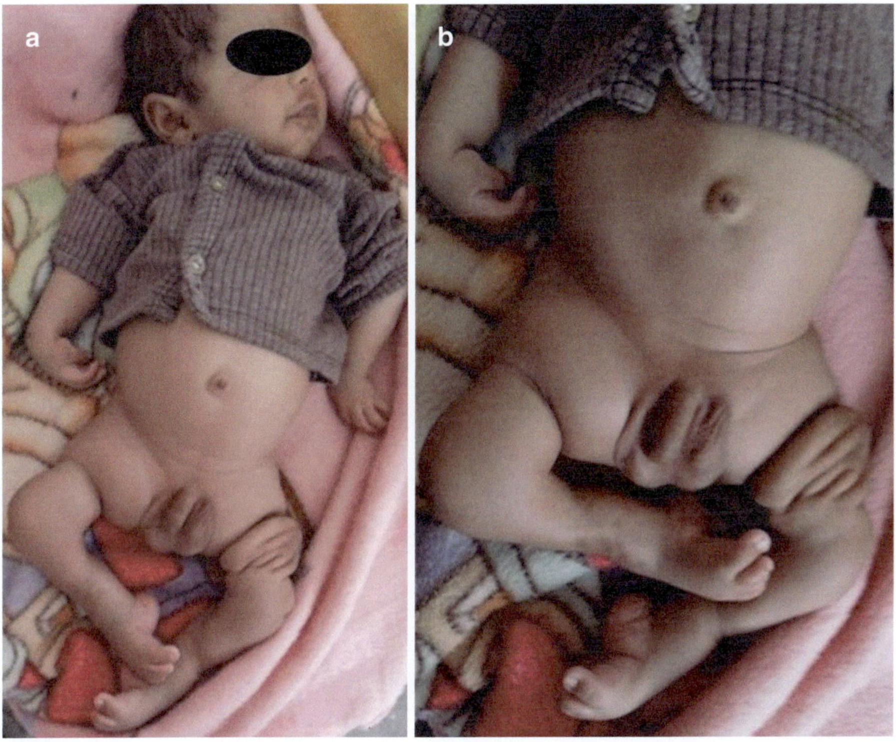

Fig. 19.126 (**a, b**) A 6-months-old child revealed severe constriction bands on both the thighs. Note the skin folds on the left thigh and flexion contracture of the right wrist. The patient also had bilateral CTEV

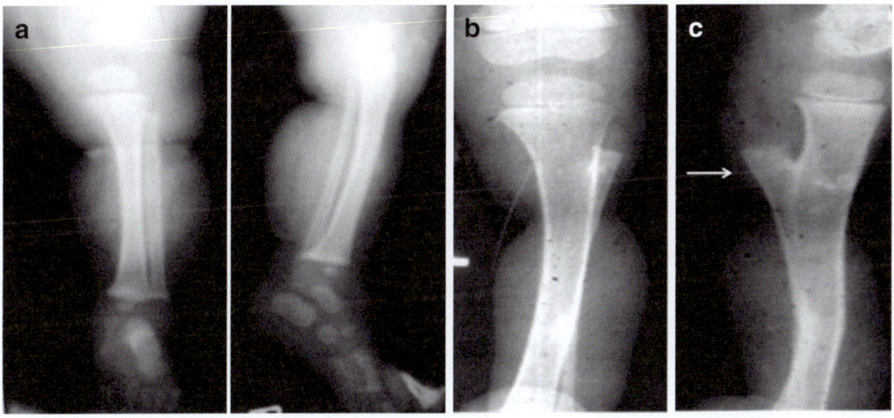

Fig. 19.127 (**a**) Roentgenogram reveals tight circular constriction band at the proximal leg with distal oedema. (**b, c**) Roentgenogram reveals severe constriction band at the proximal leg also with distal oedema. Proximal end of the fibula is visible (arrow). Note the rest of the fibula appears to be merged with mid tibial shaft which is hypertrophied-a rare anomaly

19.16.1 Congenital Hypoplastic Toes

Congenital absence of toes (aphalangia)/hypoplastic toes is a rare anomaly and is usually associated with other anomalies like—Aphalangia of the hand (Fig. 19.128). Usually it is a benign condition (Figs. 19.129a–d and 19.130a, b). It may also result in occasional pain and discomfort to the patient particularly while putting on the shoes. This may require a symptomatic treatment.

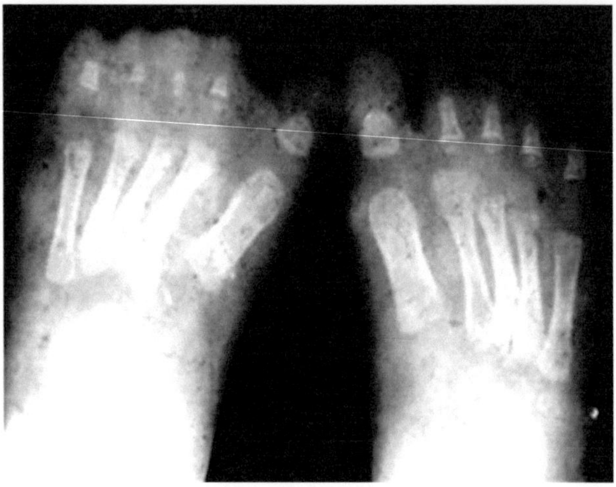

Fig. 19.128 Roentgenogram shows bilateral congenital hypoplasia with uni-phalangeal toes

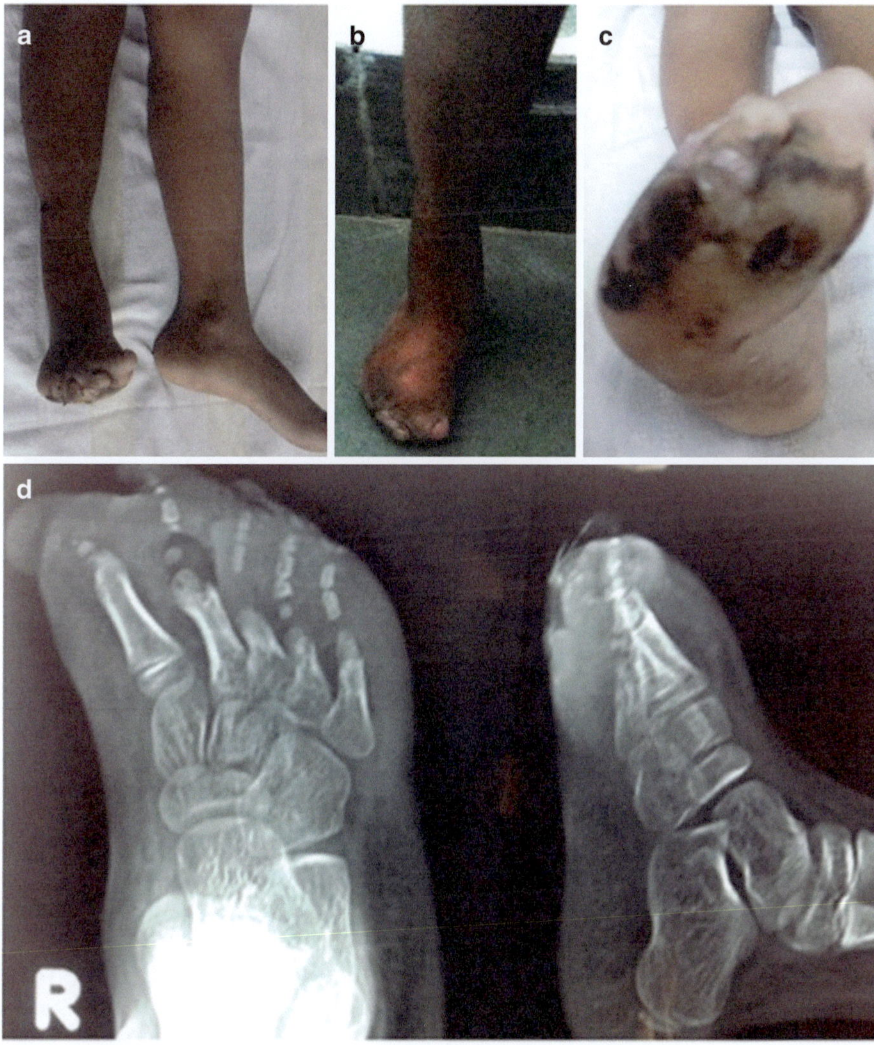

Fig. 19.129 (**a, b, c**) Clinical presentation of deformed right foot with hypoplastic toes. Note the ulcers under the forefoot. (**d**) Roentgenogram of the right foot with hypoplastic metatarsals and miniphalanges. The toes are not defined

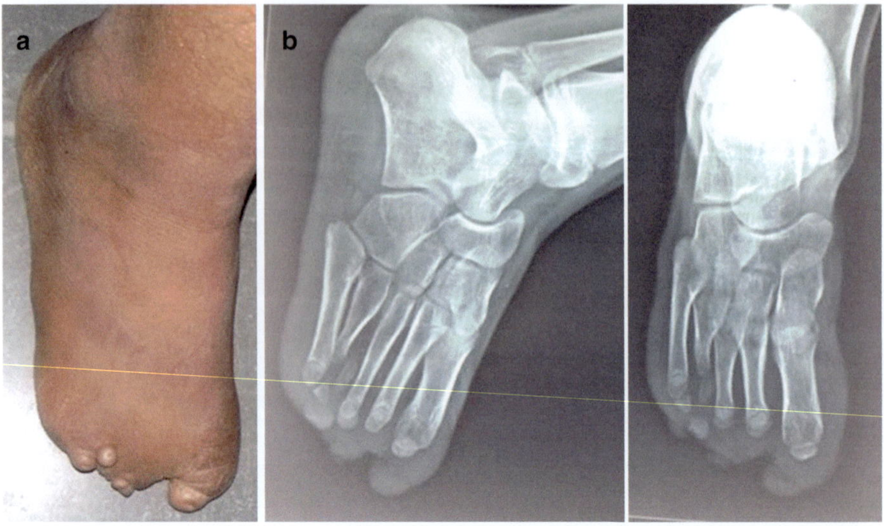

Fig. 19.130 (**a**) A 15-year-old boy presented with loss of toes on the right side (aphalangia). (**b**) Roentgenogram shows brachymetatarsia of the fourth ray and aphalangia

Suggested Readings

Barlow TG. Early diagnosis and treatment of congenital dislocation of the hip. J Bone Joint Surg Br. 1962;44:292.

Berosa GP, Feller N. Treatment of congenital dislocation of the hip. Management before walking age. Clin Orthop. 1987;225:77.

Blockey NJ. Derotation osteotomy in the management of congenital dislocation of the hip. J Bone Joint Surg Br. 1984;66:485.

Amstutz HC. Developmental (infantile) coxa vara-a distinct entity. Clin Orthop. 1970;72:242.

Babb FS, Ghormley RK, Chatterton CC. Congenital coxa-vara. J Bone Joint Surg Am. 1949;31:115.

Langenskiold F. On pseudarthrosis of the femoral neck in congenital coxa-vara. Acta Chir Scand. 1949;98:568.

Miklos S, Franklin HS. Color atlas of clinical orthopaedics. Berlin Heidelberg: Springer-Verlag; 2009.

Speen DP. Slipped capital femoral epiphysis. In: Steinberg ME, editor. In the hip and its disorders. W.B. Saunders; 1991. p. 390–418.

Ponseti IV, McLitock R. The pathology of slipping of the upper femoral epiphysis. J Bone Joint Surg. 1956;38A:71–83.

Bensahel H, Dal Monte A, Hjelmstedt A, et al. Congenital dislocation of the knee. J Pediatr Orthop. 1989;9:174.

Ferris B, Aichroth P. The treatment of congenital knee dislocation. A review of nineteen knees. Clin Orthop Rel Res. 1987;216:135.

Johnson E, Audell R, Oppenheim WL. Congenital dislocation of the knee. J Pediatr Orthop. 1987;7:194.

Gillespie R, Torode IP. Classification and management of congenital abnormalities of the femur. J Bone Joint Surg Br. 1983;65:557.

Hamanishi C. Congenital short femur. J Bone Joint Surg Br. 1980;62:307.

Murray D, Kambouroglou G, Kenwright J. One-stage lengthening for femoral shortening with associated deformity. J Bone Joint Surg Br. 1993;75:566.
Amstutz H. The natural history and treatment of congenital absence of the fibula. J Bone Joint Surg Am. 1972;54:1349.
Brown F, Pohnert W. Construction of a knee joint in meromelia tibia (congenital absence of the tibia): a 15 years follow-up study. J Bone Joint Surg Am. 1972;54:1333.
Brown LA, Osebold WR, Ponseti IV. Congenital pseudarthrosis of long bones. A clinical radiographic, histologic and ultrastructural study. Clin Orthop. 1977;122:228–42.
Crossett LS, Beaty JH, Betz RR, et al. Congenital pseudoarthrosis of the tibia: long term follow-up study. Clin Orthop. 1989;245:16.
Achterman C, Kalamchi A. Congenital deficiency of the fibula. J Bone Joint Surg Br. 1979;61:133.
Choi IH, Kumar SJ, Bowen JR. Amputation or limb-lengthening for partial or total absence of the fibula. J Bone Joint Surg Am. 1990;72:1391.
Loder R, Herring J. Fibular transfer for congenital absence of the tibia: a reassessment. J Pediatr Ortho. 1987;7:8.
Pattison RC, Fixsen JA. Management and outcome in tibial dysplasia. J Bone Joint Surg Br. 1992;74:893.
Yadav SS. Type-I tibial Hemamilia. A limb-salvage and lengthening technique. JBJS Am Open Access. 2019;18(00029)
Abel M, McFarland R. Hair and thread constriction in infants. J Bone Joint Surg Am. 1993;75:915.
Askins G, Ger E. Congenital constriction band syndrome. J Pediatr Orthop. 1988;8:461.
Bass HN, Caldwell S, Brooks BS. Michelin tire baby syndrome. Am J Med Genet. 1993;45:370.
Tada K, Yonenobu K, Swanson AB. Congenital constriction band syndrome. J Pediatr Orthop. 1984;4:726.
VanAllen MI, Siegel-Bartelt J, Dixon J, et al. Constriction bands and limb reduction defects in two newborns. Am J Med Genet. 1992;44:598.
Morrissy RT. Release of congenital constriction band. In: Morrissy RT, editor. Atlas of pediatric orthopaedic surgery. Philadelphia: JB Lippincott; 1990.
Bourne MH, Klassen RA. Congenital annular constricting bands. J Bone Joint Surg. 1987;7:218.
Greene WB. One-stage release congenital circumferential constriction bands. J Bone Joint Surg Am. 1993;75:650.
Bennett GL, Weiner DS, Leighley B. Surgical treatment of symptomatic accessory tarsal navicular. J Pediatr Orthop. 1990;10:445.
Grogan DP, Gasser SI, Ogden JA. The painful accessory navicular: a clinical and histopathological study. Foot Ankle. 1989;10:164.
Sullivan JA, Miller WA. The relationship of the accessory navicular to the development of the flatfoot. Clin Orthop. 1979;44:233.
Bonney G, Macnab I. Hallux valgus and hallux Rigidus. J Bone Joint Surg Br. 1952; 1952(34):366.
Browne D. Congenital deformities of mechanical origin. Arch Dis Child. 1955;30:37.
Conway HR, Cowell HR. Tarsal coalition: clinical significance and roentgenographic demonstration. Radiology. 1969;92:799.
Dedrick DK, Kling TJ, Smith WS. Macrodactyly of the foot reduced with ray resection. Proceedings of the American Association of Orthopaedic Surgeons, New Orleans; 1986 February.
Deutsch AL, Resnick D, Campbell G. Computed tomography and bone scintigraphy in the evaluation of tarsal coalition. Radiology. 1982;144:137.
Elkus RA. Tarsal coalition in the young athlete. Am J Sports Med. 1986;14:447.
Eyre-Brook A. Congenital vertical talus. J Bone Joint Surg Br. 1967;49:618.
Coleman SS, Stelling FH III, Jarrett JJ. Congenital vertical talus: pathomechanics and treatment. Clin Orthop. 1970;70:62.
Hall JE, Salter RB, Bhalla SK. Congenital short tendo calcaneus. J Bone Joint Surg Br. 1967;49:695.
Hark FW. Rocker-foot due to congenital subluxation of the talus. J Bone Joint Surg Am. 1950;32:344.
Howard CB, Benson MKD. Clubfoot: its pathological anatomy. J Pediatr Orthop. 1993;13:654.

Hughes JR. Pathologic anatomy and pathogenesis of congenital vertical talus and its practical significance. J Bone Joint Surg Br. 1970;52:777.

Masciocchi C, D'Archivio C, Barile A, et al. Talocalcaneal coalition: computed tomography and magnetic resonance imaging diagnosis. Eur J Radiol. 1992;15:22.

Oppenheim W, Smith C, Christie W. Congenital vertical talus. Foot Ankle. 1985;5:198.

Atar D, Lehman WB, Grant AD, Strongwater AM. Revision surgery in clubfeet. Clin Orthop Relat Res. 1992;283:223.

Piggott H. The natural history of hallux valgus in adolescent and early adult life. J Bone Joint Surg Br. 1960;42:749.

Shands AR. Congenital anomalies, accessory bones, and osteochondritis in the feet of 850 children. Surg Clin North Am. 1953;33:1643.

Venn-Watson EA. Problems in polydactyly of the foot. Orthop Clin North Am. 1976;7:909.

Tatli B, Aydinli N, Caliskan M, Ozmen M, Bilir F, Acar G. Congenital muscular tortiocollis: evaluation and classification. Pediatr Neurol. 2006;34(1):41–4.

Gigantism, Acromegaly and Dwarfism 20

20.1 Gigantism: Acromegaly

Generalized overgrowth during infancy and childhood may be due to excessive secretion of growth hormone usually the result of pituitary adenoma. If the onset of the disease is before fusion of the epiphyses, this results in gigantism which is characterized by rapid and excessive skeletal growth (Fig. 20.1a, b). Once maturity has been reached it results in acromegaly (Fig. 20.2a–d). There is bone thickening and proportionate overgrowth of the whole skeleton. The skull, mandible, spine, ribs and bones of the hand and feet are particularly affected.

In gigantism, there is proportionate overgrowth of all parts of the skeleton. There is delay in closure of the growth plate cartilage of the long bones and height is greatly increased. Features of acromegaly develop in adult life if the disease remains untreated. Acromegaly results in overgrowth of many tissues including skin, subcutaneous tissue and bone.

The vertebral bodies show cortical thickening and there may be painful spinal kyphosis. The bone overgrowth may lead to compression of the spinal nerve roots. The skeletal changes result in deformity of the articular surface and predispose to secondary osteoarthrosis. There is also fibrous overgrowth of overlying subcutaneous tissue.

S. S. Yadav, *Skeletal Dysplasias & Congenital Malformations*,
https://doi.org/10.1007/978-981-99-5679-1_20

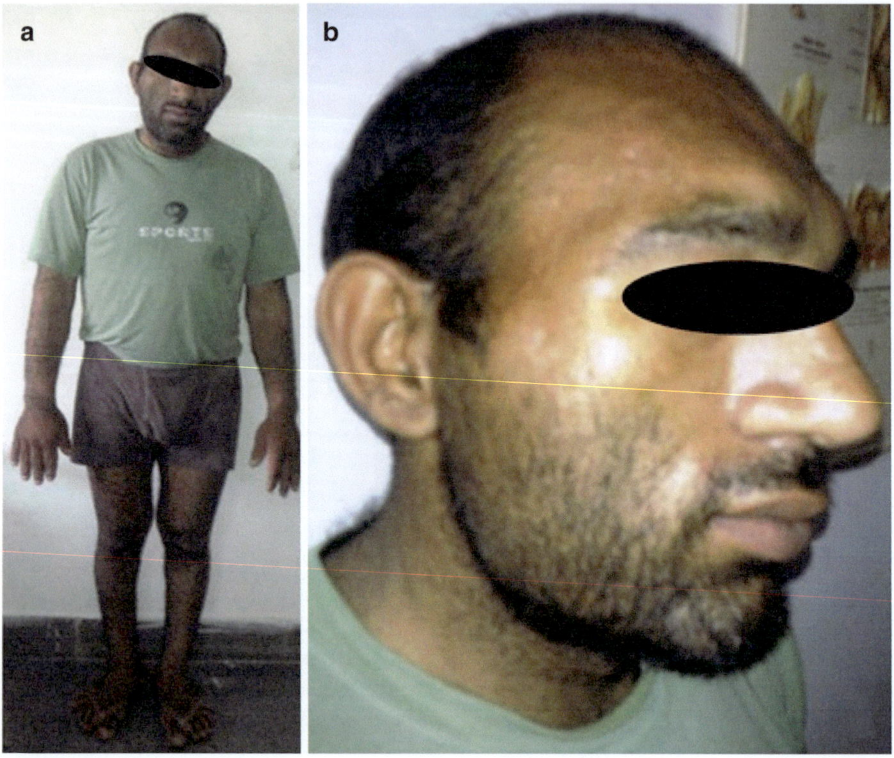

Fig. 20.1 (**a**, **b**) A 42-year-old male showing signs of acromegaly

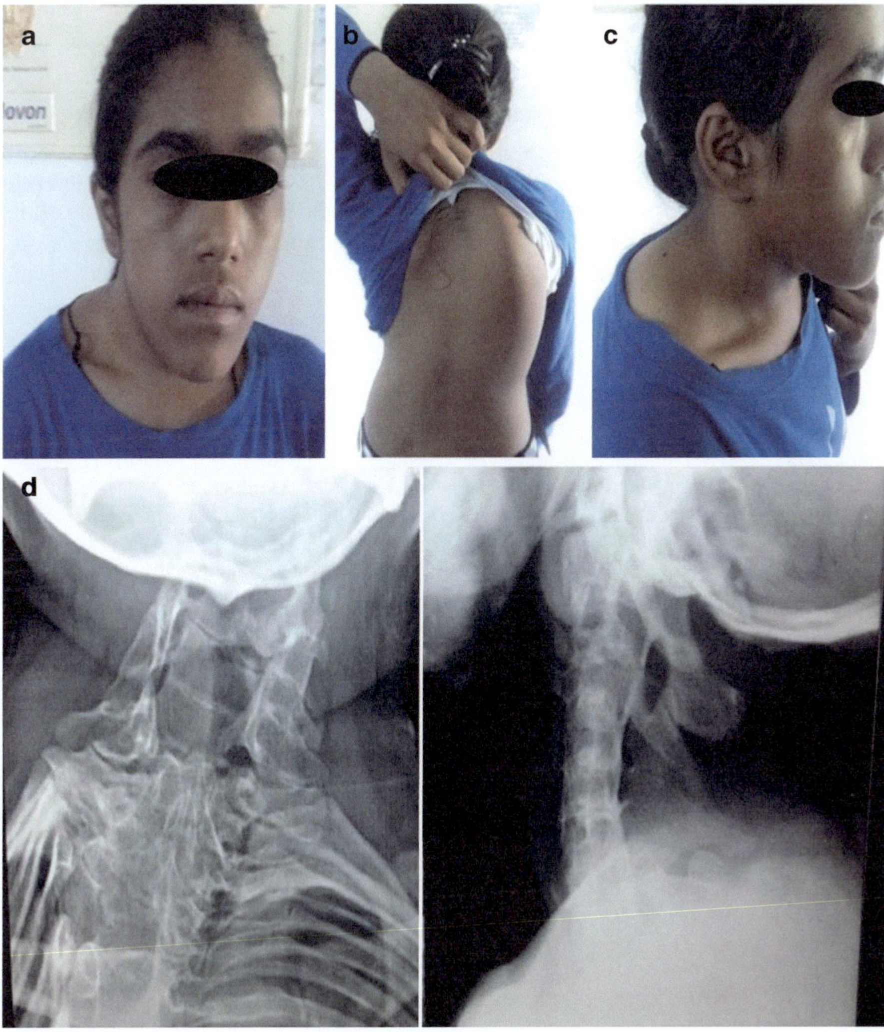

Fig. 20.2 (**a**, **b**, **c**) A 25-year-old girl revealing signs of gigantism. She also had severe congenital thoracic scoliosis. (**d**) Roentgenogram of the same patient revealed the absence of laminae and multiple anomalies of the cervical spine and in the cervico-dorsal region. She had no neurological symptoms

20.2 Dwarfism

Prepubertal panhypopituitarism may result in symmetrical dwarfism and maintenance of normal skeletal proportions with the formation of slender bones. The growth plate is very thin. The growth plate remains open and may persist into adult life. In addition to growth hormone deficiency, effects on the skeleton result from disturbances in the production of thyroid hormone and sex hormones. This results in marked atrophy of the thyroid, adrenal glands and gonads with consequent disturbances of skeletal growth and bone mass. Shortening of the spine may be the cause of both acquired and congenital dwarfism.

Suggested Readings

Thorner MO. Anterior pituitary disorders. In: Weatherall D, Ledingham J, Warrell D, editors. Oxford textbook of medicine. Oxford University Press; 1997. p. 1573–99.

Riggs BL, Randall RU, Wahner HW, et al. The nature of the metabolic bone disorder in acromegaly. J Clin Endocrinol Metab. 1972;34:911–8.

Athanasou NA. Pathological basis orthopaedic and rheumatic disease. London: Edward Arnold; 2001.

Hemihypertrophy and Hemihypotrophy 21

Patients with hemihypertrophy and hemihypotrophy are sporadic and idiopathic. Diagnosis is possible at birth or in weeks following births. Idiopathic hemihypertrophy is a condition of unknown cause affecting approximately 1 per 50,000 members of the population. There is no clear inheritance pattern. Although often primarily evident in just one limb, it may also affect upper and lower limbs, the trunk, the face and internal organs. Asymmetry remains unchanged during the growth of the body. In true hemihypertrophy, definite overgrowth of all or part of the body on one side can be found in comparison with normal standards for the age (Fig. 21.1a, b).

Hemihypertrophy generally decreases with increasing age. In hemihypotrophy, the affected side is smaller than the unaffected and is often associated with Scoliosis, Mental retardation and associated with Silver-Russel syndrome. In hemihypertrophy there is no associated scoliosis or mental retardation. Hemihypertrophy is usually not inherited. However, a case of hemihypertrophy has been described in a mother and her child.

Asymmetrical skeletal dysplasias such as enchondromatosis, dysplasia epiphysialis punctata, etc. and numerous syndromes, also contribute to localized undergrowth of portions of the skeleton. Hamartomatous malformations such as the Klippel-Trenaunay syndrome, lymphangiectasia and macrodystrophia lipomatosa are associated with localized gigantism.

Most of the cases with lower extremity discrepancy do need surgery for limb length equalization.

S. S. Yadav, *Skeletal Dysplasias & Congenital Malformations*,
https://doi.org/10.1007/978-981-99-5679-1_21

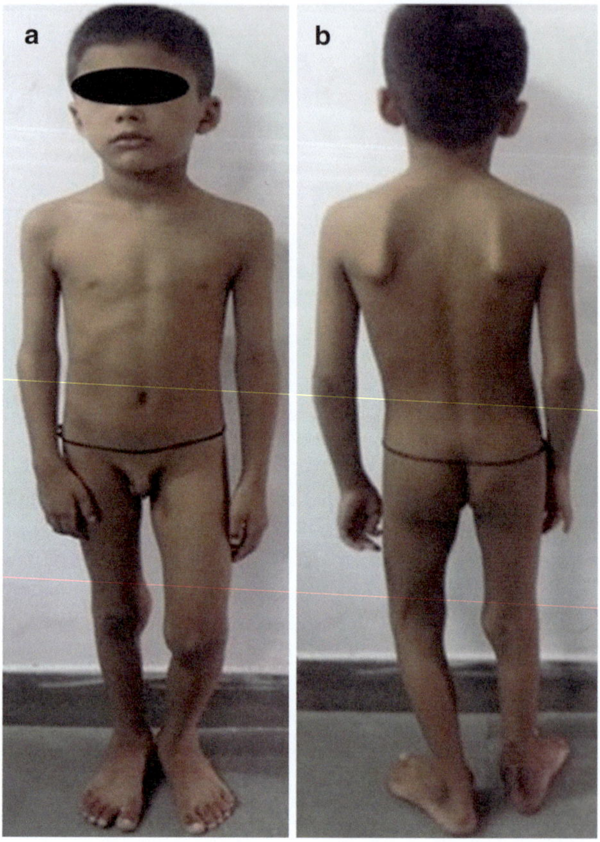

Fig. 21.1 (**a**, **b**) Clinical picture of a 6-year-old boy showing idiopathic hemi-hypertrophy on the left side. Clinical appearance with an increase in size on the left side. The left lower limb is longer as the boy stands with flexed left knee

Suggested Readings

Beals RK. Hemihyperthrophy and hemihypotrophy. Clin Orthop Relat Res. 1982;199:203.
Levine C. Imaging of body asymmetry and hemihypertrophy. Crit Rev Diagn Imaging. 1990;31:1.
Morrissy RT, Weinstein SL. Lovell and Winter's pediatric orthopaedics. 4th ed. Lippincott-Raven Publishers; 1996.

MIX
Papier aus verantwortungsvollen Quellen
Paper from responsible sources
FSC® C105338

If you have any concerns about our products,
you can contact us on
ProductSafety@springernature.com

In case Publisher is established outside the EU,
the EU authorized representative is:
Springer Nature Customer Service Center GmbH
Europaplatz 3, 69115 Heidelberg, Germany

Printed by Libri Plureos GmbH
in Hamburg, Germany